Sustainable Development and Biodiversity

Volume 46

Series Editor

Kishan Gopal Ramawat, Botany Department, Mohanlal Sukhadia University, Udaipur, India

Sustainable Development Goals are best achieved by mechanisms such as research, innovation, and knowledge sharing. This book series aims to help researchers by reporting recent progress and providing complete, comprehensive, and broad subject-based reviews about all aspects of sustainable development and ecological biodiversity.

The series explores linkages of biodiversity with delivery of various ecosystem services and it offers a discourse in understanding the biotic and abiotic interactions, ecosystem dynamics, biological invasion, ecological restoration and remediation, diversity of habitats and conservation strategies. It is a broad scoped collection of volumes, addressing relationship between ecosystem processes and biodiversity. It aims to support the global efforts towards achieving sustainable development goals by enriching the scientific literature. The books in the series brings out the latest reading material for botanists, environmentalists, marine biologists, conservationists, policy makers and NGOs working for environment protection.

We welcome volumes on the themes --

Agroecosystems, Agroforestry, Biodiversity, Biodiversity conservation, Conservation of ecosystem, Ecosystem, Endangered species, Forest conservation, Genetic diversity, Global climate change, Hotspots, Impact assessment, Invasive species, Livelihood of people, Plant biotechnology, Plant resource utilization, Sustainability of the environment, Sustainable management of forests, Sustainable use of terrestrial ecosystems and plants, Traditional methods, Urban horticulture.

Monika Thakur • Tarun Belwal
Editors

Food Biodiversity and Consumer Engagements

Pathways to Sustainable Agri Food Systems

Springer

Editors
Monika Thakur
Amity Institute of Food Technology
Amity University Uttar Pradesh
Noida, Uttar Pradesh, India

Tarun Belwal
Phytonutrients
7 Continents Natural Products LLC
Monmouth Junction, NJ, USA

ISSN 2352-474X ISSN 2352-4758 (electronic)
Sustainable Development and Biodiversity
ISBN 978-981-95-7776-7 ISBN 978-981-95-7777-4 (eBook)
https://doi.org/10.1007/978-981-95-7777-4

This Springer imprint is published by the registered company Springer Nature Singapore Pte Ltd.
The registered company address is: 152 Beach Road, #21-01/04 Gateway East, Singapore 189721, Singapore

Preface

In an era marked by rapid environmental change, shifting dietary patterns, and an urgent need for resilient food systems, *Food Biodiversity and Consumer Engagements: Pathways to Sustainable Agri-Food Systems* emerges as a timely and comprehensive exploration of the complex interplay between biodiversity, innovation, and human behaviour in contemporary food systems. This book brings together a diverse array of perspectives from scholars, academicians, food technologists, nutritionists, and industry experts who share a common commitment to advancing sustainability through research, practice, and policy.

The fundamental premise of this work is that sustainable agri-food systems are not only rooted in ecological principles but are also shaped profoundly by the choices, perceptions, and engagements of consumers. Biodiversity, encompassing the variety of species, genetic resources, and ecosystems that sustain our food supply, is a crucial yet often overlooked pillar of food security. Equally important is understanding how consumers interact with food environments, what they choose, why they choose it, and how market and cultural dynamics influence these decisions.

The twenty-first-century agri-food system stands at a crossroads, shaped by pressing global challenges including climate change, resource depletion, food insecurity, and shifting dietary patterns. In response, the pursuit of sustainability in food systems has gained momentum, driven by a synergy of technological innovation, nutritional awareness, and evolving consumer behaviour. This book, *Food Biodiversity and Consumer Engagement*, brings together diverse perspectives and pioneering research to explore how future-ready strategies can transform the way food is produced, processed, and perceived.

The contributions are grouped into four core parts, together, reflecting an interdisciplinary approach to building resilient food systems aligned with the United Nations Sustainable Development Goals (SDGs). All chapters are contributed by academicians, food scientists, technologists, and also medical practitioners with the interest of modern-day consumers in mind. By bridging science, innovation, and human behaviour, it aims to inspire transformative action towards a more sustainable food future.

The content in this book is divided into four distinct parts:

- Part I—From Field to Future: Cutting-Edge Advances in Agri-Health Systems—covered in 11 chapters, examines breakthroughs in agri-health systems, smart farming, sustainable practices, and agri-food innovations.
- Part II—Food Biodiversity: Innovations in Safe, Sustainable, Nutrient-Rich Foods—covered in 9 chapters, comprehensively explains food biodiversities, innovative food processing, trends in food products, integrating nutrition, and food sustainability.
- Part III—Eating with Intent: Sustainability in Market and Consumer Choices—covered in 4 chapters, mentions the link between marketing strategies and consumer choices.
- Part IV—Mindful Consumption: Psychology Behind Sustainable Eating—covered in 3 chapters, explores the emotional, cognitive, and social factors influencing food choices and dietary habits with their role in sustainable eating patterns.

Food Biodiversity and Consumer Engagements: Pathways to Sustainable Agri-Food Systems stands apart from other works by uniquely integrating food biodiversity with consumer engagement, bridging ecological variety and human choice in ways most books treat separately. It organizes content across four interdisciplinary sections offering a holistic perspective that combines rigorous research with practical relevance for scholars and practitioners alike, and brings together diverse disciplinary voices to explore not just what changes are needed for sustainability but how consumers actually engage with and influence food systems, making the book broadly applicable for researchers, industry professionals, policymakers, and educators.

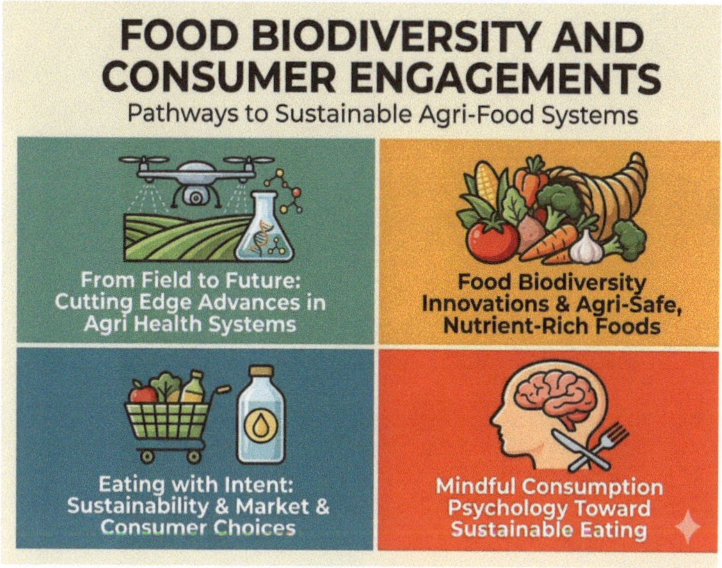

Noida, Uttar Pradesh, India
Monmouth Junction, NJ, USA

Monika Thakur
Tarun Belwal

Acknowledgements

In an era of environmental change, evolving diets, and the urgent need for resilient food systems, *Food Biodiversity and Consumer Engagements* offers a comprehensive look at how biodiversity, innovation, and human behaviour intersect in today's agri-food landscape. It unites scholars, food technologists, nutritionists, and industry experts committed to advancing sustainability through research, practice, and policy.

We are profoundly grateful to our respected Founder President, *Dr. Ashok K. Chauhan*, for his blessings and unwavering encouragement throughout this endeavour. We extend our sincere appreciation to *Dr. Atul Chauhan*, Chancellor of Amity University, Uttar Pradesh, and President of RBEF; without their inspiring guidance and support, this work would not have been possible. We also express our heartfelt thanks to *Prof. Balvinder Shukla*, Vice Chancellor, Amity University, Uttar Pradesh, for her constant motivation and support at every stage of this project.

We are deeply thankful to *Dr. V. K. Modi*, Director, Amity Institute of Food Technology (AIFT), Amity University, for his invaluable guidance, blessings, and continuous support throughout the development of this work. We also extend our gratitude to *Prof. T. N. Lakhanpal*, Himachal Pradesh University, Shimla, for his mentorship and blessings. We would like to thank all faculty and staff members of the *Amity Institute of Food Technology, Amity University Uttar Pradesh, Noida*, for their cooperation and assistance.

We are extremely indebted to all the authors who contributed chapters to this volume and graciously shared their expertise on various aspects of bioactive components, their sources, applications, and sustainability. Without their dedicated efforts, this volume would not have come to fruition.

Our sincere thanks go to Springer Publishing for their support and cooperation throughout the publication process. We are especially grateful to the Springer production team for their professionalism and efforts in bringing this book to publication. Finally, we wish to acknowledge with deep gratitude the support and encouragement of our family members, whose understanding and patience have been a source of strength throughout this journey.

Noida, Uttar Pradesh, India — Monika Thakur
Monmouth Junction, NJ, USA — Tarun Belwal

Contents

List of Figures

List of Tables

Editors and Contributors

About the Editors

Monika Thakur is working as a Professor at Amity Institute of Food Technology, Amity University, Uttar Pradesh, India. She is currently engaged in the development of functional foods, and extraction of bioactive components. She was awarded PhD in 2006 by Himachal Pradesh University. She is Double Gold Medalist (MSc and MPhil) and received several prestigious awards. Apart from her teaching assignments, she has been involved in various research and extension activities. She is a member of editorial board and reviewer in journals of national and international repute. She has published more than 70 publications, 35 book chapters, and 15 books. She has also organized various training programmes, faculty development programmes, international conferences, and national seminars. Her research interests include Plant Sciences, Functional foods and Nutraceuticals, Mycology, Food microbiology, and food safety and related areas.

Tarun Belwal is currently associated with 7 Continents Natural Products, USA, where his work focuses on food science, plant-based nutraceuticals, and their implications for human health. His earlier academic and research engagements include Texas A&M University, the University of Turin, and Zhejiang University. He earned his PhD in Plant Biotechnology from Kumaun University, Uttarakhand, India, in collaboration with the G.B. Pant National Institute of Himalayan Environment (NIHE). To date, he has edited 7 scientific books, authored over 100 international peer-reviewed research papers, and contributed 18 book chapters. His scholarly contributions have received more than 10,000 citations worldwide.

Contributors

Aadya Amity School of Economics, Amity University, Noida, Uttar Pradesh, India

Fasaha Ahmad Department of Food Technology, Laxminarayan Innovation Technological University, Nagpur, Maharashtra, India

Eid Sadun Alotaibi Department of Mathematics and Statistics, AlKhurmah University College, Taif University, Taif, Saudi Arabia

Bindvi Arora Division of Food Science and Postharvest Technology, ICAR-IARI, New Delhi, India

Mahua Bhattacharjee Foundation for Development of Rural Value Chains (FDRVC), New Delhi, India

Research and Livelihood Development, ISRA, PMU Team-NCUI, New Delhi, India

Chaiyavat Chaiyasut Innovation Center for Holistic Health, Nutraceuticals and Cosmeceuticals, Chiang Mai, Thailand

Nikhil Chandel Amity Institute of Organic Agriculture, Amity University Uttar Pradesh, Noida, Uttar Pradesh, India

Parul Chauhan Amity Institute of Food Technology, Amity University Uttar Pradesh, Noida, India

Shweta Deotale Visvesvaraya National Institute of Technology, Nagpur, Maharashtra, India

Sheetal D. Deshmukh Department of Food Technology, Laxminarayan Innovation Technological University, Nagpur, Maharashtra, India

Swarnima Dey Department of Food Technology, Faculty of Science and Humanities, SRM University Delhi- NCR, Sonepat, Haryana, India

Amity Institute of Food Technology, Amity University, Uttar Pradesh, Noida, Uttar Pradesh, India

Sayantani Dutta Grupo de Investigación en Polifenoles, Química Analítica, Nutrición y Bromatología, Universidad de Salamanca, Salamanca, Spain

Dhananjay Gaikwad Vaikunth Mehta National Institute of Cooperative Management (VAMNICOM), Pune, Maharashtra, India

Urishita Gambhir Amity School of Economics, Amity University, Noida, Uttar Pradesh, India

Manya Gupta Department of Nutrition and Dietetics, Sharda University, Greater Noida, Uttar Pradesh, India

Reeba Iqbal Amity Institute of Food Technology, Amity University Uttar Pradesh, Noida, Uttar Pradesh, India

R. Jenisha Priyadharshini Department of Food Science and Nutrition, Dr. N.G.P Arts and Science College, Coimbatore, Tamil Nadu, India

Alka Joshi Division of Food Science and Postharvest Technology, ICAR-Indian Agricultural Research Institute, New Delhi, India

Mahesh Kadam Vaikunth Mehta National Institute of Cooperative Management (VAMNICOM), Pune, Maharashtra, India

Sneha Karadbhajne Department of Food Technology, Laxminarayan Innovation Technological University, Nagpur, Maharashtra, India

Renuka Kulkarni Department of Food Technology, Lund University, Lund, Sweden

Aneesha Nalin Kumar Department of Nutrition and Dietetics, Manav Rachna International Institute of Research and Studies (Deemed to be University), Faridabad, Haryana, India

Manoj Kumar Amity Centre of Behavioural Science, Amity University, Noida, India

Rahul Kumar Department of Food and Nutritional Sciences, University of Reading, Reading, UK

Vikash Kumar Department of Agriculture, Deen Dayal Upadhyaya Gorakhpur University, Gorakhpur, Uttar Pradesh, India

Faculty of Agricultural Sciences, GLA University, Mathura, Uttar Pradesh, India

Yogesh Kumar Department of Food Technology, Faculty of Science and Humanities, SRM University Delhi- NCR, Sonepat, Haryana, India

Sushma Kumari Berlin School of Business and Innovation, Berlin, Germany

Arushi Loiwal Department of Foods and Nutrition, Faculty of Family and Community Sciences, The Maharaja Sayajirao University of Baroda, Vadodara, Gujarat, India

S. Madumitha Department of Food Science and Nutrition, Dr. N.G.P Arts and Science College, Coimbatore, Tamil Nadu, India

Peter R. Makgwane University of South Africa, Pretoria, South Africa

Corinne Malpuech-Bruger Clermont Auvergne University (UCA), Clermont-Ferrand, France

Shivani Mehta Manav Rachna International Institute of Research and Studies, Faridabad, Haryana, India

Maheshwar Mekam MYAS, ICMR-National Institute of Nutrition, Hyderabad, Telangana, India

Extension and Training Division, ICMR-National Institute of Nutrition, Hyderabad, India

Ananya Mohan Department of Food Technology, Jamia Hamdard (Deemed to be University), New Delhi, India

Ashish M. Mohite Amity Institute of Food Technology, Amity University Uttar Pradesh, Noida, India

Shireesha Mukka MYAS, ICMR-National Institute of Nutrition, Hyderabad, Telangana, India
Extension and Training Division, ICMR-National Institute of Nutrition, Hyderabad, India

Asaeli Naika Department of Applied Sciences, CETVET, Fiji National University, Suva, Fiji

Poonam Pramod Nichat MYAS, ICMR-National Institute of Nutrition, Hyderabad, Telangana, India

Valerie Orsat Department of Bioresource Engineering, McGill University, Montreal, QC, Canada

Ankit Paliwal Department of Applied Sciences, CETVET, Fiji National University, Suva, Fiji

Kshitij Parmar Amity Institute of Organic Agriculture, Amity University Uttar Pradesh, Noida, Uttar Pradesh, India

Pankaj B. Pathare Sultan Qaboos University, Al Seeb, Oman

Charlotte Peraut Université Clermont Auvergne, Clermont-Ferrand, France

Sellam Perinban Division of Food Science and Postharvest Technology, ICAR-Indian Agricultural Research Institute, New Delhi, India

Arti Pillay Department of Applied Sciences, CETVET, Fiji National University, Suva, Fiji

Nikhil Raghuvanshi Faculty of Agricultural Sciences, GLA University, Mathura, Uttar Pradesh, India

Alok Rai Department of Agriculture, Deen Dayal Upadhyaya Gorakhpur University, Gorakhpur, Uttar Pradesh, India

C. Rajna Department of Food Science and Nutrition, Dr. N.G.P Arts and Science College, Coimbatore, Tamil Nadu, India

Mayuri Rastogi Department of Nutrition and Dietetics, Sharda University, Greater Noida, Uttar Pradesh, India

Aditi Rikhari Department of Nutrition and Dietetics, Sharda University, Greater Noida, Uttar Pradesh, India

Jaydip Rokade ICAR-Central Avian Research Institute, Bareilly, Uttar Pradesh, India

Sonakshi Ruhela Rochester Institute of Technology, Dubai, United Arab Emirates

Vijayata Sengar Department of Foods and Nutrition, Faculty of Family and Community Sciences, The Maharaja Sayajirao University of Baroda, Vadodara, Gujarat, India

Neha Sharma Amity Institute of Food Technology, Amity University Uttar Pradesh, Noida, India

Vasudha Sharma Department of Food Technology, Jamia Hamdard (Deemed to be University), New Delhi, India

Shivanjali Shrivastava Amity Centre of Behavioural Science, Amity University, Noida, India

Karuna Singh Department of Nutrition and Dietetics, Sharda University, Greater Noida, Uttar Pradesh, India

Sharda school of Allied Health Sciences, Sharda University, Greater Noida, Uttar Pradesh, India

Pratibha Singh Department of Nutrition and Dietetics, Manav Rachna International Institute of Research and Studies (Deemed to be University), Faridabad, Haryana, India

Simran Singh Department of Food Technology, Laxminarayan Innovation Technological University, Nagpur, Maharashtra, India

Hardik Somaiya Laxminarayan Innovation Technological University, Nagpur, Maharashtra, India

Siddharth Soni School of Psychology and Counselling, O.P. Jindal Global University, Sonipat, Haryana, India

Chhavi Taliwal Amity Institute of Food Technology, Amity University, Noida, India

Anamika Thakur Division of Food Science and Postharvest Technology, ICAR-ICAR-Indian Agricultural Research Institute, New Delhi, India

Monika Thakur Amity Institute of Food Technology, Amity University Uttar Pradesh, Noida, Uttar Pradesh, India

Ajay Tomar Amity Institute of Organic Agriculture, Amity University Uttar Pradesh, Noida, Uttar Pradesh, India

Harika Vasa Extension and Training Division, ICMR-National Institute of Nutrition, Hyderabad, India

Arda Yıldırım Faculty of Agriculture, Department of Animal Science, Tokat Gaziosmanpasa University, Tokat, Turkey

Savita Zambre Department of Food Technology, Laxminarayan Innovation Technological University, Nagpur, Maharashtra, India

Abbreviations

%	Percent
&	And
°	Degree
μCT	Microfocus X-Ray Computed Tomography
μg	Microgram
μl	Microlitre
μm	Micrometre
AACs	Aaptamer-Affinity Columns
ACIAR	Australian Centre for International Agricultural Research
AFM	Atomic Force Microscopy
AGP	Antibiotic Growth Promoters
AI	Artificial Intelligent
ALA	Alpha-Linolenic Acid
APA	American Psychiatric Association
ASC	Added Sugar Content
ASE	Accelerated Solvent Extraction
BD	Bulk Density
BSA	Bovine Serum Albumins
CA	Controlled Atmosphere
CBT	Cognitive-Behavioural Strategies
CDC	Centers for Disease Control
CEC	Cation Exchange Capacity
CE-MS	Capillary Electrophoresis-Mass Spectrometry
CFU/ml	Colony Forming Unit per ml
CHO	Carbohydrates
cm	Centimetre
conc.	Concentrated
COP	Coefficient of Performance
CP	Cold Plasma
CREMNET	Crop and Resource Management Network

DBD	Dielectric Barrier Discharges
dil.	Dilute
DNA	Deoxy Ribonucleic Acid
DPF	Defatted Peanut Flour
e.g.	For Example
EAE	Enzyme-Assisted Extraction
EC	Evaporative cooling
Ed.	Edition
ed.	Editor
eds.	Editors
EFF	Enhanced Freshness Formulation
EFSA	European Food Safety Authority
EIA	Assurance Enzyme Immunoassay
ELISA	Enzyme-Linked Immunosorbent Assay
EO	Essential Oils
et al.	Et alia; and others
etc.	Et cetera
FAO	Food and Agriculture Organisation
FAVs	Fruits and Vegetables
FBO	Food Business Operator
FCR	Feed Conversion Ratio
Fig.	Figure
FIGs	Farmers Interest Groups
FPOs	Farmers Producers' Companies
FSMS	Food Safety Management System
FSSAI	Food Safety Standards Authority of India
FTIR	Fourier Transform Infrared Spectroscopy
g	Gram
GC	Gas Chromatography
GC-ECD/NPD	Gas Chromatography with Electron Capture
GC-MS-MS (QqQ)	Gas Chromatography-Tandem Mass Spectrometry (triple
GC-TOF-MS	Gas Chromatography-Time-of-Flight Mass spectrometry
GDP	Gross Domestic Product
GI	Glycaemic Index
GMOs	Genetically Modified Organisms
GMP	Good Manufacturing Practices
GPS	Global Positioning Systems
GRAS	Generally Recognized as Safe
HACCP	Hazard Analysis and Critical Control Points
HAZOP	Hazard and Operability Analysis
HCs	Horticulture Crops
HIS	Hyperspectral Imaging
HPLC-PDA	HPLC with Photodiode Array Detector
HPLC-UV	HPLC with Ultraviolet Detector

HPP	High Pressure Processing
hrs.	Hours
i.e.	That Is
IACs	Immuno Affinity Columns
in vitro	Within Glass
in vivo	Within Living
IR	Infrared
IRRI	International Rice Research Institute
Kg	Kilogram
l	Litre
LAR	Locally Acquired Resistance
LD	Longissimus Dorsi
LDL	Low Density Lipoprotein
LLE	Liquid-Liquid Extraction
LSR	Liquid-to-Solid Ratio
m	Metre
MAE	Microwave-Assisted Extraction
MALDI-TOF	Matrix-Assisted Laser Desorption Ionization-Time of Flight
MDA	Malondialdehyde
mg	Milligram
MH	Million Hectares
min	Minutes
MIPs	Molecularly Imprinted Polymers
MIR	Mid-Infrared
MIRs	Mid-Infrared Spectroscopy
ml	Millilitre
mln tons	Million Tonnes
mm	Millimetre
MoFPI	Ministry of Food Processing Industry
MT/ha	Metric Ton per Hectare
NIR	Near-Infrared
NIRs	Near-Infrared Spectroscopy
nm	Nanometre
NMR	Nuclear Magnetic Resonance
NTP	Non-Thermal Processing
^{o}C	Degree Celsius
PC	Phosphatidylcholine
PE	Phosphatidylethanolamine Phosphatidylinositol (PI)
PEF	Pulsed Electric Field
PG	Phosphatidylglycerol
PHL	Post-Harvest Losses
PHM	Post Harvest Management
PHT	Post-Harvest Technologies
PLD	Phospholipase D

PRPs	Prerequisite Programs
PSMs	Plant Secondary Metabolites
PZT	Lead Zirconate Titanate
QuEChERS	Quick Easy Cheap Rough and Safe
R^2	Coefficient of Regression
RJC	Real Juice Content
RMSECV	Root Mean Square Error of Cross-Validation
RMSEP	Root Mean Square Error of Prediction
RNS	Reactive Nitrogen Species
RPD	Residual Predictive Deviation
R_{pred}	Correlation Coefficient for Prediction
RRLC-MS/MS	Rapid Resolution Liquid Chromatography Tandem Mass
RTE	Ready to Eat
RTS	Ready to Serve
SAR	Systemic Acquired Resistance
SC	Carbon sequestration
SDG	Sustainable Development Goal
SE	Solid-Liquid Extraction
SE	Soxhlet Extraction
SFE	Supercritical Fluid Extraction
sp.	Species (Singular)
SPE	Solid Phase Extraction
spp.	Species (Plural)
sq Km	Square Kilometre
SSC	Soluble Solid Content
TA	Titratable Acidity
TAC	Total Anthocyanin Content
TM	Tropomyosin
TPC	Total Phenolic Content
TSEC	Two-Stage Evaporative Cooler
TSS	Total Soluble Solids
UAE	Ultrasound-Assisted Extraction
UL	Ultrasonic Treatment
UPLC-Q-TOF-MS	Ultra Performance Liquid Chromatography – Quadrupole Time-of-Flight Mass Spectrometry
UV	Ultraviolet
VALDS–ME	Vortex Assisted Low Density Solvent - Microextraction
var.	Variety
Vis/NIRs	Visible to Near-Infrared Spectroscopy
viz.	Vidalicet; Namely
VOCs	Volatile organic compounds
vol. (s)	Volume(s)
w.r.t.	With Respect To
WE	Water Extraction

WHO	World Health Organization
WPI	Whey Protein Isolate
XPS	X-ray Photoelectron Spectroscopy
XRD	X-Ray Diffraction
ZECC	Zero Energy Cool Chamber
Ψl	Leaf Water Potential

Part I
From Field to Future: Cutting-Edge Advances in Agri-Health Systems

Chapter 1
Development of Roselle (*Hibiscus sabdariffa*) Calyces Jelly: A Functional, Nutrient-Rich Food Innovation

Poonam Pramod Nichat, Chhavi Taliwal, Shireesha Mukka, and Maheshwar Mekam

Abstract *Hibiscus sabdariffa*, commonly known as roselle, is a versatile edible crop employed worldwide in food preparation and herbal remedies, where its calyces provide exceptional antioxidant capacity among leaves, fruits, roots, and seeds. This investigation formulated four types of roselle calyx jelly: dried variants 1 (76.19% calyces, 4.76% agar, 19.05% sugar) and 2 (70.80% calyces, 2.65% agar, 26.55% sugar), mirrored by fresh equivalents. Panelists rated Sample 2 highest in both dried and fresh forms for taste and consistency. These functional jellies, boast superior vitamin C, phenolic antioxidants, protein, and fiber with minimal calories and absent fat, ideal for health-focused consumers.

Keywords Antioxidant · Jellies · Roselle calyces · Standardization · Total phenol

1.1 Introduction

Roselle (*Hibiscus sabdariffa L.*) is a versatile plant from the Malvaceae family, popularly consumed in India and globally, earning the moniker "*poor man's food*" due to its affordability (Zaharah et al. 2017). The plant is adaptable and originates from tropical areas (Riaz et al. 2021) in India, the harvesting of roselle calyces can take place between the months of October and January. Calyces are handpicked.

The calyces are the succulent sepals that encircle the flower and are especially abundant in vitamins, minerals, and bioactive substances (Patel 2014). The commercially significant component of the plant is the fleshy calyx, which consists of the sepals encasing the fruit, known as capsules. There is an abundance of pigments

P. P. Nichat · S. Mukka · M. Mekam
MYAS, ICMR-National Institute of Nutrition, Hyderabad, Telangana, India

C. Taliwal (✉)
Amity Institute of Food Technology, Amity University, Noida, India
e-mail: ctaliwal@amity.edu

M. Thakur, T. Belwal (eds.), *Food Biodiversity and Consumer Engagements*, Sustainable Development and Biodiversity 46,
https://doi.org/10.1007/978-981-95-7777-4_1

Fig. 1.1 Mature, harvested, and dried roselle calyces and extract

found in roselle, which produces a red hue which act as an important provider of antioxidant properties. The phenolic compounds found in the roselle calyces naturally serve as antioxidants due to the presence of hydroxyl groups (Wu et al. 2018).

Additionally, some of these compounds can indirectly stimulate the production of endogenous antioxidants within biological systems. These compounds possess redox properties that inhibit the inactivation of metals, mitigate the scavenging action of free radicals, facilitate the decomposition of peroxides within cells, and diminish oxidative stress (Thakur and Modi 2020). These calyces additionally possess certain medicinal attributes.

The roselle calyces are recognized as a component of traditional medicine. Numerous studies have indicated that the extracts derived from these calyces possess specific compounds that exhibit antioxidant and anticancer properties. These attributes are crucial in mitigating chronic diseases such as hypertension, diabetes mellitus, cardiovascular conditions, and dyslipidemia. Consumption of these calyces has proven to enhance the immune system, decrease inflammation, and improve heart health. Additionally, Roselle is explored for its potential to enhance lactation and aid in weight loss (Hapsari et al. 2021).

According to reports, roselle calyces can be utilized as a food coloring or flavoring for sauces, jellies, marmalades, and soft beverages (Shruthi and Ramachandra 2019). Roselle is said to be a good and promising source of naturally occurring red colorants that dissolve in water. Natural colors that are soluble in water are called anthocyanins or flavonoids. Because anthocyanins are present in roselle or *Hibiscus sabdariffa* L. calyces aqueous extracts, they have a bright red color that makes them suitable for use as coloring agents in medicinal goods (Fig. 1.1). Additionally, it has been observed that the decoction of roselle calyces exhibits stability under heat. The plant's calyces can be utilized to create a beverage; the dried calyces may be steeped in water to produce a vibrant cold drink or boiled to serve as a hot infusion (Wu et al. 2018).

Furthermore, these calyces offer the benefit of being able to be integrated into various food items like syrups, jams, and jellies (Islam et al. 2016). Roselle calyx jelly has been attracting more attention due to its nutritional benefits. Rich in vitamin C, anthocyanins, and other phytochemicals, roselle calyces are responsible for their bright red color (Juhari et al. 2018).

The production of roselle calyx jelly consists of multiple stages, which include extraction, concentration, and gelling. During the extraction phase, the goal is to extract the necessary phytochemicals and nutrients from the calyces, while the concentration step is crucial for achieving the right solids content for jelly production. Finally, the gelling process usually requires the incorporation of a gelling agent, like pectin or agar, to achieve the desired texture and consistency (Halim et al. 2022).

Furthermore, the creation of roselle calyx jelly offers a chance to broaden the use of this adaptable plant and aid in the expansion of the food industry. With the ongoing increase in the global population and the need for nutrient-rich foods, the production of roselle calyx jelly can play a role in addressing these issues by offering a healthy and environmentally friendly food choice (Salami and Afolayan 2021).

The development of roselle calyces jelly represents a sustainable approach to food production by leveraging the nutritional and functional properties of this high-value crop. Roselle is rich in vitamin C, anthocyanins, and antioxidants, making it ideal for creating a healthful and flavorful jelly. The natural acidity and pectin content of the calyces reduce the need for added preservatives or gelling agents, contributing to a cleaner, more eco-friendly product. This approach supports sustainability by encouraging the cultivation of Roselle, particularly in regions where it thrives, thus reducing the carbon footprint associated with transportation and supporting local economies. Additionally, the minimal processing required to produce Roselle jelly helps preserve its natural nutrients while reducing energy consumption. With growing consumer interest in functional foods and natural ingredients, Roselle jelly appeals to health-conscious individuals seeking nutrient-rich alternatives. This product also has the potential to support small-scale farmers and create new business opportunities in rural areas, contributing to sustainable livelihoods. Beyond jelly, the versatile nature of Roselle calyces offers opportunities for innovation in other food products, enhancing their market potential.

1.2 Methodology

The production of roselle calyx jelly consists of multiple stages, which include extraction, concentration, and gelling. The production process of dried and fresh roselle calyx jelly was done. All raw materials used in these experiments were purchased from local supermarkets.

It started by weighing all the raw materials according to the formulation. The goal is to develop a healthy jelly that will contain maximum nutrients.

1.2.1 Formulation Standardization, Optimization, and Development Recipe of Both Products, i.e., Fresh and Dried Roselle Calyces Jelly

1.2.1.1 Preparation of Extract of Dried Roselle Calyces

For dried roselle calyces jelly, 150 g of water was taken, and dried roselle calyces were soaked overnight at a room temperature of about 26.0–32 °C. Next day-soaked calyces were squeezed and filtered to impart better color to roselle extract, which was used for jelly preparation. This extract was used to prepare two samples of jelly, which were named dried samples 1 and 2.

1.2.1.2 Preparation of Extract of Fresh Roselle Calyces

For fresh roselle calyces jelly. Three hundred milliliters of water were taken and 30 g of fresh roselle calyces were added to it. These calyces were boiled at 100 °C for 5 min. Later separated using a strainer (Fig. 1.2). This extract was used to prepare two samples of jelly, which were named as fresh samples 1 and 2.

1.2.1.3 Procedure of Preparation of Jelly

In Sample 1 of dried and fresh roselle calyces jelly, 5 g agar and 20 g of sugar were added, whereas in Sample 2 of dried and fresh roselle calyces jelly 3 g of agar and 30 g of sugar were added, respectively.

Fig. 1.2 Extraction process of roselle calyces: soaking, boiling, and final extract preparation

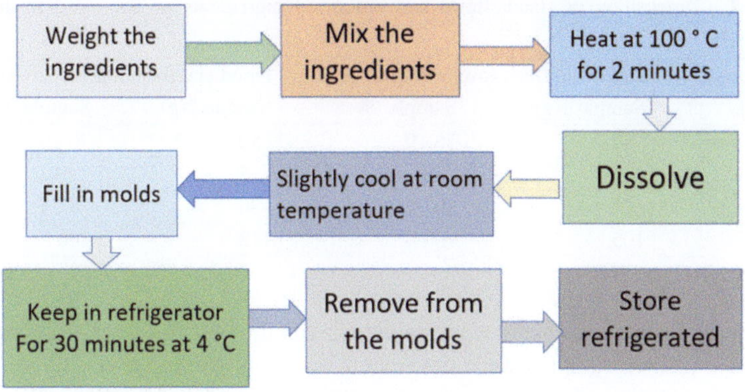

Fig. 1.3 Flow chart of development of product dried and fresh roselle calyces jelly

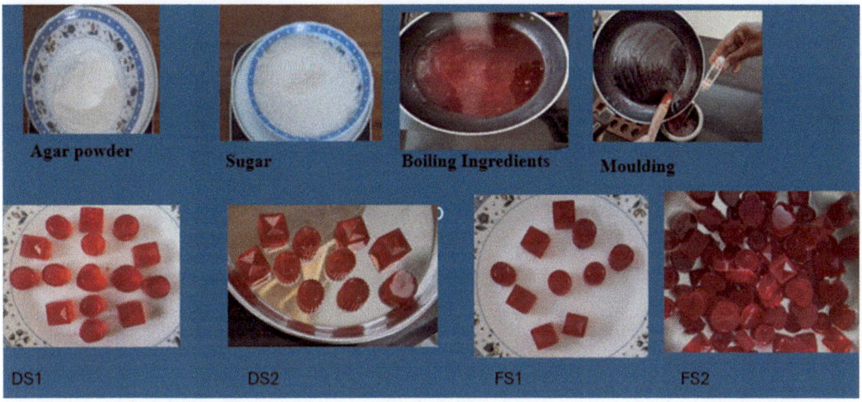

Fig. 1.4 Jelly preparation from roselle calyces: boiling with agar and sugar, molding, and final dried and fresh jelly samples

All ingredients were allowed to dissolve on heating for 2 min at 100 °C. Later, this liquid was filled into molds, cooled slightly to room temperature & stored in a refrigerator at 4 °C for 30 min. All formed jellies were removed from molds and stored in the refrigerator.

Figure 1.3 shows the procedure for jelly preparation, which is summarized as follows: All the ingredients were weighed and mixed together. This formulation was heated on the stove at a temperature of 100 °C for 2 min, long enough for the complete dissolution of the ingredients. After complete homogenization, the formulation was transferred to different shapes of tray molds, resulting in gums of approximately 5 g. These trays remained at room temperature for about 30 min until cooling to achieve an equilibrium with the kitchen temperature. Then, they were placed in the refrigerator (+4 °C) for 30 min. After 24 h, they were removed from the molds and stored in a closed container that was kept in the refrigerator until further analysis (Fig. 1.4).

Table 1.1 Composition of standardized and optimized ingredients of dried and fresh roselle calyces jelly

Ingredients	Dried and fresh roselle calyces jelly		Dried and fresh roselle calyces jelly	
	Sample 1(g)	Sample (%)	Sample 2(g)	Sample 2(%)
Rosella extract	80	76.19	80	70.80
Agar	5	4.76	3	2.65
Sugar	20	19.05	30	26.55
Total	105	100	113	100

Hence, four combinations were considered, with the best characteristics presented in Table 1.1. Two samples were developed by using dried roselle calyces (Dried Samples 1 and 2), and the other two samples were made of fresh roselle calyces (fresh Samples 1 and 2). In Samples 1 and 2, the concentration of roselle calyces was 76.19% and 70.80%, agar content was 4.76% and 2.65%, and sugar content was 19.05% and 26.55%, respectively.

The standardized and optimized ingredients composition in the dried and fresh roselle calyces jellies for Samples 1 and 2 included roselle extract at 76.19% and 70.80%, agar at 4.76% and 2.65%, and sugar at 19.05% and 26.55%, respectively.

1.2.2 *Proximate Analysis of Both Dried and Fresh Roselle Calyces Jelly*

Proximate analysis of the optimized samples was determined according to the AOAC & FSSAI Lab Manual 10. Moisture and ash contents were estimated by using the AOAC methods. Crude protein was estimated by using the Kjeldahl method, and crude fat was estimated using FSSAI guidelines. Carbohydrates were determined using the difference method. Energy values were calculated by the standard method of summing up the values of total carbohydrates, crude protein, and crude fat obtained, and multiplying the quantity of carbohydrate and protein per 100 gm by 4 kcal and that of fat per 100 gm by 9 kcal, respectively (Gupta et al. 2021a).

1.2.3 *Estimation of Ascorbic Acid Content*

Ascorbic acid is a water-soluble, fundamental, low-molecular antioxidant that operates within the human body. It plays a crucial role in managing the levels of reactive oxygen species (ROS) and enhancing the activity of other antioxidants. Ascorbic acid influences ROS levels right from the moment they are generated (Gegotek and Skrzydlewska 2022). The ascorbic acid content in the jellies and the calyxes was determined by the Dichlorophenol Indophenol Method (Tiwari 2015).

The following material is required:

- *4% Oxalic acid*: 4 gm of Oxalic acid is dissolved in 100 ml of distilled water.
- *Dye solution*: 26 mg of dye and 21 mg of sodium carbonate are individually weighed and made up to 100 ml with distilled water.
- *Ascorbic acid stock standard: (1 mg/mL)*: 100 mg of ascorbic acid is weighed into a volumetric flask and the volume is brought to 100 mL with 4% oxalic acid.
- *Ascorbic acid working standard: (100 μg/mL)*: To 10 mL of the ascorbic acid stock standard solution taken in a volumetric flask, dilute it to 100 mL with 4% oxalic acid.
- *Sample extraction*:

  Weigh 10 g of each sample individually; crush them and finely grind the sample, then make up to 100 ml with 4% oxalic acid. A 1:10 dilution of this solution is made with 4% oxalic acid. Five milliliters of this diluted sample volume is used for titration.

  Ascorbic acid reduces 2,6-dichlorophenol indophenols, a blue colored dye, into a colorless leuco-dye under acidic conditions. Ascorbic acid undergoes oxidation to dehydro-ascorbic acid, and the dye gets reduced. Although the dye is blue, the appearance of pale pink indicates the endpoint of titration. The DCPIP dye itself acts as a self-indicator; there is no need to add any additional indicators (Tiwari 2015).

  To 5 ml of ascorbic acid working standard (0.1 mg/ml) taken in a conical flask, 10 ml of oxalic acid was added, vortexed, and the contents were titrated against DCPIP dye solution taken in a burette until the pink color endpoint is observed. The color persists for a few minutes. Immediately record the titer value. Similarly, 5 ml of each test sample was titrated against the DCPIP solution, and burette readings were recorded.

Calculation

$$\text{Amount of ascorbic acid content (mg/100 g)}$$
$$= \frac{(V_2)\ \text{Titer value of the test} \times \text{concentration of standard} \times 10 \times 10 \times 100}{(V_1)\ \text{Titer value of the standard} \times 5}$$

where

V_2 = volume of dye consumed by 5 ml of test sample
V_1 = volume of dye consumed by 5 ml of standard ascorbic acid
500 = concentration of standard ascorbic acid in μg
10 = 10 gm of the sample
10 = 1:10 dilutions
100 = ascorbic acid content/100 gm of sample
5 = volume of test sample taken for titration

1.2.3.1 Estimation of Total Phenol Content

TPC activity involves determining the level of phenolic content present in samples. The phenolic compounds found in plants possess redox properties, enabling them to function as antioxidants (Johari and Khong 2019).

The TPC of jelly of roselle calyces and the extracts of fresh and dried calyces are determined by using the Folin–Ciocalteau procedure (Dragovic-Uzelac et al. 2010).

Reagents Required

1. Gallic acid (stock solution-1 mg/ml): weigh 100 mg of gallic acid into a volumetric flask and dissolve it in 40 ml of distilled water, then make up the volume to 100 ml.
2. Gallic acid (working standard- 400 μg/ml): transfer 40 ml of gallic acid stock solution and make up the volume to 100 ml with distilled water.
3. Folin–Ciocalteau (2N): 1:1 dilution: 50 ml of commercially available Folin–Ciocalteau, add an equal volume of distilled water, and make up the volume to 100 ml. This is to be prepared freshly.
4. Sodium carbonate anhydrous (10%w/v): weigh 10 gm of anhydrous sodium carbonate and dissolve it in 100 ml of distilled water.
5. Methanol (99.6%).

Roselle Samples Fresh roselle calyces, jelly, dried roselle calyces, fresh roselle calyces, dried roselle calyces, dried roselle calyces powder.

Sample Extraction

Weigh 10 gm of each of the samples and finely grind them, then the contents are transferred into a dark colored flask and mix in 200 ml of Methanol and vortexed at room temperature. After 24 h, the extracts were filtered with Whatmann No. 1 filter paper, and the residues were re-extracted with 200 ml of methanol. After 48 h. The supernatant was collected, and both supernatants were combined. The obtained extracts were appropriately diluted to 1 mg/ml and used to determine the total phenol content.

Procedure The Phenolic content reacts with Folin–Ciocalteu reagent to form a blue colored complex whose absorbance is measured at 765 nm. The Phenolic hydroxyl group of Gallic acid reacts with Folin-Ciocalteu reagent and reduces the heteropolyacid (phosphomolybdic-phosphotungstic acid) under alkaline conditions to form a blue colored molybdenum and tungsten complex, which can be measured at 765 nm.

For the determination of total phenolic content (TPC), gallic acid is one of the best standards among other phenolic standards in the Folin–Ciocalteu method as it contains three phenolic groups. The standard solution of gallic acid with a concentration ranging from 0.08 to 0.4 mg/ml was aliquoted into a series of tubes. Followed by the addition of 2.5 ml of Folin–Ciocalteu reagent, and the contents were mixed. Then, 2.5 ml of 10% anhydrous sodium carbonate solution was added to the tubes, and then vortexed the content of the test-tubes. The test tubes were incubated at room temperature for 5 min, and then the absorbance was read at 765 nm.

Similarly, the diluted methanol extracts of roselle jellies and the fresh and dried calyces with a concentration of 1 mg/ml were analyzed with the above reagents, and the absorbance was measured. To determine the phenolic content in the extracts, a standard curve was constructed with a concentration of gallic acid on the *x*-axis and absorbance on the *y*-axis, and the total phenol content in 100 gm of extract was calculated and expressed in gm GAE/100 gm of extract.

1.2.4 Sensory Evaluation

The sensory evaluation was conducted by using the nine-point hedonic test for the overall acceptability, appearance, color, texture, odor, and palatability of the four formulations of roselle jellies, which were presented coded so as not to influence their judgment. A scale of 1–9 was used in the hedonic test, with 1 being dislike extremely and 9 being like extremely (Gupta et al. 2021b). A total of 50 panel members, including students and faculty from the ET Division of ICMR-NIN, took part in the product evaluation process.

1.3 Result and Discussion

Jellies have gained widespread popularity across all age groups, and when crafted with wholesome ingredients, they not only deliver a delightful taste but also offer health advantages. Therefore, this current investigation involved the development of four recipes incorporating both dried and fresh calyces.

Aim to offer healthful jellies derived from natural plant sources, providing nutritional richness and health benefits without the use of chemical artificial colors, flavors, or preservatives, which can lead to various adverse effects

This chapter summarizes the standardized recipes, findings of the Study and discusses the implications of those findings. The chapter ends with suggested future research directions and concluding remarks.

1.3.1 Optimization and Standardization of Ingredients of Dried and Fresh Roselle Calyces Jelly

The formulation of standardized ingredients for jelly made from both dried and fresh roselle calyces in Samples 1 and 2 was optimized. In Sample 1, roselle extract constituted 76.19%, agar comprised 4.76%, and sugar made up 19.05%. In Sample

2, roselle extract constituted 70.80%, agar comprised 2.65%, and sugar made up 26.55%.

In a study of the development of roselle (*Hibiscus sabdariffa* L.) calyx jelly candies of different concentrations were studied. The concentrations of sucrose used were 63.56%, 70.62%, 77.68%, and 84.75%. The concentrations of dried roselle calyx used were 6.67%, 10.00%, 13.33%, and 16.67%, respectively (Halim et al. 2022).

This indicates the different concentrations of ingredients like roselle calyx extract, agar & sugar in roselle calyx jelly and different concentrations of ingredients such as sucrose and roselle calyx in roselle calyx candy, which were optimized and standardized in the processing of jelly developed different characteristics, which are crucial for sensory evaluation and acceptance.

1.3.2 Mean Sensory Scores of Rosella Calyces Jelly

The results of the hedonic test for a sensorial profile of the dried roselle calyces jellies are presented in Fig. 1.4, and they correspond to the average of the scores attributed by the 50 panelists to the different attributes in each sample.

Table 1.2 depicts the sensory features of the dried rosella calyces jellies developed. Sample 2 received higher ratings for taste attributes compared to sample 1. The absence of chemical flavoring agents resulted in a lower odor rating for sample 1 compared to sample 2. Regarding textural attributes, the jellies in sample 2 exhibited a more enhanced jelly-like texture compared to sample 1. The color of the jelly obtained high approval in both samples. Overall, the acceptability of the jellies was moderate for sample 2 compared to sample 1.

A study which was conducted on Monascus-fermented durian seed jelly drink: effect of roselle extract. Taste, color, and mouthfeel were evaluated. The average sensory evaluation scores for color were found to be 3.35–5.58, 2.10–5.05 for mouthfeel, and 2.60–5.37 for taste (Angela et al. 2021).

This indicates the different sensory characteristics such as taste, odor, color, texture, and overall acceptability in dried roselle calyces jelly between samples 1 and 2 and the preceding study in which taste, color, and mouth feel were evaluated

Table 1.2 Sensory profile of developed dried rosella calyces jelly of samples 1 and 2

	DS1	DS2
Parameters	Mean ± SD	Mean ± SD
Taste	5.8 ± 1.7	7.1 ± 1.2
Odor	5.5 ± 1.7	6.02 ± 1.3
Color	7.44 ± 0.9	7.8 ± 0.9
Texture	5.32 ± 1.7	7.12 ± 1.3
Overall acceptability	5.85 ± 1.5	7.02 ± 1.1

DS1 dried rosella calyces jelly sample 1, *DS2* dried rosella calyces jelly sample 2

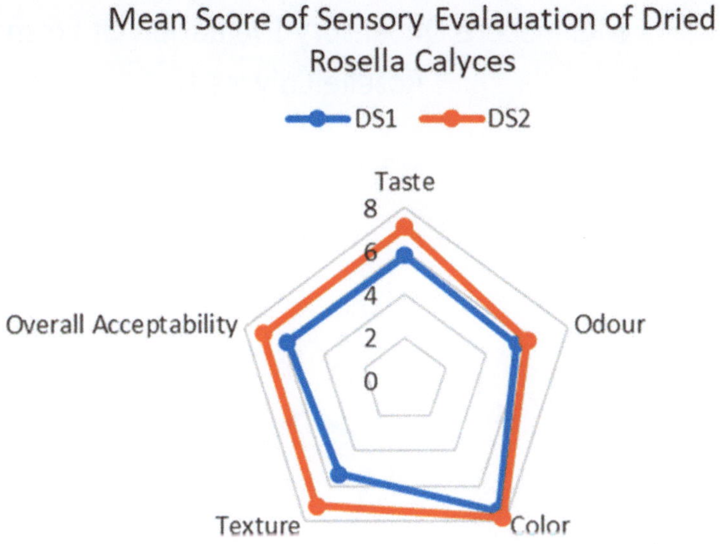

Fig. 1.5 Descriptive sensorial profiles of the two developed dried roselle jellies

Table 1.3 Sensory profile of developed fresh roselle calyces jelly of samples 1 and 2

	FS1	FS2
Parameters	Mean ± SD	Mean ± SD
Taste	7.02 ± 1.03	7.44 ± 1.07
Odor	7.02 ± 1.03	7.08 ± 1.27
Color	7.26 ± 1.00	7.64 ± 0.92
Texture	7.08 ± 1.49	7.24 ± 1.36
Overall acceptability	7.08 ± 1.00	7.28 ± 1.01

FS1 fresh rosella calyces jelly sample 1, *FS2* fresh rosella calyces jelly sample 2

in seed drink. All characteristics were highly accepted than odor in roselle calyces jelly otherwise overall acceptance in both studies was highly accepted.

The radar chart illustrates the sensory evaluation attributes of jelly made from dried roselle calyces. N = 50, Hedonic 9-point scale method (Fig. 1.5) (Table 1.3).

The characteristics were consistently rated for the two formulations under consideration. Both samples 1 and 2 of fresh roselle calyces jellies were moderately sweet. However, sample 2 exhibited a more pronounced jelly-like texture compared to sample 1 in terms of textural attributes. Since no chemical flavoring agents were incorporated, the product had no discernible odor. The color of the jelly received high approval. Overall, consumers found both samples 1 and 2 of fresh roselle calyces jellies to be acceptable (Fig. 1.6).

A study was conducted on the development of rose jelly. The parameters for sensory tests were color, aroma, texture, sweet taste, and overall acceptance. The average sensory evaluation scores for color were found to be 5.94, aroma was 5.06,

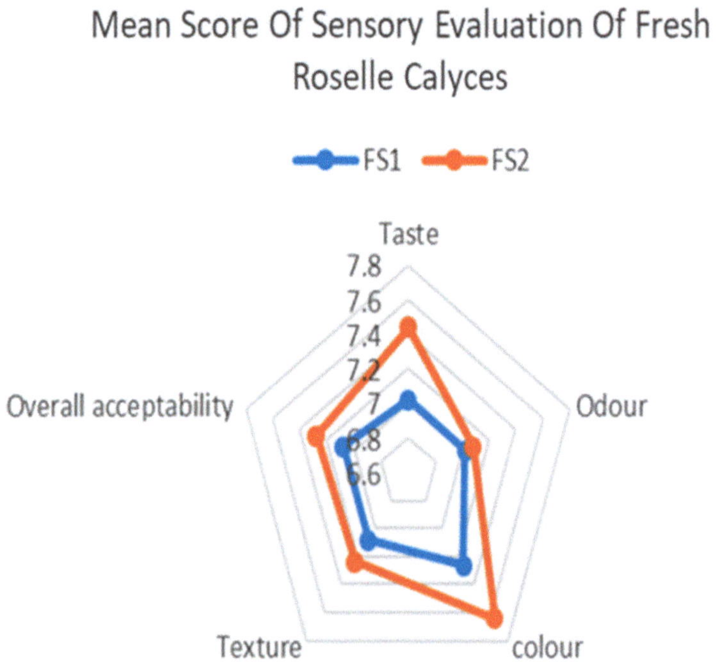

Fig. 1.6 Descriptive sensorial profiles of the two developed fresh roselle jellies

taste was 5.58, texture was 4.83, and overall acceptance was 5.60 (Franzen et al. 2020).

This indicates that roselle calyces jelly had better acceptability in all sensory characteristics than rose jelly.

1.3.3 Proximate Analysis of Roselle Calyces Jelly

The following tables show the proximate analysis of both dried and fresh calyce jelly samples (Table 1.4).

The moisture content of dried rosella calyces jelly in sample 2 measured 57.01 per 100 g, protein content was 1.02 per 100 g, ash content was 0.65 per 100 g, and total fat was absent. The determined dietary fiber was 0.44 per 100 g, carbohydrate content stood at 41.32 per 100 g, and the energy content was recorded at 169.36 kcal per 100 g. The absence of fat makes this jelly healthy for consumption.

An investigation was carried out using sun-dried roselle calyces. It was discovered that the moisture content was 8.6% & protein was roughly 5%. While ash and crude fat were found to be 5.3% and 1.6%, respectively. There was a significant amount of crude fiber, which was roughly 17.6%, and carbohydrates were found to be 60% (Amoasah et al. 2018).

Table 1.4 Proximate analysis of dried roselle jelly of sample 2

S No.	Parameters	Units	Mean ±SD
1	Moisture	g/100	57.01 ± 0.27[a]
2	Protein	g/100	1.02 ± 0.12[a]
3	Ash	g/100	0.65 ± 0.14[a]
4	Total fat	g/100	0
5	Dietary fiber	g/100	0.44 ± 0.16[a]
6	Carbohydrate	g/100	41.32 ± 0.22[a]
7	Energy	Kcal/100 g	169.36 ± 0.44[a]

Mean ± SD values of each parameter analyzed individually in triplicate
[a]Denotes a non-significant difference

Table 1.5 Proximate analysis of fresh roselle jelly of sample 2

S No.	Parameters	Units	Mean ±SD
1	Moisture	g/100	47.36 ± 0.24[a]
2	Protein	g/100	1.13 ± 0.17[a]
3	Ash	g/100	0.56 ± 0.12[a]
4	Total fat	g/100	0
5	Dietary fiber	g/100	0.47 ± 0.16[a]
6	Carbohydrate	g/100	50.95 ± 0.29[a]
7	Energy	Kcal/100 g	208.32 ± 0.29[a]

Mean ± SD values of each parameter analyzed individually in triplicate
[a]Denotes a non-significant difference

This indicates that the dried calyces had a higher protein content and a lower moisture content than the jelly. Dried roselle calyces had a higher ash content than jelly. On the other hand, fat was found to be lesser in dried roselle calyces and completely absent in jelly. The amount of fiber in dried calyces was higher than in its jelly. Additionally, dried roselle calyces had higher carbohydrate contents than jelly, and jelly had higher energy contents than dried calyces because sugar was added during the jelly's processing.

Table 1.5 shows a proximate analysis of fresh roselle jelly. The moisture content was 47.36 per 100 g, the protein content was 1.13 per 100 g, the ash content was 0.56 per 100 g, and total fat was absent. The determined dietary fiber was 0.47 per 100 g, carbohydrate content stood at 50.95 per 100 g, and the energy content was recorded at 208.32 kcal per 100 g.

A similar study found that the nutritional composition of *H. sabdariffa* fresh calyces was it showed that the moisture content of the calyces is 86.8%. The ash content revealed that the calyces had 1.09% ash. Crude lipid of the calyces was 0.90%. From the analysis, the fiber content of the calyces was 2.59%. The crude protein found in this research was 2.95%, obtained from the calyces (Shruthi et al. 2017).

This reveals that fresh calyces had higher amounts of moisture and ash content than roselle calyces jelly. Crude protein and fiber were found to be more in fresh calyces than in its jelly. Whereas fat was totally absent in jelly and present in lesser quantity in fresh calyces. Also, fresh roselle calyces jelly had high amounts of carbohydrates and energy due to the sugar content, which was added during the processing of the jelly.

1.3.4 Vitamin C Estimation

The vitamin C levels in fresh roselle calyces were measured at 66.17 mg per 100 g, which was the highest content compared to dried roselle calyces, which had 60.29 mg per 100 g. Dried roselle powder, on the other hand, contained 62.00 mg of vitamin C per 100 g (Fig. 1.7).

Table 1.6 shows that the jelly made from fresh roselle calyces had a vitamin C content of 58.82 mg per 100 g. In contrast, the jelly prepared from dried roselle calyces displayed a vitamin C content of 51.47 mg per 100 g.

According to Fig. 1.8, it was observed that the level of ascorbic acid concentration was greater in jellies made from fresh roselle calyces than in those prepared from dried roselle calyces. Additionally, fresh roselle calyces exhibited a higher

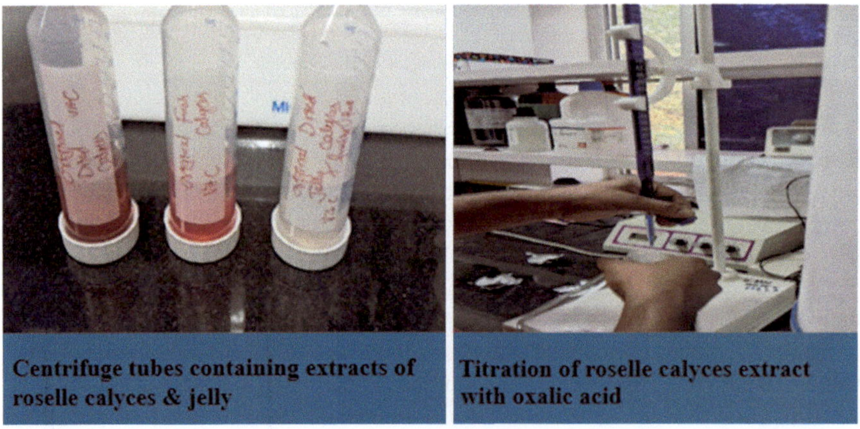

Fig. 1.7 Roselle calyces extract and jelly analysis: sample preparation and titration

Table 1.6 Ascorbic content mg/100 g

S.no	Roselle sample	Mean value (mg/100 g)[a]
1	Fresh roselle calyces	66.17 mg/100 g
2	Dried roselle calyces	60.29 mg/100 g
3	Dried roselle calyces powder	62.00 mg/100 g
4	Fresh roselle calyces jelly	58.82 mg/100 g
5	Dried roselle calyces jelly	51.47 mg/100 g

[a]Mean values of each parameter were analyzed individually in triplicate

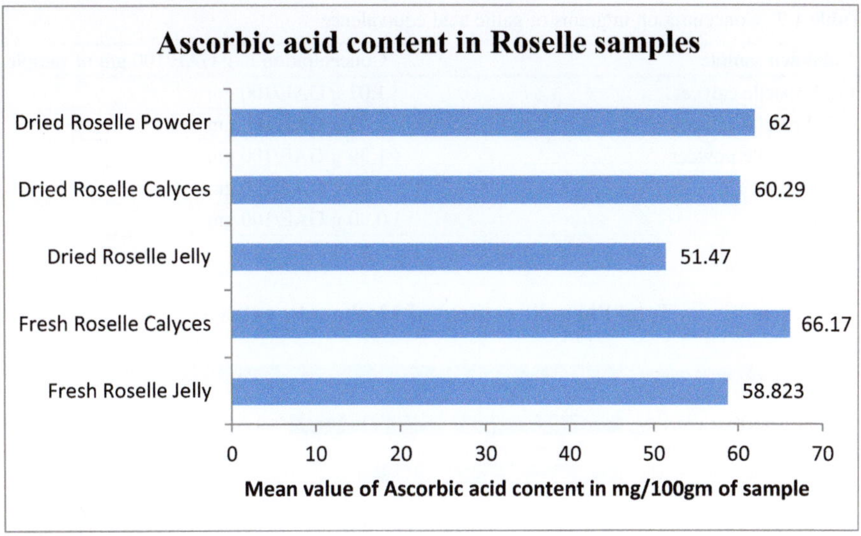

Fig. 1.8 Content of vitamin C presented through graph

ascorbic acid content compared to both dried roselle calyces and dried roselle calyces powder.

In the previous study done by Babalola on roselle calyces, the ascorbic acid content was found to be 54.8 mg 100 g (Babalola et al. 2001).

In the previous study done by Puro et al. (2017) in roselle calyces, the ascorbic acid content was found to be 0.801 mg g^{-1}.

This indicates a moderate amount of ascorbic acid is found in fresh roselle calyces, whereas a high amount of ascorbic acid is found in guava jelly, which is comparatively higher than roselle calyces.

1.3.5 Total Phenol Content

Analysis of total phenol content was carried out on jelly made from both fresh and dried roselle calyces. Additionally, the total phenol content in g GAE of fresh roselle calyces and dried calyces was examined. The greatest concentration of total phenol content in g GAE/100 gm of sample was observed in the dried roselle calyces.

Table 1.7 indicates the concentration in grams of gallic acid equivalence per 100 g of sample. The total phenol content in fresh and dried roselle calyces was determined to be 1.07 g GAE/100 gm and 1.25 g GAE/100 gm of sample, respectively, whereas dried roselle calyces powder exhibited a total phenol content of1.39 g GAE/100 gm of sample. Fresh roselle jelly displayed a total phenol content of 0.86 g GAE/100 gm of sample, while dried roselle calyces jelly had 0.90 g GAE/100 gm of sample.

Table 1.7 Concentration in grams of gallic acid equivalence

Unknown sample	Concentration in g GAE/100 gm of sample
Fresh roselle calyces	1.07 g GAE/100 gm
Dried roselle calyces	1.25 g GAE/100 gm
Dried roselle powder	1.39 g GAE/100 gm
Fresh jelly	0.86 g GAE/100 gm
Dried jelly	0.90 g GAE/100 gm

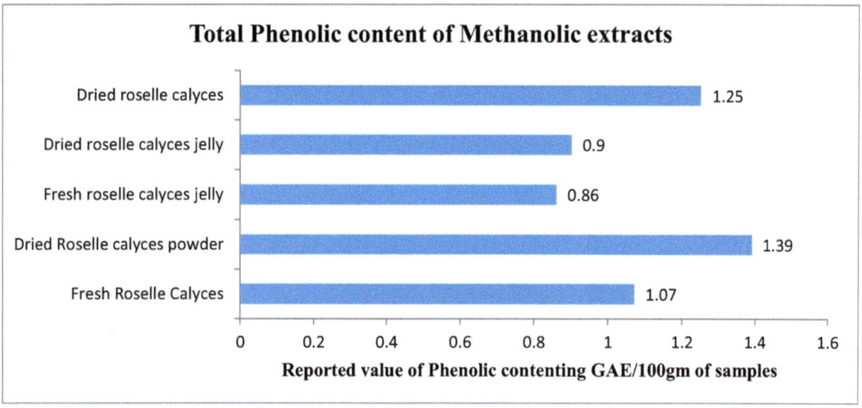

Fig. 1.9 Total phenolic content of methanolic extracts

According to Fig. 1.9, it was noted that the total phenolic content (TPC) was slightly higher in the jelly made from dried roselle calyces compared to the jelly prepared from fresh roselle calyces. Moreover, the TPC in dried roselle calyces was found to be higher than that in fresh roselle calyces, indicating an overall higher TPC in the dried samples compared to the fresh ones.

A similar study of the total phenolic content of roselle calyces was conducted by a similar method. The total phenolic content of *H. sabdariffa* calyx was found to be 41.07 mg gallic acid/g (Ahmed et al. 2014). The study conducted on the usage of red beet and roselle jelly showed a content of 520 mg (GAE/100 g) of total phenol content.

In the previous study done by (Altun and Aydemir 2022). It was also revealed that TPC ranged from 0.6 to 38.2 mg GAE/100 g, and the mean value was determined as 11.08 mg GAE/100 g in royal jelly. Puro et al. (2017) also analyzed the biochemical parameters of the roselle calyces. They found that the total phenol content of methanolic extract was measured at 7.12 mg GAEg-1. In the study of Ben Rejeb et al. (2020), the TPC content in the citrus jellies ranged between 123.16 and 192.76 mg GAE/100 g.

1.3.6 Nutritional Value of Developed Roselle Jellies Compared to Their Commercially Available Counterparts

Table 1.8 represents the nutritional value of developed roselle calyces jelly and commercial jellies. The nutritional content of developed roselle jellies was compared to that of commercially available jellies per 100 g. The moisture content in dried roselle calyces jelly was 57.01 g, whereas in fresh roselle calyces jelly, it was 47.36 g. In comparison, commercially available jelly had a moisture content of 47.36 g. The protein content in dried roselle calyces jelly was 1.02 g, and in fresh roselle calyces jelly, it was 1.13 g. Commercially available jelly, on the other hand, had no detectable protein content. Fat was not present in any of the jelly types. The dietary fiber in dried roselle calyces jelly was 0.44 g, and in fresh roselle calyces jelly, it was 0.47 g. The carbohydrate content in dried roselle calyces jelly was 41.32 g, while in fresh roselle calyces jelly, it was 50.95 g. Commercial jellies, however, had a high carbohydrate content of 85 g. The energy content in dried roselle calyces jelly was 169.36 kcal, in fresh calyces jelly it was 208.32 kcal, and in commercial jellies, it was significantly higher at 346 kcal.

This showed that dried roselle calyces jelly had more amount of moisture than fresh roselle calyces jelly and commercially available jelly. Protein and dietary fiber content were present in both dried and fresh calyces jelly, whereas it was totally absent in commercially available jelly. Fat was absent in all types of jelly. Carbohydrate and energy content was highest in commercially available jelly than both dried and fresh calyces jelly.

Table 1.8 Nutritional value of developed roselle calyces jelly compared with commercial jelly

Nutritional value of developed roselle jellies compared to their commercially available counterparts			
Nutrients per 100 g			
	Dried roselle calyces jelly	Fresh roselle calyces jelly	Commercially available jelly
Moisture (g)	57.01	47.36	47.36
Protein (g)	1.02	1.13	0
Total fat (g)	0	0	0
Dietary fiber (g)	0.44	0.47	0
Carbohydrate (g)	41.32	50.95	85
Energy (kcal)	169.36	208.32	346

1.3.7 Observation of Storage Stability of Roselle Calyces Jelly at Room Temperature

Table 1.9 indicates the storage stability of roselle calyces jelly at room temperature for 30 days. During storage of roselle calyces jellies, the changes in odor, pH, color, flavor, and fungal growth of roselle jellies were observed at 10-day intervals up to 30 days at room temperature. It was observed that the roselle jellies did not show any remarkable change up to 30 days of storage. After a period of 30 days, the sample calyces jelly sample (FS1) developed an undesirable flavor and odor. Fungal growth was also seen.

Studies on the processing and preservation of dragon fruit jelly were conducted. Storage stability was experimented on dragon fruit jelly. Color, flavor, and turbidity were observed. It was found out that at the fifth month, color and off flavor was observed (Islam et al. 2012). Another study on the preparation and storage of jelly from dragon fruit (Hylocereusundatus) was conducted. Parameters like moisture, acidity, reducing sugar, total sugar, phosphorus, and ascorbic content were monitored for a 3-month duration. Microbial growth (103 cfu/ml) was observed from the few days of production, and it was observed till the end of third month duration. It was observed that microbial growth was at its peak level both for ambient and refrigerated temperatures at the end of third month.

This indicates that dragon fruit jelly and roselle calyces jelly had better storage stability, which was conducted for an interval of 5 months and 3 months, respectively, than the preceding dragon fruit jelly study, which was conducted for 3 months.

Table 1.9 Observation of storage stability of roselle calyces jelly at room temperature

Storage period (days)	Sample	Color	Flavor	Odor	Fungal growth
0	DS1	Dark red	Pleasant	No bad smell	Not visible
	DS2	Dark red	Pleasant	No bad smell	Not visible
	FS1	Dark red	Pleasant	No bad smell	Not visible
	Fs2	Dark red	Pleasant	No bad smell	Not visible
10	DS1	Dark red	Pleasant	No bad smell	Not visible
	DS2	Dark red	Pleasant	No bad smell	Not visible
	FS1	Dark red	Pleasant	No bad smell	Not visible
	Fs2	Dark red	Pleasant	No bad smell	Not visible
20	DS1	Dark red	Pleasant	No bad smell	Not visible
	DS2	Dark red	Pleasant	No bad smell	Not visible
	FS1	Dark red	Pleasant	No bad smell	Not visible
	Fs2	Dark red	Pleasant	No bad smell	Not visible
30	DS1	Dark red	Pleasant	No bad smell	Not visible
	DS2	Dark red	Pleasant	No bad smell	Not visible
	FS1	Dark red	Unpleasant	Bad smell	Visible
	Fs2	Dark red	Pleasant	No bad smell	Not visible

1.4 Conclusion

Conventional jellies available in the market typically provide limited nutritional benefits and are characterized by their high energy density, primarily due to elevated sugar content. In contrast, jellies made from roselle calyces are derived from plants and demonstrate lower energy and fat levels while also boasting higher concentrations of vitamin C and total phenolic compounds. Consequently, the innovation of jellies utilizing roselle calyces represents a promising alternative, presenting consumers with a healthier choice in comparison to traditional jelly candies.

The innovation of jelly made from roselle calyces represents a significant opportunity for harnessing the benefits of this nutrient-dense plant. By capitalizing on the distinctive attributes of roselle calyces, food producers can develop a flavorful and healthful product that may provide various health advantages for consumers.

References

Ahmed E, Hassan HM, Sirag N, Mm E, Algaili AM, Ohaj M (2014) Determination of total phenolic content and antioxidant activity of Roselle (*Hibiscus sabdariffa* L.) calyx ethanolic extract. Stand Res J Pharm Pharmacol 1(2):034–039

Altun SK, Aydemir ME (2022) Determination of antioxidant capacities with the phenolic and flavonoid contents of royal jelly mixtures. Bee Stud ARI 14(1):9–15

Amoasah B, Appiah F, Kumah P (2018) Effects of different drying methods on proximate composition of three accessions of Roselle (*Hibiscus sabdariffa*) calyces. Int J Plant Soil Sci 21(1):1–8

Angela J, Ristiarini S, Nugerahani I, Srianta I (2021) Development of Monascus-fermented durian seed jelly drink: effect of roselle extract concentration on physicochemical and organoleptic properties of the jelly drink. IOP Conf Ser Earth Environ Sci 794(1):012136

Babalola SO, Babalola AO, Aworh OC (2001) Compositional attributes of the calyces of roselle (*Hibiscus sabdariffa* L.). J Food Tech Africa 6(4):133–134

Ben Rejeb I, Dhen N, Kassebi S, Gargouri M (2020) Quality evaluation and functional properties of reduced sugar jellies formulated from citrus fruits. J Chem 2020:1–8

Dragovic-Uzelac V, Savic Z, Brala A, Levaj B, Kovacevic DB, Bisko A (2010) Evaluation of phenolic content and antioxidant capacity of blueberry cultivars (Vaccinium corymbosum L.) grown in the Northwest Croatia. Food Technol Biotechnol 48(2):214–221

Franzen FDL, Oliveira MSR, Menegaes JF, Gusso AP, Silva MN, Richards NSPDS (2020) Physico-chemical, microbiological and sensory characteristics of jellies made with rose and *Hibiscus* flowers. Braz J Dev 6(3):14828–14845

Gegotek A, Skrzydlewska E (2022) Antioxidative and anti-inflammatory activity of ascorbic acid. Antioxidants 11(10):1993

Gupta C, Khedkar R, Negi K, Singh K (2021a) Calcium enriched nutrimix flour supplement for lactating mothers: optimized by response surface methodology (RSM). Curr Res Nutr Food Sci 9(1):267–279

Gupta C, Khedkar R, Negi K, Singh K (2021b) Optimization and nutritional evaluation of traditional instant pancake flour for lactating mothers. Plant Arch 21(1):2112–2116

Halim Y, Evelyne C, Rosa D, Ramli S (2022) Development of Roselle (Hibiscus sabdariffa L.) calyx jelly candy. Caraka Tani J Sustainable Agric 37(2):357–372

Hapsari BW, Manikharda, Setyaningsih W (2021) Methodologies in the analysis of phenolic compounds in roselle (Hibiscus sabdariffa L.): composition, biological activity, and beneficial effects on human health. Horticulturae 7(2):1–36

Islam M, Khan M, Hoque M, Rahman M (2012) Studies on the processing and preservation of dragon fruit (Hylocereus undatus) jelly. The Agriculturists 10(2):29–35

Islam A, Jamini TS, Islam M, Yeasmin S, Hasan AK (2016) Fundamental and applied agriculture Roselle: a functional food with high nutritional and medicinal values. Fundam Appl Agric 1(2): 44–49

Johari MA, Khong HY (2019) Total phenolic content and antioxidant and antibacterial activities of *Pereskia bleo*. Adv Pharmacol Sci 2(2019):7428593

Juhari NH, Bredie WLP, Toldam-Andersen TB, Petersen MA (2018) Characterization of Roselle calyx from different geographical origins. Food Res Int 112:378–389

Patel S (2014) Hibiscus sabdariffa: an ideal yet under-exploited candidate for nutraceutical applications. Biomed Prev Nutr 4(1):23–27

Puro K, Aochen C, Ghatak S, Das S, Sanjukta R, Mahapatra KP, Jha AK, Shakuntala I, Sen A (2017) Studies on the therapeutic properties of Roselle (Hibiscus sabdariffa) calyx: a popular ingredient in the cuisine of North East India. Int J Food Sci Nutr 2(5):1–6

Riaz G, Naik SN, Garg M, Chopra R (2021) Phytochemical composition of an underutilized plant Sorrel/Roselle (Hibiscus sabdariffa L.) cultivated in India. Lett Appl NanoBioScience 10(2): 2138–2147

Salami SO, Afolayan AJ (2021) Evaluation of nutritional and elemental compositions of green and red cultivars of roselle: Hibiscus sabdariffa L. Sci Rep 11(1030):1–13

Shruthi VH, Ramachandra CT (2019) Roselle (Hibiscus sabdariffa L.) calyces: a potential source of natural color and its health benefits. Food Bioactives 2019:169–190

Shruthi VH, Ramachandra CT, Nidoni U, Hiregoudar S, Naik N, Kurubar AR (2017) Physicochemical, nutritional and functional properties of Roselle (Hibiscus sabdariffa L.). Int J Curr Microbiol App Sci 6(12):2976–2982

Thakur M, Modi VK (2020) Emerging technologies in food science—focus on the developing world. Springer Nature

Tiwari A (2015) Practical biochemistry: a student companion. LAP

Wu HY, Yang KM, Chiang PY (2018) Roselle anthocyanins: antioxidant properties and stability to heat and pH. Molecules 23(6):1367

Zaharah SS, Suhana PS, Shukor JA, Alam A, Aslani F (2017) Growth performance of roselle (*Hibiscus sabdariffa* L.) in response to oil palm by-product media and controlled release fertilizer. Bangladesh J Bot 46(1):427–431

Chapter 2
From Field to Future: The "Powerful Bar" as Future Agri-Health Innovation

Hardik Somaiya, Sneha Karadbhajne, Renuka Kulkarni, and Savita Zambre

Abstract This research introduces Powerful Bar, an innovative plant-based protein bar designed for a healthy and active lifestyle. Powerful Bar boasts a complete protein source derived from a unique blend of nature's powerhouses: pumpkin, sesame, chia, sunflower, and flax seeds. Each 50 g serving packs a powerful punch with 5.32 g of protein, 6.23 g of total fat, 10.23 g of dietary fibre, 200 kcal of energy, and a well-balanced profile of essential minerals, including 1.2 mg of iron, 550 mg of calcium, and 235 mg of phosphorus. Powerful Bar is meticulously formulated to cater to individuals across all age groups, addressing the growing demand for convenient yet nutritious on-the-go options. The full nutritional potential of these seeds, which are rich in all essential amino acids, vitamins, and minerals, is utilised. Unlike sugary snacks dominating the market, Powerful Bar boasts zero added sugars, making it a healthier and guilt-free indulgence. The convenient bar format enhances accessibility, allowing consumers to incorporate healthy choices into their busy lives without compromising on taste or satisfaction. Sensory evaluation using a 9-point hedonic scale with a scale ranging from 1 to 9, with 1 representing strong dislike, 5 indicating a neutral stance (neither like nor dislike), and 9 denoting strong liking, revealed high overall acceptability: a score of 8.58 even after 8 days of refrigerated storage, and a score of 8.19 without refrigeration. The bar upon texture analysis proved to be softer and easier to chew after storage, unlike major protein bars in the market that harden upon storage. This paper delves deeper into the nutritional composition, outlining the sustainable and responsible manufacturing processes behind Powerful Bar. The paper explores the potential

H. Somaiya
Laxminarayan Innovation Technological University, Nagpur, Maharashtra, India

S. Karadbhajne · S. Zambre (✉)
Department of Food Technology, Laxminarayan Innovation Technological University, Nagpur, Maharashtra, India

R. Kulkarni
Department of Food Technology, Lund University, Lund, Sweden
e-mail: renuka_sanjay.kulkarni.8448@student.lu.se

M. Thakur, T. Belwal (eds.), *Food Biodiversity and Consumer Engagements*, Sustainable Development and Biodiversity 46,
https://doi.org/10.1007/978-981-95-7777-4_2

health benefits associated with its unique ingredient blend, positioning Powerful Bar as a viable and delicious solution for a balanced and nutritious diet in today's fast-paced world.

Keywords Plant protein source · Complete protein · Naturally sweeten · Nutritional source · Convenience food · Sensory analysis

2.1 Introduction

The continually expanding global population and our fast-paced lifestyles, characterised by prolonged working hours and evolving dietary habits, have notably diminished the frequency and quality of traditional meals we partake in. This fosters an ongoing demand for convenient and sustainable food choices that are not only nourishing but also practical (Singh et al. 2024). A sustainable food system can be defined as one that ensures the provision of nourishing sustenance to fulfil present dietary requirements while concurrently safeguarding robust ecosystems capable of sustaining forthcoming generations, all with minimal adverse ecological consequences. This system fosters local production and distribution infrastructures, endeavours to make nutrient-rich nourishment universally attainable, accessible, and economically feasible, and operates in an equitable and ethical manner, safeguarding the well-being of farmers, workers, consumers, and communities (Thakur and Modi 2020). The intricate fabric of the food system is substantially intricate, driven by a multitude of economic, sociocultural, and environmental factors that extend both within and beyond its confines. The interwoven nature of these interactions accentuates the necessity for systematic approaches and comprehensive evaluation tools to steer transformative change. Furthermore, the surge in physical activity and engagement in sports as a means of upholding health necessitates safeguarding our bodies against the mechanical strain and oxidative stress induced by exercise. These free radicals, whose production escalates during physical exertion, can precipitate oxidative damage to crucial molecules within our bodies. Consequently, the market witnesses a surge in products such as high-protein bars, initially crafted to bolster muscle growth in athletes and later used by common people to achieve their daily protein goal (Marques et al. 2015). These bars furnish a convenient avenue for promptly replenishing energy and nutrients post-exercise, thus conserving invaluable time often lacking for meal preparation. The utilisation of cereal products in food formulations can increase the content of total dietary fibre and bioactive compounds such as phenolic acids and flavonoids, oligosaccharides, proteins, folates, sterols, vitamins, and minerals. Literature data have shown that oat bran is rich in fibre (6–8 g/100 g total dietary fibres). Also, among dietary fibre sources widely used for food enrichment, inulin represents an essential ingredient with a potentially beneficial effect on human health, mainly related to its ability to counteract constipation and promote microflora growth in the digestive tract. However, notwithstanding the extensive array of protein bars on offer, their formulations typically encompass 5–50 g of premium protein per 100 g of product alongside sugars, sugar alcohols such as glycerol and sorbitol, fats like palm oil, and a meagre quantity of water, complemented by a smattering of vitamins and minerals. Irrespective of the specific recipe, crafting protein bars remains a challenge due to intricate interactions among ingredients during blending (Lu and Tipper 2009). These interactions may precipitate a swiftly fluctuating sensory profile, yielding a bar with a fleeting shelf life. This poses a quandary when the bar solidifies or adopts a tough texture that is unpalatable to consumers. Such solidification may arise from a myriad of physical or chemical transformations during storage, including Maillard reactions (reactions between proteins and sugars that yield new flavours and aromas), sugar crystallisation, and molecular migration within the bar. Consequently, researchers and manufacturers must pivot towards modifying protein bar formulas to surmount these obstacles and ensure a product that retains its gustatory allure and desirability throughout its shelf life. But through this paper, a protein bar that does not adopt a tough texture upon storage but becomes increasingly likely with 8 days of storage was developed.

2.2 Preparation of Powerful Bar

The ingredients—flax seeds, pumpkin seeds, sunflower seeds, chia seeds, sesame seeds, rolled oats, dates, and unsweetened honey were bought from local markets in Nagpur, India. The dates were deseeded and then made into a fine paste in a grinding machine. The seeds and oats were roasted over a medium flame in a pan to remove their rawness and improve their flavours. The seeds, after roasting, were cooled and ground into coarse particles for ease in mixing. In a mixing bowl, roasted oats, honey, date paste, and coarse seeds were mixed to form a dough. On a stainless-steel plate, greased with a little coconut oil, the dough was placed and rolled over from the top to get a uniform-sized serving. The mixture was cooled at 6 °C to set the bar. With the help of a knife, cuts were made to get servings of 3 cm width and 6 cm length. Various variations of ingredients were used to get the desired product. Each serving weighed 50 g and had the following composition (Table 2.1).

Table 2.1 Composition of Powerful Bar

Ingredients	%
Dates	40–42
Oats	22–24
Pumpkin seeds	8–10
Sunflower seeds	8–10
Flax seeds	2–4
Sesame seeds	2–4
Chia seeds	2–4
Honey	2–4

2.2.1 Nutritional Analysis

Standard protocols were utilised to analyse the proximate and mineral compositions of all the variants.

2.2.1.1 Proximate Composition

The moisture content of each sample was assessed by drying 5 g of the sample in an oven at 105 °C for 16 h and then weighing it. The ash content in the various samples was determined by incinerating the samples in a muffle furnace at a temperature of 550 °C for 5 h. Carbohydrate content was calculated using the difference method (Pomeranz and Meloan 1994). The protein content was determined using the Kjeldahl method. Fat content was determined by extraction with petroleum ether using the Soxhlet apparatus, and the further free fatty acid value was calculated (Hyvönen 1996). Total fibre was estimated by digesting the fat-free samples in 1.25% sulphuric acid followed by 1.25% sodium hydroxide solution (Prosky et al. 1985). Water activity of the sample was measured by AMTAST Water Activity Meter (Rödel 2001). The total energy provided by Powerful Bar was assessed using a combustion chamber apparatus, employing the principle of direct determination of the volume of oxygen necessary to combust a specified weight of the substance (Jiang et al. 2014). This method involves calculating the potential energy of the substance based on established factors for the calorific value of a litre of oxygen. These factors are derived from previous combustion experiments conducted with similar materials in a bomb calorimeter.

2.2.1.2 Mineral Content

The iron, calcium, and phosphorus levels in the sample were determined by subjecting a measured quantity of the sample to ashing at 550 °C. Mineral content was evaluated using colorimetry using LMSPUV1200—Single Beam Spectrophotometer—LABMAN, and the titrimetric method. The colour development was facilitated by the addition of ammonium molybdate solution following the phosphovanado-molybdate colorimetric procedure for phosphorus. Iron content was assessed employing hydroxylamine hydrochloride, and the Fe (II) was subsequently measured spectrophotometrically as its coloured complex with α-α-dipyridyl, the solution being buffered with acetate buffer solution (Adams 1995). To ascertain the calcium content in the sample, the ash was treated with concentrated hydrochloric acid, then transferred to a volumetric flask and adjusted to a final volume of 50 ml. A portion of ash solution was utilised for determining calcium levels using the titrimetric method (Titchenal and Dobbs 2007).

2.2.1.3 Organoleptic Attributes

The sensory characteristics of Powerful Bar were assessed at ambient temperature in a well-illuminated, hygienic, and adequately ventilated environment. A panel of 11 semi-trained assessors scrutinised and appraised the product using a 9-point hedonic rating scale (Wichchukit and O'Mahony 2015). The degree of preference was articulated on a numerical scale ranging from 1 to 9, with 1 representing strong dislike, 5 indicating a neutral stance (neither like nor dislike), and 9 denoting strong liking. Participants were instructed to evaluate the Powerful Bar on its appearance, colour, taste, aftertaste, texture, and overall acceptability.

2.2.1.4 Texture Analysis

Utilising instrumentation for texture analysis represents an economically efficient and innovative methodology for elucidating the textural properties of a food item (Day and Golding 2016). The texture assessments of the three variants were conducted employing the TA. XTplus Texture Analyser, scrutinising parameters such as Hardness (g), Chewiness (g), and Cutting (g). Calibration settings employed encompassed the 5 kg load cell and a 25 mm cylinder probe. The measurement mode configurations for double-cycle compression were established with a test speed of 5.0 mm/s, while the probe height was meticulously adjusted to 5.0 mm above the sample. A trigger force of 5.01 g-f was meticulously chosen, and the samples were subjected to compression up to 50% of their original size.

2.2.1.5 Water Sorption Curve

Understanding the water sorption curve of protein bars is crucial in assessing their stability and shelf life. This curve represents the relationship between the equilibrium moisture content (EMC) of the bar and the surrounding relative humidity (RH). By plotting the EMC against RH, valuable insights into the bar's moisture uptake behaviour can be gleaned. The importance of this curve lies in its ability to predict the bar's response to changes in environmental conditions, such as storage humidity levels (Al-Muhtaseb et al. 2002). A well-defined water sorption curve aids in determining the bar's susceptibility to moisture-related issues, including textural changes, microbial growth, and flavour degradation. The interpretation of the curve involves analysing its shape, slope, and inflection points. A steep slope at low RH indicates rapid moisture absorption, while a plateau at high RH signifies saturation. Understanding these characteristics enables manufacturers to formulate protein bars with optimal moisture content for the desired texture and stability. Additionally, the water sorption curve assists in selecting appropriate packaging materials that can maintain the desired moisture levels and prevent moisture-related issues such as staleness, mould growth, or texture changes. For instance, if the water sorption curve indicates high moisture uptake at moderate humidity levels, packaging with low moisture permeability would be necessary to prevent moisture ingress. Overall, the water sorption curve serves as a valuable tool in protein bar formulation, quality control, and product development, ultimately contributing to consumer satisfaction and market success.

2.2.1.6 Microbial Analysis

The microbiological integrity of the protein bars underwent assessment utilising the serial dilution method to ascertain the total bacterial count on both the day of production and after an 8-day storage period. This meticulous technique entails generating a succession of increasingly diluted solutions derived from an initial sample. In our study, a dilution factor of 1:10 was implemented. Typically, a precisely measured volume, such as 1 mL, of the sample is aseptically transferred (devoid of any contamination) into a receptacle containing 9 mL of sterile diluent, often a buffered saline solution. This action yields a 1:10 dilution. The procedure is then iterated, with 1 mL from the preceding dilution being transferred to a fresh container holding 9 mL of diluent, thereby resulting in a 1:100 dilution, and so forth. This sequential dilution process facilitates the examination of a broad spectrum of bacterial concentrations present in the sample. After an incubation period conducted under appropriate conditions, the enumeration of colony forming units (CFU) ensues on agar plates previously inoculated with specific dilutions (Ben-David and Davidson 2014). The selection of dilutions is contingent upon the anticipated bacterial burden. Through the comparative analysis of CFU counts across dilutions, an approximation of the total bacterial count in the original sample can be derived. This method assumes paramount importance in safeguarding the microbiological integrity of comestible items such as protein bars. It affords the capability to detect and quantify potential bacterial contamination, which has the propensity to compromise product longevity and pose a risk to consumer well-being.

2.3 Results and Discussions

The samples are named as S1, S2, S3, S4, and S5. S1 refers to the sample that was freshly made. S2 and S4 refer to samples kept at room temperature (28 °C) and evaluated on days 3 and 8 of being made. S3 and S5 refer to samples kept in a refrigerator (6 °C) and evaluated on days 3 and 8 of being made. All the samples, while being stored, were carefully wrapped in aluminium foil.

2.3.1 Proximate Composition

A detailed evaluation of the Powerful Bar reveals it delivers 200 kcal per serving (50 g per serving), with a macronutrient profile of 5.32 g protein, 27.595 g carbohydrates (including 10.23 g dietary fibre and 17.365 g sugars), and 6.23 g fat. This composition characterises the bar as one of moderate protein content, with a primary focus on energy provision. Carbohydrates are the dominant source of calories, likely originating from ingredients such as oats, alongside natural sugars potentially derived from honey and dates. While the protein content offers a supply of essential amino acids, presumably sourced from a composite blend of various seeds and oats, it establishes the bar as a reasonably effective option for a protein-rich snack. The fat content, the lowest amongst the macronutrients, might suggest a prioritisation of calorie control. However, it likely still incorporates beneficial fats rich in polyunsaturated fatty acids (PUFAs) and monounsaturated fatty acids (MUFAs), while maintaining low levels of saturated fats, potentially derived from the inclusion of seeds. The exiguous free fatty acid (FFA) content of 0.27% measured in the Powerful Bar stands as a scientifically validated indicator of its superior compositional quality. From a biochemical standpoint, FFA levels serve as a critical barometer of lipid degradation and the overall product's freshness. A demonstrably low FFA value such as this signifies minimal breakdown of lipids within the bar, strongly suggesting that the constituent ingredients were exceptionally fresh at the time of incorporation and have undergone negligible oxidation or spoilage throughout the production and storage processes. This translates to a product that demonstrably adheres to stringent quality standards and regulations, ultimately ensuring that consumers can confidently expect a protein bar that delivers an optimal sensory experience in terms of taste and aroma, while simultaneously safeguarding its intended nutritional value. Examining Powerful Bar within the framework of a 2000 kcal daily diet recommended for Indian adults reveals its role in fulfilling macronutrient requirements. The bar provides 5.32 g of protein, meeting approximately 10.64% of the Recommended Dietary Allowance (RDA) based on the standard recommendation of 0.8 g of protein per kilogram of body weight for an average adult, and serves as a convenient protein boost throughout the day. The carbohydrate content stands at 27.595 g, equating to around 13.79% of the daily recommendation for a 2000 kcal diet. With fat content at its lowest, measuring 6.23 g, it contributes roughly 7% to the daily RDA. This hints at a potential emphasis on calorie control, and the fat sources being beneficial, such as those rich in polyunsaturated fatty acids (PUFAs) and monounsaturated fatty acids (MUFAs) provides good fats. Overall, the protein bar provides a moderate contribution to macronutrients within a 2000 kcal diet and serves as a useful snack choice for protein and carbohydrate intake. The evaluation additionally unveils a moisture content of 14.98% and a water activity (aw) of 0.61 at 28.5 °C. These values act as pivotal indicators of the product's stability, shelf-life, and overall quality. The moisture content signifies the water present in the bar, with 14.98% falling within acceptable ranges to maintain desirable texture and freshness. Meanwhile, the water activity of

Table 2.2 Shelf-life analysis of Powerful Bar

Parameters	Sample 1	Sample 2	Sample 3	Sample 4	Sample 5
Moisture (%)	14.98	15.17	15.02	15.38	15.19
Water activity (aW)	0.61	0.62	0.61	0.64	0.62

0.61 indicates the unbound water available for microbial growth, with lower values suggesting reduced microbial activity and a protracted shelf-life. While moisture and water activity play a pivotal role in assessing the quality of the Powerful Bar, it is notable that the moisture content of the bar has experienced a marginal increase while the water activity has remained more or less the same, indicating its stable shelf life of at least 8 days. The specifics of the moisture and water activity at 28 °C are outlined in Table 2.2.

2.3.2 *Mineral Composition*

The analysis of the protein bar reveals a total ash content of 2.07% and a notable mineral profile within the context of a 2000 kcal diet recommended for Indian adults. The results stated further are based on values provided by one serving of 50 g. It provides 1.2 mg of iron, meeting 15% of the Recommended Dietary Allowance (RDA). This indicates a moderate contribution towards daily iron needs, particularly beneficial for individuals vulnerable to iron deficiency, such as menstruating women or those with dietary restrictions limiting iron-rich sources like red meat. Additionally, the bar delivers a substantial 550 mg of calcium, satisfying 55% of the RDA. This is significant as calcium is pivotal for bone health and density. Furthermore, the bar offers 235 mg of phosphorus, fulfilling 33.57% of the RDA. Phosphorus works in tandem with calcium for bone health and is involved in various biological processes. While the protein bar doesn't single-handedly meet all daily mineral requirements, its provision of iron, calcium, and phosphorus is noteworthy, especially when integrated into a balanced diet. This mineral profile suggests the bar could be a valuable component of a dietary strategy to bolster bone health, prevent mineral deficiencies, and ensure optimal physiological function.

2.3.3 *Texture Analysis*

Texture analysis of the Powerful Bar was conducted for each sample, evaluating parameters such as hardness, chewiness, and cutting characteristics. The tests were performed in duplicate, along with ANOVA, Normality Shapiro–Wilk Test, and the results were meticulously recorded in Table 2.3 and Fig. 2.1 as the average of two readings. Observations revealed a gradual decrease in all parameters over the storage period of up to 8 days, indicating ongoing changes in texture, but W and p-value

Table 2.3 Texture analysis of Powerful Bar

Texture parameters	Sample 1	Sample 2	Sample 3	Sample 4	Sample 5	W	p-value
Hardness (g)	914.878	878.963	889.854	724.311	746.678	0.96	0.786
Chewiness (g)	−79.656	−71.368	−69.365	−49.586	−50.327	0.932	0.471
Cutting (g)	3.467	2.963	3.125	2.057	2.687	0.94	0.556

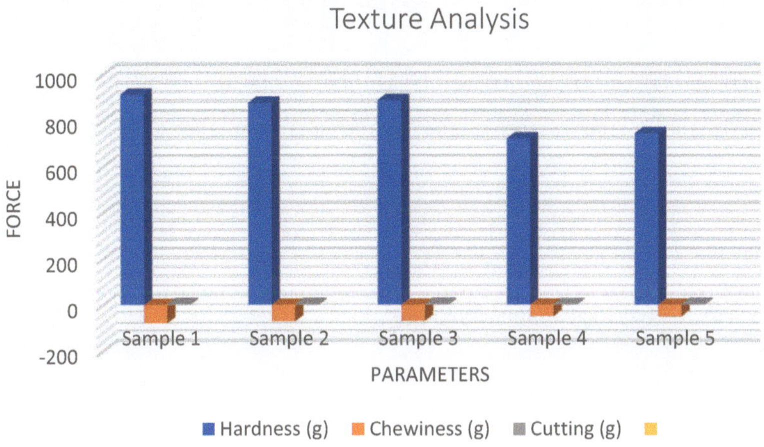

Fig. 2.1 Texture analysis comparison of Powerful Bar

indicate that the texture of the product did not deviate from normality. Notably, the overall acceptability of the Powerful Bar remained consistent throughout the 8-day storage period. Unlike many other protein bars, the Powerful Bar did not exhibit significant hardening over time, thereby achieving the intended goal of formulating a protein bar resistant to texture deterioration during storage. Samples 4 & 5 were particularly notable for their ease of chewing and absence of textural hardening. Furthermore, sensory evaluators reported a slight increase in sweetness, possibly attributed to the breakdown of complex sugars into glucose and its even distribution throughout the bar, thereby improving mouthfeel of Powerful Bar.

2.3.4 *Organoleptic Attributes*

An organoleptic evaluation was undertaken to assess consumer preference for the Powerful Bar across various storage conditions, with a focus on an 8-day timeframe. The results demonstrates average of generally favourable sensory attributes across all samples evaluated by the panellists. Sample 1 (S1) exhibited an overall acceptability score of 7.98, with comments highlighting a perceived hardness in texture. Interestingly, samples 2 (S2) and 3 (S3) displayed a marked improvement in overall

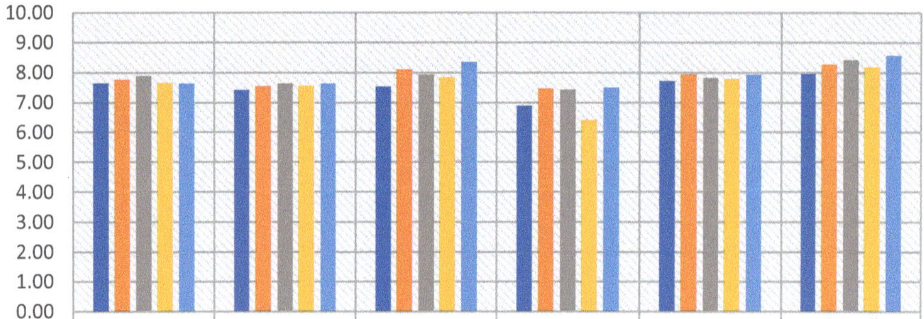

	Appearance	Colour	Taste	After Taste	Texture	Overall Acceptability
S1	7.66	7.44	7.55	6.91	7.74	7.98
S2	7.78	7.56	8.12	7.48	7.96	8.28
S3	7.89	7.63	7.93	7.42	7.82	8.41
S4	7.68	7.58	7.86	6.43	7.81	8.19
S5	7.65	7.65	8.36	7.52	7.95	8.58

S1 S2 S3 S4 S5

Fig. 2.2 Organoleptic analysis of Powerful Bar

acceptability, reaching scores of 8.28 and 8.41, respectively. Panellists attributed this enhancement to an improvement in mouthfeel, translating to a more enjoyable chewing experience and a heightened perception of taste. Notably, after 3 days of storage under both conditions, these positive changes were observed. Sample 4 (S4) demonstrated a slight decrease in overall acceptability, scoring 8.19. Here, comments pointed towards the presence of an undesirable aftertaste. However, it is noteworthy that S4 still offered a more favourable chewing experience compared to S2. Finally, sample 5 (S5) displayed the most significant positive shift, achieving an overall acceptability score of 8.58. Panellists remarked on improvements in mouthfeel, taste, and chewiness for S5. A similar ANOVA test, when conducted on organoleptic properties, provided with p-values as 0.482, 0.514, 0.528, 0.342, and 0.314, respectively, for S1, S2, S3, S4, and S5 (Fig. 2.2). These findings suggest that storage conditions can influence consumer perception of the Powerful Bar, with some variations emerging after 8 days.

2.3.5 *Costing Analysis*

A cost analysis was conducted to determine the production cost per serving of a 50 g Powerful Bar (Marouli and Maroulis 2005). The analysis utilized retail pricing data for raw materials as of February 2024 at Nagpur, Maharashtra. To account for processing costs, an additional 30% markup was applied to the total raw material cost. The raw materials themselves amounted to a cost of ₹9.33 per serving. Subsequently, factoring in the 30% processing cost markup, the final production cost per 50 g serving of the Powerful Bar reaches ₹12.13. This cost analysis provides valuable insights into the economic viability of the product at its current formulation and production methodology.

2.3.6 *Water Sorption Curve Analysis*

The water sorption curve, Fig. 2.3, was plotted for Powerful Bar, where initial and critical points were identified as yellow and green colour points, respectively. The danger point was found at 71.345% relative humidity. Since a steep increase is seen after 50% relative humidity, it can be concluded that Powerful Bar is a hygroscopic substance. The water vapour transmission of packaging material was $98.3 * 10^{-8}$ cc/cm^2/s/cm of Hg at 35 °C, hence high-density polyethylene with aluminium can be used as a packaging material with a large safety margin.

2.3.7 *Microbial Analysis*

Upon microbial analysis of S1 of Powerful Bar, just one count was observed in 10^{-1} dilution plate, which had a count of $1 * 10^1$ cfu. The analysis for S4 and S5 had counts of $3 * 10^2$ and $2 * 10^2$ cfu, respectively. The low microbial count after 8 days of storage in both conditions shows that Powerful Bar can be considered fit for human consumption.

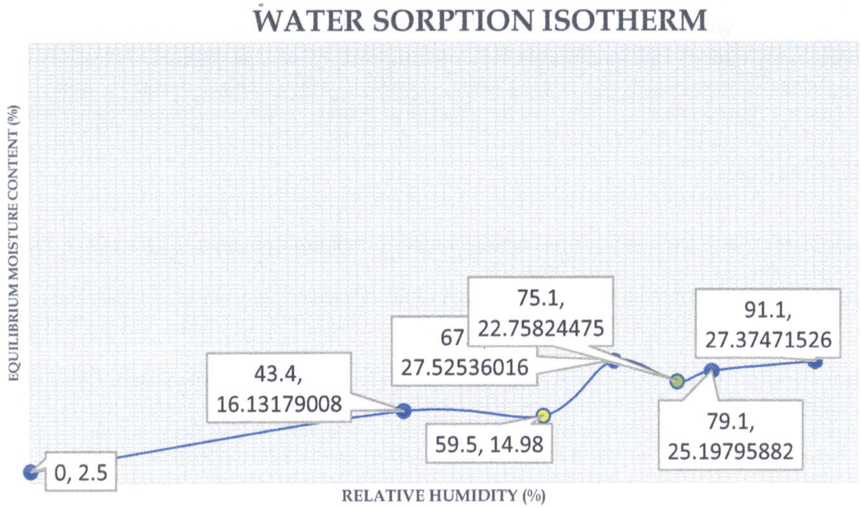

Fig. 2.3 Water sorption curve for Powerful Bar

2.4 Conclusion

A detailed analysis of the Powerful Bar reveals a uniquely formulated protein snack delivering a moderate protein content per serving. This protein boasts a complete essential amino acid profile attributable to a rich blend of various seeds incorporated into the bar's formulation. Additionally, the Powerful Bar provides a substantial amount of dietary fibre, clocking in at 10.23 g per serving. The seeds employed contribute a commendable quantity of polyunsaturated and monounsaturated fatty acids (PUFAs and MUFAs), classified as beneficial fats. These contribute to promoting cardiovascular health and offer a range of other health advantages. These combined attributes position the Powerful Bar as an ideal nutritional snack option suitable for all age groups. Microbiological analysis conducted on the eighth day of storage demonstrated excellent shelf life under both standard and refrigerated storage conditions. This suggests the Powerful Bar remains viable for consumption for up to 8 days. Furthermore, the implementation of improved packaging materials alongside refrigerated storage has the potential to extend this shelf life even further. The bar's mineral profile, exhibiting a decent presence of iron, calcium, and phosphorus, further bolsters its nutritional value. A unique characteristic of the Powerful Bar is its texture, which demonstrably softens and becomes easier to chew with storage duration. This stands in stark contrast to many commercially available protein bars, which tend to develop a hardened texture upon storage, leading to consumer dissatisfaction regarding texture and mouthfeel. Interestingly, the overall consumer acceptability of the Powerful Bar was observed to increase with storage time, signifying its potential for strong market adoption within both consumer and nutraceutical sectors. From a unit economics perspective, the production cost per 50 g serving is calculated at ₹12.13, making it a cost-effective product for manufacturing and distribution. In light of the findings from proximate analysis and mineral evaluation, the Powerful Bar can be confidently positioned as an exceptional protein-rich snack option suitable for consumption by all age groups.

References

Adams PE (1995) Determining iron content in foods by spectrophotometry. J Chem Educ 72(7): 649. https://doi.org/10.1021/ed072p649

Al-Muhtaseb AH, McMinn WAM, Magee TRA (2002) Moisture sorption isotherm characteristics of food products: a review. Food Bioprod Process 80(2):118–128. https://doi.org/10.1205/09603080252938753

Ben-David A, Davidson CE (2014) Estimation method for serial dilution experiments. J Microbiol Methods 107:214–221. https://doi.org/10.1016/j.mimet.2014.08.023

Day L, Golding M (2016) Food structure, rheology, and texture. In: Encyclopedia of food chemistry. Elsevier, pp 125–129. https://doi.org/10.1016/B978-0-08-100596-5.03412-0

Hyvönen L (1996) Approach to fat analysis of foods. Food Chem 57(1):23–26. https://doi.org/10.1016/0308-8146(96)00058-1

Jiang B, Tsao R, Li Y, Miao M (2014) Food safety: food analysis technologies/techniques. In: Encyclopedia of agriculture and food systems. Elsevier, pp 273–288. https://doi.org/10.1016/B978-0-444-52512-3.00052-8

Lu R, Tipper NC (2009) A portable device for the bioyield detection to measure apple firmness. Appl Eng Agric 25(4):517–523. https://doi.org/10.13031/2013.27455

Marouli AZ, Maroulis ZB (2005) Cost data analysis for the food industry. J Food Eng 67(3): 289–299. https://doi.org/10.1016/j.jfoodeng.2004.04.031

Marques TR, Corrêa AD, de Carvalho Alves AP, Simão AA, Pinheiro ACM, de Oliveira Ramos V (2015) Cereal bars enriched with antioxidant substances and rich in fiber, prepared with flours of acerola residues. J Food Sci Technol 52(8):5084–5092. https://doi.org/10.1007/s13197-014-1585-2

Pomeranz Y, Meloan CE (1994) Determination of moisture. In: Food analysis. Springer US, pp 575–601. https://doi.org/10.1007/978-1-4615-6998-5_34

Prosky L, Asp N-G, Furda I, Devries JW, Schweizer TF, Harland BF (1985) Determination of total dietary fiber in foods and food products: collaborative study. J AOAC Int 68(4):677–679. https://doi.org/10.1093/jaoac/68.4.677

Rödel W (2001) Water activity and its measurement in food. In: Instrumentation and sensors for the food industry. Elsevier, pp 453–483. https://doi.org/10.1533/9781855736481.2.453

Singh A, Sharma V, Grenier M, Singh V, Thakur M (2024) Nourishing the future: introduction to sustainable food systems with concepts and framework. Springer, Cham, pp 3–24. https://doi.org/10.1007/978-3-031-47122-3_1

Thakur M, Modi VK (eds) (2020) Emerging Technologies in Food Science. Springer Singapore. https://doi.org/10.1007/978-981-15-2556-8

Titchenal CA, Dobbs J (2007) A system to assess the quality of food sources of calcium. J Food Compos Anal 20(8):717–724. https://doi.org/10.1016/j.jfca.2006.04.013

Wichchukit S, O’Mahony M (2015) The 9-point hedonic scale and hedonic ranking in food science: some reappraisals and alternatives. J Sci Food Agric 95(11):2167–2178. https://doi.org/10.1002/jsfa.6993

Chapter 3
Harnessing Electronic Nose Technology for Quality and Safety in Sustainable Food Processing

Anamika Thakur, Alka Joshi, Sellam Perinban, and Valerie Orsat

Abstract Food plays a vital role in human life, and as times change, people's expectations and preferences regarding food have evolved significantly. Today, consumers are more conscious about aspects such as food safety, nutritional value, ingredients, brand, origin, and how the food is processed. All of these factors highlight the importance of reliable food testing technologies. Among the many available methods, the electronic nose (eNose) has emerged as a highly efficient, rapid, non-destructive, and promising technology in the field of food analysis. This paper begins by explaining the fundamental working principles and components of the eNose. It then delves into important technical aspects such as the choice of gas sensors, design of the sampling process, methods of data collection, and approaches to information analysis. The paper further reviews recent and notable uses of eNose technology across the food industry, focusing on six main areas: evaluating freshness, monitoring production processes, analyzing flavor, verifying authenticity, ensuring quality control, tracing the origin of products, and detecting pesticide residues. In conclusion, the paper addresses some of the major challenges that still hinder the wider adoption of eNose technology in food applications and outlines future research directions that could help overcome these issues and expand the technology's use in the industry.

Keywords Food safety · Non-destructive · Freshness · Monitoring production processes · Analyzing flavour

A. Thakur (✉) · A. Joshi · S. Perinban
ICAR-Indian Agricultural Research Institute, New Delhi, India

V. Orsat
Department of Bioresource Engineering, McGill University, Montreal, Canada

M. Thakur, T. Belwal (eds.), *Food Biodiversity and Consumer Engagements*, Sustainable Development and Biodiversity 46,
https://doi.org/10.1007/978-981-95-7777-4_3

3.1 Introduction

Human survival is fundamentally dependent on food. As the global population continues to grow rapidly, the demand for food has surged, driving significant expansion in the food industry. However, alongside economic progress and a thriving market, people's expectations regarding food quality have undergone notable changes. Consumers are now more vigilant about the presence of harmful substances in what they eat, leading to heightened concern for food safety and authenticity. Today, the food industry faces several critical challenges. These include ensuring the authenticity and safety of food products, enabling real-time monitoring throughout processing stages, evaluating the flavor of food items, and accurately assessing food quality. Addressing these concerns has become essential for maintaining public health and meeting modern consumer standards. The presence of problematic or contaminated food in the market can lead to serious health risks and economic consequences (Li et al. 2023). To prevent such outcomes, it is crucial to implement detection technologies that enable rapid, accurate, and real-time monitoring of food products. These technologies play a vital role in ensuring food safety, maintaining quality standards, and protecting consumer health, making their application highly significant in today's food industry (Loutfi et al. 2015). Food testing today generally falls into two main categories: subjective sensory evaluation and objective instrumental analysis. Subjective sensory evaluation involves assessing food quality based on human senses—such as sight, touch, and smell—typically carried out by trained panelists or experts with a refined sense of smell and taste. This method relies on personal perception and experience. On the other hand, objective instrumental testing uses scientific instruments to analyze the physical and chemical properties of food in a more standardized and accurate manner. One commonly used technique in this category is gas chromatography, which helps in identifying and quantifying the chemical compounds present in food products. This approach minimizes human bias and provides precise, reproducible results (Cascos et al. 2023; Wei et al. 2022).

During the mid-twentieth century, significant advancements were made in the theoretical understanding and practical application of various chemical sensors. These developments laid the foundation for the emergence of electronic nose (eNose) technology. By mimicking the human sense of smell using sensor arrays and pattern recognition systems, eNose technology evolved into a powerful tool for detecting and analyzing complex odors and volatile compounds, especially in the food industry (Fei et al. 2021). Fruits, as a key category of food, emit distinctive aromas formed by specific combinations and concentrations of volatile organic compounds (VOCs) (El Hadi et al. 2013; Le Phuc et al. 2020; Spadafora et al. 2016). These VOCs are influenced by various factors such as enzyme activity, substrate availability, temperature, and more (Zhang et al. 2011). Because of this, the VOC profile of a fruit can serve as a reliable indicator of its type, ripeness, geographic origin, and storage conditions (Shen et al. 2016). By analyzing the types and levels of these emitted compounds, it is possible to accurately identify the fruit and monitor its freshness or spoilage status (Hohmann et al. 2015). Earlier studies using eNose technology to analyze fruit VOCs typically employed traditional machine learning techniques, such as principal component analysis (PCA), linear discriminant analysis (LDA), and support vector machines (SVM), for feature extraction and classification (Chen et al. 2018). More recently, one-dimensional convolutional neural networks (1D-CNNs) have been introduced for VOC analysis (Khorramifar et al. 2021). These models use 1D convolutional layers to extract features and dense layers for classification, often achieving high accuracy (Xiong et al. 2014). However, there is ongoing debate in the research community regarding whether the convolutional layers of 1D-CNNs truly offer superior feature extraction compared to traditional, manual methods. In fact, artificial neural networks (ANNs), especially those using multi-layer perceptrons (MLPs) similar to the dense layers in CNNs, have already demonstrated high classification performance in eNose applications. Therefore, further investigation is needed to confirm if convolutional feature extraction adds significant advantages in this context (Esteki et al. 2018).

3.1.1 eNose System

An electronic nose (eNose) is an electronic sensing device specifically engineered to detect and differentiate complex odors or flavors, thereby emulating the human olfactory system. The core idea behind "electronic sensing" is the reproduction of human sensory capabilities through the synergistic use of chemical sensor arrays and sophisticated pattern recognition systems. These instruments are designed to perceive volatile organic compounds (VOCs), which are the chemical constituents responsible for odors and flavors. The conceptual groundwork for e-Nose technology dates back as early as 1982, marking the beginning of dedicated research efforts to develop systems capable of detecting and recognizing odors and flavors instrumentally. The first rudimentary E-nose, developed by Persaud and Dodd in 1982 at the University of Warwick, utilized three metal-oxide sensors that reacted to odor-producing substances, mirroring the activation of olfactory cells in the human nose, with results derived from the resulting signal patterns. An early practical application was seen in 1984 when the US Coast Guard employed eNose devices to identify volatile substances in emergency scenarios. The electronic nose (eNose) is a type of bionic instrument that has seen rapid development over the past 30 years (Munekata et al. 2023). It was originally designed to replicate the olfactory system of mammals, aiming to identify and distinguish between different gases and odors. The eNose works by detecting specific odor molecules dispersed in the air, using a sensor array that mimics the sensory receptors in a biological nose (Abu-Khalaf and Masoud 2022; Carrillo-Gómez et al. 2021). By analyzing the unique patterns generated by these sensors in response to various volatile compounds, the eNose can effectively recognize, classify, and monitor different gases, making it a powerful tool in applications ranging from food quality control to environmental monitoring. An eNose system (Electronic Nose) is a sophisticated device designed to detect and identify complex odors or volatile compounds, mimicking the human sense of smell. It plays a significant role in fields such as food quality control, environmental monitoring, medical diagnostics, and more. Over the subsequent decades, the field has witnessed substantial advancements, particularly in sensor materials, sensor array design, and, critically, the integration of artificial intelligence (AI) and machine learning (ML) algorithms. These computational enhancements have significantly improved the accuracy, reliability, and range of applications for eNose systems, transforming them into powerful analytical tools for complex matrices like food. The evolution reflects a journey from basic detection to nuanced interpretation and learning. Early eNose systems could primarily indicate the presence of certain volatiles. However, modern iterations, augmented by advanced algorithms, are capable of not only detecting but also classifying, identifying, and even predicting odor characteristics based on learned patterns. This enhanced interpretative power makes them far more versatile and robust for the subtle and varied demands of food quality assessment, moving beyond controlled laboratory conditions to real-world industrial environments.

3.1.2 Basic Components of an eNose System

(i) Sampling system: Collects and delivers the gas or odor sample to the sensor array in a controlled and consistent manner. It ensures sample stability and prevents contamination.
(ii) Sensor array: The core component that detects volatile organic compounds (VOCs). It typically consists of a group of non-specific chemical sensors (e.g., metal oxide sensors, polymer sensors, quartz crystal microbalance sensors), each reacting differently to various compounds. It generates a unique response pattern or "smellprint" for each odor.
(iii) Signal Conditioning and Data Acquisition Unit: It converts raw sensor outputs into digital signals. It filters and processes the signals to reduce noise and improve accuracy.
(iv) Data Processing and Pattern Recognition Module: It analyzes the sensor data using computational models. It also employs statistical or machine learning techniques such as PCA, LDA, SVM, ANN, or 1D-CNN to recognize patterns and classify odors.
(v) Output and Display Unit: It presents the analysis results in a readable format (e.g., classification of food freshness, type, or spoilage level) (Fig. 3.1).

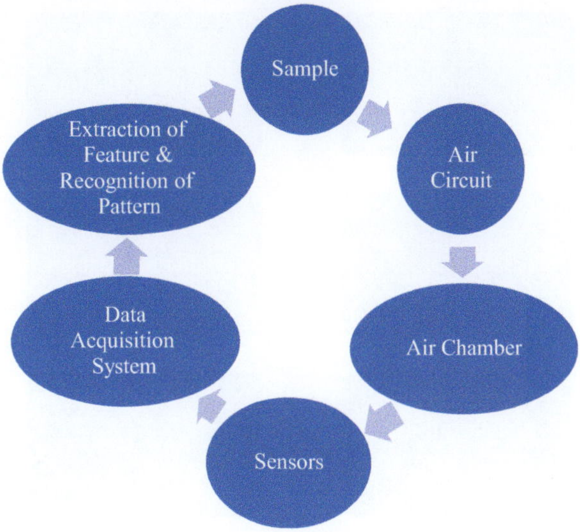

Fig. 3.1 eNose system

3.1.3 Working Principle

The eNose operates by drawing in the sample gas, which interacts with the sensor array. Each sensor in the array responds to specific chemical compounds, altering its electrical properties (such as resistance). The combined response of all sensors generates a distinctive signal pattern for the sample, which is then processed and compared to known reference patterns for identification or classification.

3.1.4 Data Analysis Method

In an eNose system, the process begins with the sampling system, which collects the volatile compounds released by a sample. These compounds are transported through a gas pathway to the gas chamber, where they interact with a sensor array. When the VOCs come into contact with the sensors, chemical reactions occur on the sensor surfaces, generating a natural response unique to the composition of the sample. This response is then sent to the data acquisition system, which converts the sensor signals into electrical data readable by a computer. The resulting data undergo further processing and analysis through suitable data processing techniques (Tatli et al. 2022).

The data analysis typically involves three key stages:

1. Pre-processing: Raw electrical signal data are cleaned and refined to eliminate noise and enhance data quality.

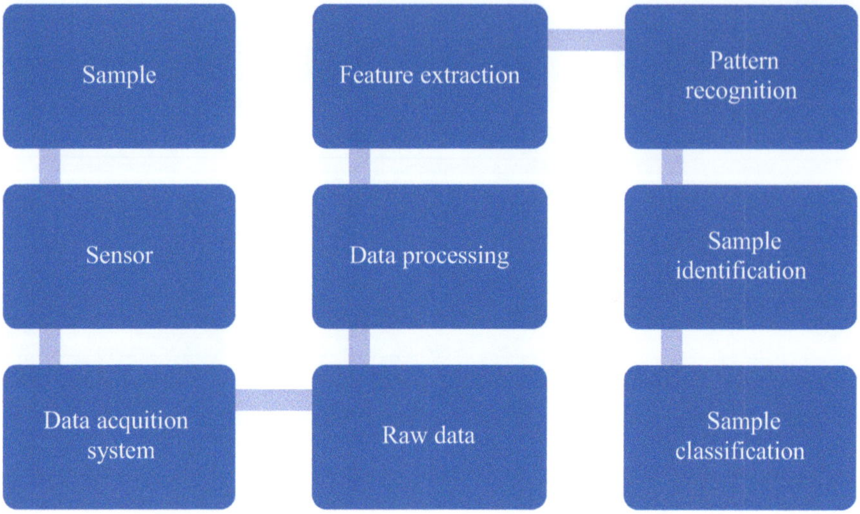

Fig. 3.2 Data analysis flow of eNose system

2. Feature Extraction: Key characteristics or patterns in the pre-processed data, which represent the specific VOCs of the sample, are identified and extracted (Fig. 3.2).
3. Pattern Recognition: Advanced algorithms or statistical methods are applied to classify or quantify the sample, enabling qualitative identification or quantitative analysis.

3.1.5 Application of eNose in the Food Industry

In recent years, the rapid advancement of eNose technology has attracted significant attention from researchers exploring its potential across various sectors of the food industry. Numerous studies have demonstrated the effectiveness of eNoses in detecting, evaluating, and classifying food products, highlighting the technology's broad and promising applications in ensuring food quality and safety. eNoses have shown strong potential in six key areas:

(i) Food Freshness Assessment: eNoses are used to detect spoilage indicators by analyzing changes in VOCs over time, enabling early identification of perishable food deterioration.
(ii) Process Monitoring: The technology allows for real-time monitoring during food processing, helping maintain consistent product quality and detecting deviations during manufacturing.

(iii) Flavor Evaluation: eNoses can objectively assess and compare flavor profiles, aiding in product development and quality standardization by analyzing aroma components.
(iv) Authenticity Evaluation: The system helps verify the authenticity of food items, distinguishing between genuine and adulterated or mislabeled products.
(v) Quality Control: eNoses are employed in routine quality assurance to maintain desired standards by rapidly screening large volumes of products.
(vi) Origin Traceability and Pesticide Residue Detection: By identifying specific VOC patterns, eNoses can trace the geographic origin of food and detect the presence of chemical residues, contributing to food safety and supply chain transparency.

The accumulation of research in these areas underscores the growing value of eNose technology as an efficient, non-destructive, and intelligent tool for modern food quality management.

3.1.6 Real-Time Freshness Monitoring in Uncontrolled Environments

In earlier studies, researchers primarily focused on evaluating the freshness of meat and aquatic products under frozen storage conditions. However, more recent investigations have shifted towards assessing freshness under natural (ambient) storage conditions. Kim et al. in South Korea conducted a study using an eNose system to evaluate the freshness of meat stored under such conditions. Their research emphasized the development of high-performance, highly sensitive biological gas sensors, and the design of two specialized eNose instruments: the CRGeNose and TAAR13NDsBeNose. These devices were modeled after previous work by Xu et al. and their effectiveness was validated through successful freshness assessment of meat. From a different angle, Liu et al. approached the problem by designing an Input Modified Convolutional Neural Network (IMCNN) algorithm. Using an eNose to capture volatile compounds released from lamb, they were able to accurately predict the total volatile basic nitrogen (TVB-N) content—an established indicator of meat freshness. Similarly, Zhang et al. employed correlation analysis and partial least squares regression (PLSR) to link the response intensity (RI) of various sensors in the eNose system with the TVB-N values of goat and duck meat. This allowed them to effectively estimate the freshness level based on sensor data. Together, these studies demonstrate the growing capability of eNose technology, not only in controlled environments but also in real-world, ambient storage conditions, highlighting its utility in practical food freshness evaluation.

3.1.7 *Processed Foods and Other Applications*

The applications of eNose technology extend to a wide variety of processed foods and other quality control scenarios within the food industry.

3.1.7.1 Monitoring Shelf-Life and Packaging Integrity

eNoses are valuable tools for investigating and monitoring the shelf-life of processed food products. For example, they have been used to monitor quality decay in processed meats like bacon during storage, distinguishing between samples stored for different durations based on changes in their VOC profiles.

Another important aspect is the evaluation of food packaging materials to ensure they do not impart undesirable odors or flavors to the packaged food. eNoses can assess the odor quality of the packaging itself. An emerging trend is the integration of eNose sensor technology with smart packaging systems. Such systems aim to provide real-time monitoring of food freshness within the package, potentially alerting consumers or supply chain partners to spoilage.

3.1.7.2 Ensuring Batch-to-Batch Consistency

Maintaining consistent product quality from one production run to the next is a hallmark of industrial food manufacturing. eNoses are employed in process and production departments to help ensure this batch-to-batch consistency by monitoring the aroma profile of products. Deviations from a "golden batch" standard can indicate variations in raw materials, processing parameters, or other factors that might affect consumer perception and brand reputation.

3.1.7.3 Detection of Contamination, Spoilage, and Adulteration

Across the broad category of processed foods, eNoses serve as general-purpose tools for detecting various quality and safety issues, including contamination (e.g., from cleaning agents, foreign materials), spoilage (microbial or chemical), and adulteration. They are utilized in quality control laboratories to assess incoming raw materials, intermediate products during processing, and final packaged goods before release.

3.1.7.4 Analysis of Edible Oils and Other Foodstuffs

The versatility of eNose technology is further demonstrated by its application to other food matrices. eNoses have been used for monitoring the quality of edible oils, where oxidative rancidity is a common concern. They have also been applied to the classification of various liquids, including different types of beer, alcohol, and sesame oil. Furthermore, eNoses are employed in the analysis of aroma profiles in mushrooms and truffles to assess quality, monitor shelf-life, and determine optimal maturation stages. Even the detection of off-flavors in drinking water, which can arise from microbial contamination or chemical taints, has been successfully addressed using eNose systems.

The application of eNoses to assess packaging materials and the forward-looking integration into smart packaging concepts signify a proactive shift in quality assurance strategies. Instead of solely evaluating the food product itself at discrete points in time, this approach extends quality monitoring to the food's immediate environment—the package. The eNose can help ensure that the packaging material is inert and does not negatively impact the food's sensory qualities by leaching undesirable volatiles. Moreover, by embedding miniaturized eNose sensors into the packaging structure, it becomes possible to actively and continuously monitor the food's condition throughout its shelf-life, from the factory to the consumer. This extends quality control beyond the confines of the manufacturing facility, potentially providing real-time freshness indicators to consumers or dynamic shelf-life information to retailers and distributors, thereby improving inventory management and reducing food waste.

The use of eNoses for ensuring batch-to-batch consistency in processed foods, which often involve intricate formulations and multi-stage manufacturing processes, implies a significant role for this technology within the framework of Process Analytical Technology (PAT). PAT emphasizes understanding and controlling manufacturing processes through timely measurements of critical quality and performance attributes of raw and in-process materials. By providing rapid, objective feedback on aroma profiles—a key quality indicator directly influenced by raw material quality and processing parameters—eNoses can help identify and correct process deviations in near real-time. If the aroma profile of a current batch begins to diverge from the established standard of a "golden batch," it can trigger alerts or even automated adjustments to process parameters. This capability moves eNose applications beyond simple end-product QC checks towards active process optimization and control, a core tenet of PAT, leading to more efficient production, reduced product variability, and consistently higher quality.

3.1.8 *Comparative Analysis: eNose vs. Traditional Analytical Methods*

The adoption of eNose technology in the food industry is often considered in the context of existing analytical methods. A comparative analysis against human sensory panels and instrumental techniques like gas chromatography (GC) and

GC–mass spectrometry (GC–MS) helps to position eNoses appropriately and understand their specific advantages and limitations.

3.1.8.1 Human Sensory Panels

Human sensory panels, comprising trained testers or consumers, are traditionally employed to evaluate the organoleptic properties of food, including aroma, flavor, and texture. A key difference lies in subjectivity versus objectivity. Human sensory analysis, while indispensable for gauging consumer acceptance and perceived quality, is inherently subjective. Perceptions can vary significantly between individuals due to genetic differences in sensory receptors, personal preferences, cultural backgrounds, and even transient physiological states. Factors like sensory fatigue (adaptation to an odor after prolonged exposure) or context effects can also influence judgments. In contrast, eNoses are designed to provide objective measurements based on the chemical composition of VOCs, yielding reproducible data that is less susceptible to human biases.

In terms of cost and time, maintaining a trained human sensory panel can be expensive, involving costs for recruitment, training, panel maintenance, and the execution of sensory sessions, which can be time-consuming. eNose analysis, once the system is calibrated, can often provide faster results, potentially at a lower per-sample cost for routine analyses.

Regarding performance in discriminative tests, such as triangle tests (where panelists identify the odd sample among three), some studies have indicated that eNoses can outperform human panels. For instance, one study reported that an eNose correctly classified all food samples under investigation in a triangle test, whereas a human sensory panel was only able to perform the test correctly in 50% of the cases. However, a crucial capability of humans that eNoses inherently lack is the ability to describe smells using a rich vocabulary (e.g., "fruity," "floral," "burnt") and to assess hedonic aspects (liking or disliking). While eNoses can be trained to correlate their patterns with such descriptors, they do not possess intrinsic descriptive or hedonic assessment capabilities.

Often, the roles are seen as complementary. Efforts are increasingly being made to supplement, rather than entirely replace, human sensory analysis with instrumental methods like eNoses. eNoses can be used as effective screening tools to pre-select samples that require more detailed sensory evaluation, to monitor consistency where human panels are not always available, or to provide objective data to support or investigate findings from sensory panels.

3.1.8.2 Gas Chromatography (GC) and GC–Mass Spectrometry (GC–MS)

GC and particularly GC–MS are powerful analytical techniques widely used for the separation, identification, and quantification of individual volatile compounds in complex mixtures.

A primary distinction relates to speed, cost, and complexity. GC/GC–MS analyses are typically laboratory-based, requiring sophisticated instrumentation, skilled operators, and often involving time-consuming sample preparation and chromatographic run times. eNoses, on the other hand, are generally designed to be faster, more cost-effective for high-throughput screening, and easier to operate, with many systems offering portability for on-site or at-line analysis.

The level of information detail also differs. GC–MS provides a detailed chemical breakdown of an aroma, identifying and quantifying its constituent VOCs. However, it does not inherently indicate whether a specific identified compound contributes significantly to the perceived odor, nor can it describe the overall smell. eNoses, in their basic form, provide a holistic "smellprint" or pattern representing the entire mixture of volatiles. They classify or differentiate samples based on these overall patterns without necessarily identifying or quantifying every individual compound. It is worth noting, however, that some advanced eNose designs are evolving to incorporate GC principles for pre-separation or use MS as a highly sensitive detector (e.g., the SPME-MS-MVA system described earlier), blurring the lines somewhat.

A technique that bridges the gap between instrumental analysis and human perception is GC–Olfactometry (GC–O). In GC–O, the effluent from the GC column is split, with one portion going to a standard GC detector (like FID or MS) and the other to a "sniffer port" where a human assessor can smell the individual compounds as they elute from the column. This allows for the identification of which specific compounds in a complex mixture are odor-active and contribute to the overall aroma. Despite their different approaches, studies have shown that significant correlations can often be established between the signals generated by eNose (and eTongue) systems, the volatile component data obtained from GC–MS, and the sensory attributes assessed by human panels. This suggests that these different methods, while providing distinct types of information, are often measuring related underlying chemical and sensory phenomena. eNoses are not positioned as a universal replacement for either human sensory panels or detailed chromatographic methods like GC–MS. Instead, they occupy a complementary niche, excelling in applications that require rapid, objective, and potentially on-line or at-line screening and monitoring of complex aroma profiles. The strength of an eNose lies in its ability to recognize patterns in these complex mixtures quickly and consistently. In contrast, GC-MS excels at providing a detailed chemical inventory of those mixtures, and human panels remain the ultimate arbiters of perceived aroma quality and hedonic response. An effective analytical strategy in the food industry often involves leveraging the strengths of each method: eNoses can handle high-throughput screening of many samples, flagging those that are anomalous or problematic. These flagged samples can then be subjected to more detailed GC-MS analysis for chemical identification or to sensory panel evaluation for a deeper understanding of their organoleptic characteristics. The ongoing efforts to correlate eNose data with both the specific chemical profiles generated by GC–MS and the descriptive sensory language used by human panels reflect a crucial endeavor: to "translate" the somewhat abstract outputs of eNose systems (i.e., sensor response patterns) into terms that are more universally understood, scientifically grounded, and directly relevant to food technologists and sensory scientists. This translation is vital for building broader trust and confidence in eNose technology and for facilitating its wider adoption by industries that are traditionally accustomed to, and reliant upon, detailed chemical data and human sensory assessments. Successfully establishing these correlations and developing robust "translation" models requires sophisticated chemometrics, a deep understanding of the complex relationships between volatile chemical composition, sensor responses, and the intricacies of human sensory perception. This will be key to moving eNoses from specialized research tools to mainstream industrial instruments whose outputs are readily interpretable and actionable.

3.1.9 Conclusion

Electronic Nose technology has demonstrably transitioned from a niche research concept to a viable and increasingly valuable analytical tool for the food processing industry. Its core strength lies in its ability to rapidly and objectively assess complex aroma and volatile organic compound profiles, offering a non-destructive alternative or complement to traditional sensory and chemical analyses. Throughout various sectors—from ensuring the freshness of perishable meats and fish, to grading the ripeness of fruits, detecting off-flavors in dairy products, classifying the nuanced aromas of beverages like coffee and wine, and screening grains for contamination—eNoses are providing actionable insights that contribute to enhanced quality control, improved food safety, and optimized processing efficiency. The operational principles, mimicking the mammalian olfactory system through sensor arrays and sophisticated pattern recognition, are now well-established. A diverse range of sensor technologies, each with its own set of advantages and limitations, allows for a degree of customization to specific analytical challenges. However, the true power of modern eNose systems is increasingly unlocked by the integration of artificial intelligence and machine learning, which enable the interpretation of complex sensor data, compensation for inherent instrumental challenges like sensor drift, and the development of robust predictive models. Despite its significant potential, the widespread adoption of eNose technology is contingent upon addressing several key considerations. The challenges of sensor stability, calibration rigor, environmental sensitivity, and the need for standardization remain pertinent. Successful implementation demands a strategic approach that involves careful selection of suitable applications, thorough validation against reference methods, comprehensive operator training, and thoughtful integration into existing quality management frameworks. The trend towards miniaturization, IoT connectivity, and advanced AI algorithms promises a future where eNoses are even more powerful, portable, and interconnected, potentially becoming integral components of smart food production and supply chain systems. These systems could offer predictive analytics for shelf-life, real-time process adjustments, and enhanced traceability, contributing to reduced waste and greater consumer confidence. In conclusion, the eNose represents a significant technological advancement with the capacity to transform aspects of quality and safety assurance in food processing. While not a panacea, its continued development, coupled with a pragmatic and informed approach to its application, will undoubtedly solidify its role as an indispensable instrument in the pursuit of higher quality, safer, and more consistent food products. The journey ahead will likely focus on further enhancing reliability, reducing costs, and fostering greater understanding and trust in the data these intelligent sensory systems provide.

References

Abu-Khalaf N, Masoud W (2022) Electronic nose for differentiation and quantification of yeast species in white fresh soft cheese. Appl Bionics Biomech 2022:8472661

Carrillo-Gómez JK, Durán Acevedo CM, García-Rico RO (2021) Detection of the bacteria concentration level in pasteurized milk by using two different artificial multisensory methods. Sens Bio-Sens Res 33:100428. [Online]. Available: https://www.sciencedirect.com/science/article/pii/S2214180421000337

Cascos G, Lozano J, Arroyo P, Ruiz-Canales A, Oates MJ, Martín-Vertedor D (2023) Fusion data of digital olfation devises for the evaluation of the quality of fresh coffee beans. https://doi.org/10.21203/rs.3.rs-2408518/v1

Chen Q, Song J, Bi J, Meng X, Wu X (2018) Characterization of volatile profile from ten different varieties of Chinese jujubes by HS-SPME/GC–MS coupled with eNose. Food Res Int 105:605–615

El Hadi MA, Zhang FJ, Wu FF, Zhou CH, Tao J (2013) Advances in fruit aroma volatile research. Molecules 18:8200–8229

Esteki M, Shahsavari Z, Simal-Gandara J (2018) Use of spectroscopic methods in combination with linear discriminant analysis for authentication of food products (Spectroscopy–linear discriminant analysis for authenticating food products). Food Control 91:100–112

Fei C, Ren C, Wang Y, Li L, Li W, Yin F, Lu T, Yin W (2021) Identification of the raw and processed crataegi fructus based on the electronic nose coupled with chemometric methods. Sci Rep 11(1):1849

Hohmann M, Monakhova Y, Erich S, Christoph N, Wachter H, Holzgrabe U (2015) Differentiation of organically and conventionally grown tomatoes by chemometric analysis of combined data from proton nuclear magnetic resonance and mid-infrared spectroscopy and stable isotope analysis. J Agric Food Chem 63:9666–9675

Khorramifar A, Rasekh M, Karami H, Malaga-Toboła U, Gancarz MA (2021) Machine learning method for classification and identification of potato cultivars based on the reaction of MOS type sensor-array. Sensors 21:5836

Le Phuc H, Tran D, Hertog M, Nicolaï B (2020) Effect of controlled atmosphere storage on the quality attributes and volatile organic compounds profile of dragon fruit (*Hylocereus undatus*). Postharvest Biol Technol 173:111406

Li Y, Ye Z, Li Q (2023) Precise identification of food smells to enable human–computer interface for digital smells. Electronics 12(2):418. [Online]. Available: https://www.mdpi.com/2079-9292/12/2/418

Loutfi A, Coradeschi S, Mani GK, Shankar P, Rayappan JBB (2015) Electronic noses for food quality: a review. J Food Eng 144:103–111. [Online]. Available: https://www.sciencedirect.com/science/article/pii/S0260877414003276

Munekata PES, Finardi S, de Souza CK, Meinert C, Pateiro M, Hoffmann TG, Domínguez R, Bertoli SL, Kumar M, Lorenzo JM (2023) Applications of electronic nose, electronic eye and electronic tongue in quality, safety and shelf life of meat and meat products: a review. Sensors (Basel) 23(2):672. [Online]. Available: https://www.mdpi.com/1424-8220/23/2/672

Shen F, Wu Q, Su A, Tang P, Shao X, Liu B (2016) Detection of adulteration in freshly squeezed orange juice by electronic nose and infrared spectroscopy. Czech J Food Sci 34:224–232

Spadafora ND, Paramithiotis S, Drosinos EH, Cammarisano L, Rogers HJ, Müller CT (2016) Detection of *Listeria monocytogenes* in cut melon fruit using analysis of volatile organic compounds. Food Microbiol 54:52–59

Tatli S, Mirzaee-Ghaleh E, Rabbani H, Karami H, Wilson AD (2022) Prediction of residual NPK levels in crop fruits by electronic-nose VOC analysis following application of multiple fertilizer rates. Appl Sci 12(21):11263. [Online]. Available: https://www.mdpi.com/2076-3417/12/21/11263

Wei M, Liu X, Xie P, Lei Y, Yu H, Han A, Xie L, Jia H, Lin S, Bai Y, Sun B, Zhang S (2022) Characterization of volatile profiles and correlated contributing compounds in pan-fried steaks from different chinese yellow cattle breeds through GC-Q-orbitrap, E-nose, and sensory evaluation. Molecules 27(11):3593. [Online]. Available: https://www.mdpi.com/1420-3049/27/11/3593

Xiong Y, Xiao X, Yang X, Yan D, Zhang C, Zou H, Lin H, Peng L, Xiao X, Yan Y (2014) Quality control of Lonicera japonica stored for different months by electronic nose. J Pharm Biomed Anal 91:68–72

Zhang B, Xi WP, Wei WW, Shen JY, Ferguson I, Chen KS (2011) Changes in aroma-related volatiles and gene expression during low temperature storage and subsequent shelf-life of peach fruit. Postharvest Biol Technol 60:7–16

Chapter 4
Plant-Based Saponin as Feed Additives in Poultry: Revolutionizing Sustainable Agri-Food Systems

Dhananjay Gaikwad, Arda Yıldırım, Jaydip Rokade, and Mahesh Kadam

Abstract Poultry has been an important component of agriculture in recent years, and the most expensive aspect of raising poultry is the cost of feed. As a result, producers are looking for ways to reduce feed costs while avoiding the use of antibiotic growth promoters (banned in many countries including India). Some naturally available compounds, such as feed additives, herbal products, and bioactive agents, are used worldwide. These products are safe for the environment and human health and aid in improving bird characteristics such as weight, feeding efficiency, blood quality, and carcass quality in layers and broilers. Plant-derived saponins serve as bioactive feed additives in poultry and monogastric nutrition, offering a sustainable alternative to antibiotic growth promoters. They enhance membrane functionality, exert immunomodulatory effects, and contribute to lowering total cholesterol concentrations in both serum and muscle tissue. It has also demonstrated substantial changes in animal weight gain and feed efficiency.

Keywords Bioactive · Saponin · Poultry · Growth · Feed additives and health

4.1 Introduction

The increasing global demand for safe, sustainable, and antibiotic-free poultry production has intensified the search for natural feed additives with multifunctional properties. Among these, phytogenic feed additives, particularly plant-derived

D. Gaikwad (✉) · M. Kadam
Vaikunth Mehta National Institute of Cooperative Management (VAMNICOM), Pune, Maharashtra, India

A. Yıldırım
Faculty of Agriculture, Department of Animal Science, Tokat Gaziosmanpasa University, Tokat, Turkey
e-mail: arda.yildirim@gop.edu.tr

J. Rokade
ICAR-Central Avian Research Institute, Bareilly, Uttar Pradesh, India

M. Thakur, T. Belwal (eds.), *Food Biodiversity and Consumer Engagements*,
Sustainable Development and Biodiversity 46,
https://doi.org/10.1007/978-981-95-7777-4_4

saponins, have garnered attention for their potential to enhance growth performance, immune function, and gut health in poultry (Thakur et al. 2020; Saddoris-Clemons et al. 2024). The role of phytogenic feed additives in poultry nutrition has shown positive effects on the performance of birds as well as on their overall health and welfare (Keshri et al. 2018; Dai et al. 2023; Luo et al. 2022). In poultry enterprises, feed cost is the most significant component of production expenses, often accounting for 60–70% of the total cost (Youssef et al. 2020). Therefore, strategies that improve feed conversion efficiency and nutrient utilization are crucial for economic sustainability. People prefer chicken because of its favorable taste, color, and leanness, along with its well-defined carcass characteristics. Broiler meat is a vital source of affordable, high-quality protein, especially in developing nations, and its consumption continues to rise globally (Fan et al. 2018). However, disease outbreaks remain a major challenge, often resulting from poor biosecurity and intensive production systems. Historically, antibiotic growth promoters (AGPs) were used to manage these issues, but their ban in many countries due to antimicrobial resistance concerns has driven interest in alternative bioactive compounds (Sun et al. 2009). Saponins, natural glycosides with amphiphilic properties, are considered one of the most promising plant-derived alternatives to AGPs (Das et al. 2012; Kregiel et al. 2017). Their structure, composed of a hydrophobic aglycone (steroidal or triterpenoid) linked to hydrophilic sugar chains, imparts detergent-like characteristics, enabling them to act as emulsifiers and surfactants (Sparg et al. 2004). Saponins are widely distributed secondary metabolites in plants and are known for their hemolytic, antimicrobial, anti-inflammatory, antioxidant, and immune-modulatory properties (Francis et al. 2002). Many plants synthesize saponins as protective agents against microbial infections and herbivory, and these compounds are increasingly explored for their potential as feed additives. These compounds exhibit diverse bioactivities, including antioxidant, anti-inflammatory, hypocholesterolemic, and immunomodulatory effects (Bafundo et al. 2021; Dai et al. 2023). Saponins also play a role as vaccine adjuvants and can enhance the absorption of calcium and other nutrients, thereby supporting digestion and growth performance (Sun et al. 2009). Recent studies have shown that saponins derived from plants such as *Yucca schidigera, Quillaja saponaria, soapnut* (*Sapindus mukorossi*), and fenugreek improve nutrient absorption, reduce ammonia emissions, modulate gut microbiota, and enhance antioxidant defenses (Luo et al. 2022). For example, *Yucca* and *Quillaja* saponins have been reported to improve villus height and nutrient digestibility in broilers, leading to better growth performance (Bafundo et al. 2021; Youssef et al. 2020). In addition, saponins exert hypocholesterolemic effects by reducing serum cholesterol and lipid content in meat and eggs, which enhances the nutritional value of poultry products (Afrose et al. 2010; Zhang et al. 2016). Their ability to reduce oxidative stress and enhance immune responses makes them valuable for poultry raised under challenging conditions, such as heat stress or pathogen pressure (Luo et al. 2022). Moreover, the current research trend emphasizes the shift towards natural alternatives to AGPs not only for performance improvement but also for reducing environmental pollutants such as ammonia and methane, contributing to a more sustainable poultry industry (Wina et al. 2017).

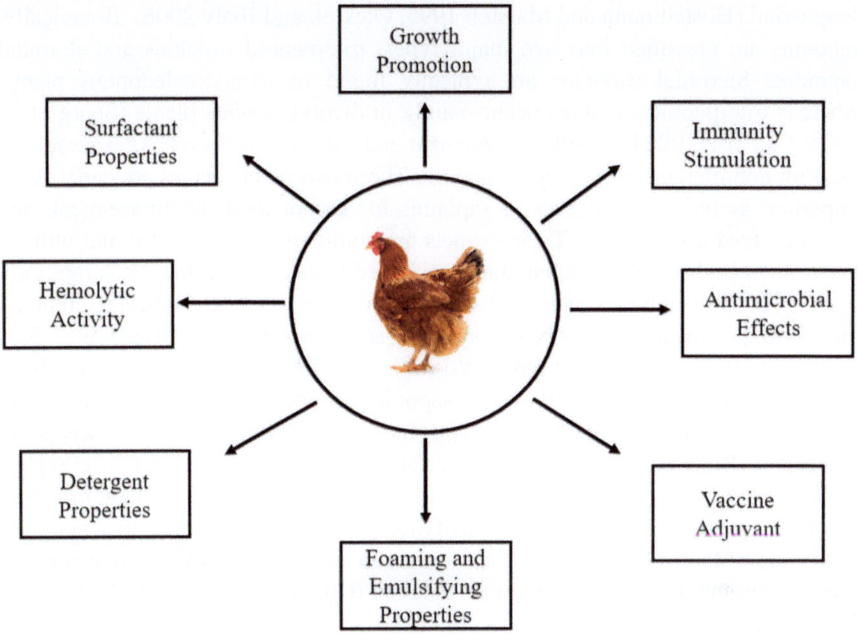

Fig. 4.1 Multifunctional role of saponins in poultry nutrition

Therefore, the aim of this chapter is to provide a comprehensive overview of saponins as functional feed additives in poultry diets, focusing on their sources, chemistry, biological activities, optimal inclusion levels, and roles in enhancing growth, immunity, and reproductive performance (Fig. 4.1). By integrating recent findings (2020–2024), this section underscores the potential of saponins as a sustainable and natural alternative to antibiotics in poultry production.

4.2 Sources of Saponin

Saponins are widely distributed secondary metabolites found predominantly in higher plants and, to a lesser extent, in certain marine organisms such as starfish and sea cucumbers (Sparg et al. 2004; Cao et al. 2024). Widely, saponin is contained in vegetables, herbs, and beans. Common dietary sources include legumes (soybean and peas), oats, tea, and yams, which contribute to the human intake of saponins through regular consumption. In addition to food plants, various medicinal herbs and forage crops—such as ginseng (*Panax* spp.), alfalfa (*Medicago sativa*), Panax notoginseng, Bupleurum, and Saponaria are rich in bioactive saponins that have attracted interest for their pharmaceutical and industrial applications (Güçlü-Üstündağ and Mazza 2007). Soybean, peas, and herbs are important sources having foaming properties, such as soapwort, soapberry, soapbark, soap root, and soap wood (Hostettmann and Marston 1995; Oleszek and Bialy 2006). Botanically, saponins are classified into two major types: triterpenoid saponins and steroidal saponins. Steroidal saponins are typically found in monocotyledonous plants, whereas triterpenoid saponins occur mainly in dicotyledonous plants (Sparg et al. 2004; Cao et al. 2024). *Quillaja saponaria* and *Yucca schidigera* (Agavaceae) are used for commercial extraction of saponin. These two plant species are particularly important as industrial sources of saponins for use in food, pharmaceutical, and livestock feed applications. Their extracts are commonly standardized and utilized due to their high saponin content and established bioactivity profiles. It is rare for a plant to contain only a single type of saponin; most species synthesize complex mixtures. For example, soybean saponins are generally categorized into four groups, A, B, E, and DDMP types (Yoshiki et al. 1998), while alfalfa roots have been shown to contain over 20 distinct saponin compounds, including medicagenic acid, zanhic acid, hederagenin, soyasapogenol A, B, and C, as well as bayogenin glycosides (Bialy et al. 1999). Due to their amphiphilic chemical structure, comprising a hydrophobic aglycone (triterpenoid or steroid) and one or more sugar moieties, saponins exhibit surfactant and emulsifying properties. These features not only contribute to their characteristic foam formation but also underpin their functional roles in nutrient absorption and gut modulation (Güçlü-Üstündağ and Mazza 2007; Sparg et al. 2004; Cao et al. 2024).

4.3 Chemistry of Saponin

Saponins are widely distributed across numerous plants and plant-derived products and play a significant role in both human and animal nutrition. These amphiphilic compounds consist of a hydrophilic sugar moiety and a lipophilic aglycone (sapogenin), which can be classified as either triterpenoid or steroidal in nature (Bialy et al. 1999; Cao et al. 2024). Triterpenoid saponins typically originate from dicotyledonous plants and include structures such as oleananes, ursanes, and lupanes, while steroidal saponins, commonly found in monocotyledons, are characterized by spirostanol, furostanol, and related skeletons (Yang et al. 2014). These compounds exhibit various biological activities, including membrane-permeabilizing, immunostimulatory, and hypocholesterolemic effects, and have been reported to influence growth performance and feed intake in animals. Additionally, saponins possess antiprotozoal properties, can interfere with protein digestion and the absorption of vitamins and minerals in the gastrointestinal tract, and may also exert hypoglycemic effects. Consequently, their impact on animals can be both beneficial and detrimental, depending on the context of their use and concentration in the diet (Das et al. 2012). Furthermore, their antiviral, antifungal, antioxidant, and anticancer effects have been documented in multiple studies, contributing to their broad pharmacological profile (Güçlü-Üstündağ and Mazza 2007; Cao et al. 2024). From a physicochemical perspective, most saponins appear as white amorphous powders, although some, such as hederagenin, may form needle-like crystals. Due to their

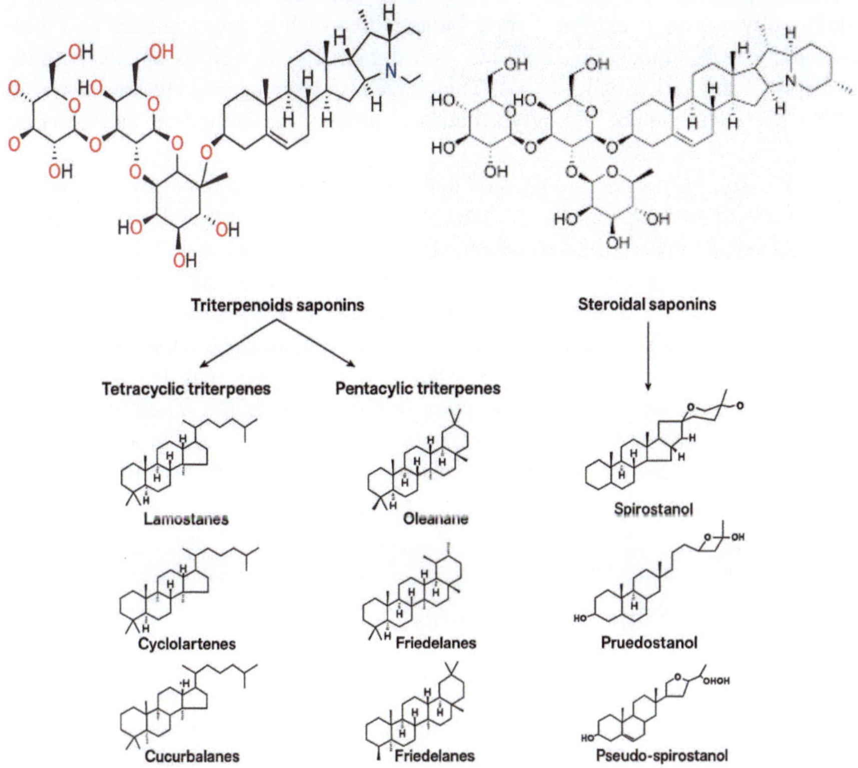

Fig. 4.2 The molecular structure of saponins comprises two main components: a glycosidic portion and a sugar moiety. Based on their chemical structure, saponins are generally classified into two types: triterpenoid and steroidal saponins. Among triterpenoid saponins, further distinction is made between tetracyclic and pentacyclic forms

amphipathic nature, they act as natural surfactants, readily soluble in water and aqueous alcohols but poorly soluble in organic solvents like ether and acetone. This property underpins their ability to form stable foams, as seen in traditional uses of *Quillaja saponaria* by Chilean natives for cleansing purposes (Reichert et al. 2019; Fink and Filip 2023).

The core structures of triterpenoid and steroidal saponins are typically oleanane and either spirostane or furostane, respectively (Sparg et al. 2004; Siegler 1998; Fig. 4.2). These aglycones are glycosidically linked to sugar residues such as glucose, rhamnose, arabinose, xylose, galactose, or glucuronic acid, influencing the saponin's solubility, biological activity, and toxicity (Yang et al. 2014; Cao et al. 2024). Saponins, which are glycosides of either steroidal or triterpenoid origin, are widely present in various plants and plant-derived products and play an important role in both human and animal nutrition. These compounds are associated with a broad spectrum of biological activities. Research has extensively documented their membrane-permeabilizing, immunostimulatory, hypocholesterolemic, and anticancer properties, along with their notable influence on animal reproduction. Additionally, saponins exhibit diverse bioactivities such as protozoacidal and molluscicidal effects, impacts on cold-blooded organisms, as well as analgesic, antinociceptive, and antioxidant actions. They have also been reported to interfere with protein digestion, induce hypoglycemia, and function as antifungal and antiviral agents (Desai et al. 2009).

Importantly, saponins may modulate metabolic pathways by activating AMP-activated protein kinase (AMPK), regulating transcription factors such as PPARs, C/EBP, and SREBPs, which control lipid oxidation and biosynthesis. Such mechanisms not only influence energy balance but also make saponins potential therapeutic agents for metabolic disorders (Liu et al. 2019; Cao et al. 2024). The pharmacological effects of saponins depend on their chemical structure and concentration, and while they offer numerous benefits, high doses may lead to adverse effects, such as reduced nutrient absorption or mucosal irritation. Therefore, understanding their chemistry is essential for optimizing their inclusion in animal diets. Table 4.1 summarizes the reported effects of various saponins on animal performance, growth, and nutrient utilization.

4.4 Effects of Saponin on Poultry

Saponins, naturally occurring plant glycosides, have shown promising effects on poultry performance and health, contributing to sustainable food production (Thakur et al. 2020; Fig. 4.3). They enhance nutrient absorption and feed efficiency by modulating gut microbiota, improving digestion, and reducing harmful pathogens. In addition, saponins possess antioxidant and anti-inflammatory properties that strengthen immune function and reduce disease incidence, minimizing the need for antibiotics. Their use in poultry diets can also lower cholesterol levels in meat and eggs, adding value to the final product. Overall, incorporating saponins supports healthier poultry and more sustainable, antibiotic-free production systems.

4.5 Safety and Optimal Inclusion Levels of Saponins in Poultry Diets

Saponins are amphiphilic glycosides derived from plants, known for their immunomodulatory, hypocholesterolemic, and antimicrobial properties. Although their potential as natural alternatives to antibiotic growth promoters (AGPs) in poultry nutrition is promising, their effects and safety vary greatly depending on the administered dose, plant source, structural composition, and duration of supplementation. At low to moderate inclusion levels (e.g., 50–150 mg/kg feed), saponins can improve gut health, immunity, and lipid metabolism. However, higher

Table 4.1 Effect of saponin on animal performance

Source	Levels	Effects/findings	References
Yucca schidigera extract	0.01%	Gain in weight and feed conversion ratio	Alfaro et al. (2007)
Ginseng leaves and stem	0.40%	The growth rate was 16.6% higher in the treatment group compared to the control group.	Park et al. (2015)
Mango saponin (MS)	0.14% (MS 0.14%), or 0.28% (MS 0.28%)	Observed gain in weight.	Zhang et al. (2016)
Alfalfa saponins (AS)	60, 120, 240, and 480 mg/kg of AS (Alfalfa saponins)	Effectively reduce egg cholesterol concentrations	Fan et al. (2018)
Acacia concinna (Dietary Peptasan®)	500, 750 and 1000 ppm	*Acacia concinna* supplementation has been shown to enhance performance, reduce mortality, and effectively prevent coccidiosis in broiler chickens.	Sánchez-Hernández et al. (2019)
Soapnut (*Sapindus mukorossi*) shell powder	100, 150 and 200 mg/kg diet)	Serum cholesterol, abdominal fat, *Escherichia coli,* and triglyceride were reduced significantly by the inclusion of saponin.	Bera et al. (2019)
Quillaja saponin blend (saponins)	46.0 mg/kg	There was an improvement in protein digestibility and an enhanced absorption of dietary nutrients.	Youssef et al. (2020)
Yucca schidigera and Trigonella foenum-graecum mixture (Norponin XO2)	250 ppm	Supplementation was as effective as monensin in reducing intestinal lesions occurrence caused by experimental *Eimeria* spp. challenge, and numerically compensated for the reduction in body weight due to infection.	Benarbia et al. (2022)
Quillaja saponaria saponins (Q. saponaria bark extract)	250 and 500 ppm	Improved growth performance, feed efficiency, ileal morphology, cecal microbial diversity, and immune gene expression (IL-1β, IFN-γ, IL-10); reduced *Clostridium perfringens* abundance and mortality in necrotic enteritis challenge model	Blue et al. (2023)
Yucca schidigera saponins	250 mg/kg in feed for 42 days	In free-range broilers under *Eimeria* spp. challenge, Y. schidigera saponins improved body weight gain, reduced oocyst excretion, decreased lesion scores in intestines, lowered inflammatory response, and preserved intestinal architecture	Guerrini et al. (2024)

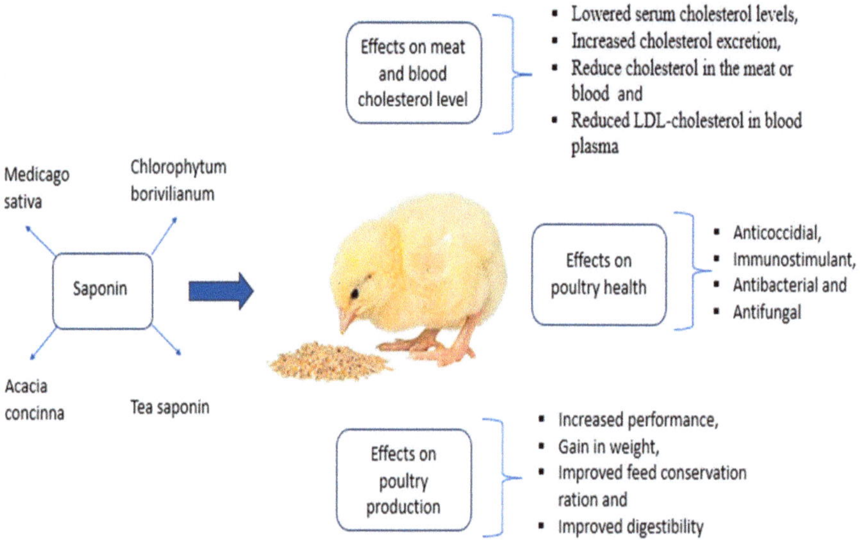

Fig. 4.3 Effects of saponin on poultry performance and health for sustainable food production

concentrations (>250–300 mg/kg), especially in monogastric species, may cause reduced feed intake, gastrointestinal irritation, and hemolytic effects. Therefore, careful formulation and dose–response evaluation are essential to ensure efficacy and safety.

Studies have demonstrated that dietary inclusion of saponins in broiler diets within the range of 50–150 mg/kg can support a variety of beneficial outcomes. For example, supplementation with saponins in the form of soapnut (*Sapindus mukorossi*) shell powder at 100–150 mg/kg has significantly improved both humoral and cell-mediated immunity, reduced abdominal fat deposition, and modulated gut microbiota without negatively affecting growth or feed intake (Bera et al. 2019; Chaudhary et al. 2018a, b). Similarly, Quillaja saponins administered at ~46 mg/kg, either alone or in combination with essential oils, have been shown to enhance intestinal villus height and proteolytic enzyme activities, thereby improving protein digestion and growth performance (Youssef et al. 2020). However, higher saponin concentrations (≥200 mg/kg) have been associated with adverse effects, including decreased feed intake, reduced weight gain, and signs of gastrointestinal irritation (Bera et al. 2019; Chaudhary et al. 2018a, b). These negative outcomes are largely attributed to the detergent-like action of saponins, which can compromise epithelial integrity and induce hemolysis at high systemic concentrations.

Yucca schidigera extract (YSE), another well-known saponin source, has been found to enhance antioxidant enzyme activities (SOD, CAT, GPx) and improve feed conversion ratio (FCR) at inclusion levels of 100–200 mg/kg. However, at or above 300 mg/kg, feed intake declined without further performance benefits, indicating a narrow safety margin (Sun et al. 2019).

A combination of *Quillaja* and *Yucca* saponins has been proposed to exploit synergistic effects while minimizing individual dosage levels. Such combinations at moderate inclusion levels (e.g., 30–50 mg/kg each) have been reported to improve gut permeability, ileal villus architecture, and nutrient absorption, thereby enhancing overall performance and minimizing toxicity risks (Saddoris-Clemons et al. 2024).

Recent findings also highlight the potential of fenugreek (*Trigonella foenum-graecum*) saponins. Inclusion of fenugreek distillates at approximately 100 mg/kg has shown to support immune responses and improve serum lipid profiles without eliciting anti-nutritional effectsu. The supplementation with fenugreek powder at levels of 0.5–1.0% in the diet has led to improvements in protein digestion, body weight gain, and feed efficiency. However, higher levels (1.5–3.0%) may reduce feed intake and negatively affect growth. Therefore, strategic inclusion of saponins in poultry diets at 100–150 mg/kg appears optimal in terms of both efficacy and safety. These levels offer benefits such as enhanced immune modulation, improved gut integrity, lipid metabolism, and productive performance. Nevertheless, the toxic potential at high concentrations underscores the importance of gradual introduction, standardization of botanical sources, and use of synergistic combinations to achieve maximum efficacy with minimal risk.

Plant-based saponins, including soapnut, *Quillaja, Yucca*, and fenugreek—are potent natural additives that, when appropriately applied, can enhance growth, health, and welfare in poultry. However, dose–response studies, biomarker-based toxicity monitoring, and long-term safety evaluations are essential, particularly when considering species-specific responses and diverse production systems. Future research should aim to refine dose–response relationships and integrate saponins into precision nutrition strategies for sustainable poultry production.

4.6 Effect of Saponins on the Immune System

Saddoris-Clemons reported that the adverse effects of coccidial infection persist beyond the peak phase and extend into the recovery period. Biomarkers such as pro-inflammatory cytokines and gut permeability (FITC-d levels) remained elevated and did not return to unchallenged control levels even 12 days post-infection. In the present study, our findings further indicate that MPU supplementation supports intestinal barrier function and modulates immune responses during coccidial challenge. Both SAL and MPU treatments demonstrated benefits not only during the acute phase but also throughout recovery, contributing to improved intestinal integrity and growth performance. These observations add to the growing body of evidence supporting the use of saponins from *Quillaja saponaria* and *Yucca schidigera* as viable alternatives to conventional antibiotics for promoting gut health and immune modulation.

Furthermore, studies on GSL (ginseng stem and leaf) extract have shown notable improvements in broiler performance. Supplementation with varying doses of GSL enhanced body weight, feed conversion ratio (F/G), average daily feed intake

(ADFI), average daily gain (ADG), and meat quality compared to controls. Significant improvements ($p < 0.05$) in antioxidant status and immune function were observed, particularly in the high-dose (GSL-H) group. Additionally, medium and high doses of GSL extract significantly reduced serum triglycerides (TG) and total cholesterol (TC) levels ($p < 0.05$), indicating beneficial effects on blood lipid profiles.

Similarly, Krishnan et al. (2024) reported interactive effects on early growth parameters of Sentul chickens during a 4-week study. Their findings suggest that vaccination alone is more effective than when combined with saponins, whereas administering saponins independently especially at higher doses (4.5 μg) yields superior weight gain and feed efficiency outcomes.

4.7 Effect of Semen Quality and Sex Hormones

Over the past few decades, there has been a growing interest in the use of plant-derived bioactive compounds as alternatives to synthetic additives in animal nutrition. This shift is largely driven by increased consumer awareness regarding the need for clean, safe, and residue-free animal products (Chaudhary et al. 2018a, b). Among these bioactive compounds, plant secondary metabolites (PSMs), including essential oils, organic acids, flavonoids, and saponins have gained considerable attention due to their multifunctional properties. Unlike primary plant metabolites, PSMs do not serve a direct nutritional role; however, they exhibit significant pharmacological and physiological effects that can enhance animal health and productivity.

Saponins, a class of high-molecular-weight glycosides known for their foam-forming ability in aqueous solutions, are widely studied for their diverse biological effects. These compounds are associated with hypocholesterolemic, antimicrobial, anti-inflammatory, antioxidant, anti-carcinogenic, and immunomodulatory activities. Their presence in various plant parts, including seeds, roots, stems, bark, pericarps, and shells, makes them a versatile feed additive for poultry and monogastric animals. Optimal supplementation levels of saponins in feed have been reported to improve overall growth and reproductive performance (Chaudhary et al. 2018a, b). Notably, dietary inclusion of saponin-rich plant materials at 0.026% (equivalent to 75 ppm) has been linked to enhanced reproductive performance in broiler breeders.

Further experimental studies have supported the positive influence of plant extracts on semen quality parameters. Suripta et al. (2023) demonstrated that supplementation with Putri Mal leaf extract (0.50–2.0 cc/day per bird) significantly improved semen attributes, including sperm count, motility, viability, thickness, and color density in free-range male chickens. Interestingly, semen volume and pH remained unaffected, indicating that the improvements were primarily related to sperm quality rather than semen quantity.

Similarly, Farag (2019) reported that dietary supplementation of Gimmizah chickens with ginseng or ginsenosides at 300 mg/kg resulted in enhanced egg production, egg mass, feed conversion ratio, and fertility rates. The inclusion of these bioactive compounds was also associated with improved antioxidant status and elevated levels of reproductive hormones such as estrogen and progesterone, particularly during the laying period.

In mammals, comparable findings have been documented. Agwa et al. (2020) observed that oral administration of EMS at varying doses (900–1500 mg per buck) improved key reproductive parameters in rabbit bucks. Positive responses included increased sperm motility, viability, membrane integrity, sperm concentration, and total sperm output. Enhanced antioxidant capacity and higher levels of reproductive hormones, including testosterone, follicle-stimulating hormone (FSH), and luteinizing hormone (LH) were also recorded. Moreover, reductions in lipid profiles, abnormal sperm morphology, and oxidative stress markers such as malondialdehyde (MDA) were observed. Among the tested doses, 900 mg/buck emerged as the most effective, also contributing to improved pregnancy rates, litter sizes, and kit viability at birth and weaning stages.

Collectively, these studies underscore the potential of plant-derived metabolites and extracts as natural enhancers of semen quality and reproductive hormone profiles in poultry and livestock species. The findings provide a strong foundation for developing sustainable and antibiotic-free strategies to improve reproductive efficiency and productivity in modern animal farming systems.

4.8 Conclusion

Plant-based saponins offer a natural, effective alternative to synthetic additives, enhancing poultry health and productivity while reducing environmental impact. Their integration into feed systems marks a significant step toward more sustainable, eco-friendly agri-food practices that align with global food security and health goals. It is concluded that saponin has been playing an important role in the poultry diet; it has positive effects on body weight gain, feed palatability, and immunity. As well as saponin plays an important role as an antimicrobial effect in poultry. However, saponin effects are seen on the rate of inclusion in the poultry diet.

References

Afrose S, Hossain MS, Maki T, Tsujii H (2010) Hypocholesterolemic response to karaya saponin and *Rhodobacter capsulatus* in broiler chickens. Asian Australas J Anim Sci 23(6):733–741

Agwa O, Sherif KE, Fouda SF (2020) Improving reproductive hormones, semen quality, and antioxidative capacity of rabbit bucks administrated with moringa seeds extract. J Anim Poult Prod 11(11):449–455

Alfaro DM, Silva AVF, Borges SA, Maiorka FA, Vargas S, Santin E (2007) Use of *Yucca schidigera* extract in broiler diets and its effects on performance results obtained with different coccidiosis control methods. J Appl Poult Res 16:248–254

Bafundo KW, Männer K, Duerr I (2021) The combination of quillaja and yucca saponins in broilers: effects on performance, nutrient digestibility, and ileal morphometrics. Br Poult Sci 62(4):589–595

Benarbia MEA, Gaignon P, Manoli C, Chicoteau P (2022) Saponin-rich plant premixture supplementation is as efficient as Ionophore Monensin supplementation under experimental *Eimeria* spp. challenge in broiler chicken. Front Vet Sci 9:946576. https://doi.org/10.3389/fvets.2022.946576

Bera I, Tyagi PK, Mir NA, Begum J, Dev K, Tyagi PK, Biswas A, Sharma D, Mandal AB (2019) Effect of dietary saponin-rich soapnut (Sapindus mukorossi) shell powder on growth performance, immunity, serum biochemistry and gut health of broiler chickens. J Anim Physiol Anim Nutr 103(6):1800–1809

Bialy Z, Jurzysta M, Oleszek W, Piacente S, Pizza C (1999) Saponins in alfalfa (Medicago sativa L.) root and their structural elucidation. J Agric Food Chem 47(8):3185–3192

Blue CEC, Emami NK, White MB, Cantley S, Dalloul RA (2023) Inclusion of *Quillaja saponin* Clarity Q manages growth performance, immune response, and nutrient transport of broilers during subclinical necrotic enteritis. Microorganisms 11(8):1894. https://doi.org/10.3390/microorganisms11081894

Cao S, Liu M, Han Y, Li S, Zhu X, Li D, Shi Y, Liu B (2024) Effects of saponins on lipid metabolism: the gut–liver axis plays a key role. Nutrients 16(10):1514

Chaudhary SK, Mandal AB, Bhar R, Gopi M, Kannan A, Jadhav SE, Rokade JJ (2018a) Effect of graded levels of soapnut (*Sapindus mukorossi*) shell powder on reproductive performance in broiler breeders. Asian Australas J Anim Sci 32(1):118

Chaudhary SK, Rokade JJ, Aderao GN, Singh A, Gopi M, Mishra A, Raje K (2018b) Saponin in poultry and monogastric animals: a review. Int J Curr Microbiol Appl Sci 7(7):3218–3225

Dai Z, Wang H, Liu J, Zhang H, Li Q, Yu X, Yang C (2023) Comparison of the effects of Yucca saponin, Yucca schidigera, and Quillaja saponaria on growth performance, immunity, antioxidant capability, and intestinal flora in broilers. Animals 13(9):1447

Das TK, Banerjee D, Chakraborty D, Pakhira MC, Shrivastava B, Kuhad RC (2012) Saponin: role in animal system. Vet World 5(4):248–254. https://doi.org/10.5455/vetworld.2012.248-254

Desai SD, Desai DG, Kaur H (2009) Saponins and their biological activities. Pharma times 41(3): 13–16

Fan W, Zhang X, Shi P, Li J, Wang C, Li D, Zhu X (2018) Effects of dietary alfalfa saponins on laying performance, egg cholesterol concentration, and ATP-binding cassette transporters G5 and G8 expression in laying hens. J Appl Anim Res 46(1):1051–1058

Farag M (2019) Effect of dietary ginsenosides and ginseng supplementation on productive and reproductive performance in Gimmizah chickens: 2. During laying period. Egypt Poult Sci J 39(4):993–1014. https://doi.org/10.21608/epsj.2019.67518

Fink R, Filip S (2023) Surface-active natural saponins. Properties, safety, and efficacy. Int J Environ Health Res 33(7):639–648

Francis G, Kerem Z, Makkar HP, Becker K (2002) The biological action of saponins in animal systems: a review. Br J Nutr 88(6):587–605

Güçlü-Üstündağ Ö, Mazza G (2007) Saponins: properties, applications and processing. Crit Rev Food Sci Nutr 47(3):231–258

Guerrini A, Zago M, Avallone G, Brigandì E, Tedesco DEA (2024) Yucca schidigera saponins efficacy on broiler chickens health and growth performance during coccidiosis infection in rural free-range breeding system. Vet Anim Sci 26:100611. https://doi.org/10.1016/j.vas.2024.100611

Hostettmann K, Marston A (1995) Saponins. Cambridge University Press, Cambridge

Keshri A, Singh D, Leitanthem VK, Mir SH, Ojha L, Agrawal A, Tarun A (2018) Role of saponin as feed additive for poultry, *Pashu Sandesh*, 2018. https://pashusandesh.com/saponin. Accessed 3 Aug 2025

Kregiel D, Berlowska J, Witonska I, Antolak H, Proestos C, Babic M et al (2017) Saponin-based, biological-active surfactants from plants. In: Application and characterization of surfactants. intechopen. https://doi.org/10.5772/68062

Krisnan R, Hidayat C, Unadi A, Isbandi I, Rusdiana S, Ginting SP et al (2024) The effectiveness of using a combination of Saponin and vaccine given in Ovo on hatchability and early growth performance of local chickens. In: BIO web of conferences, vol 88. EDP Sciences, p 00020

Liu Y, Deng J, Fan D (2019) Ginsenoside Rk3 ameliorates high-fat-diet/streptozocin induced type 2 diabetes mellitus in mice via the AMPK/Akt signaling pathway. Food Funct 10(5):2538–2551

Luo JJ, Chen W, Qu H, Liu YQ, Luo CL, Ji J, Shu DM, Wang J (2022) Dietary supplementation with Yucca alleviates heat stress in growing broilers exposed to high ambient temperature. Front Vet Sci 9:850715

Oleszek W, Bialy Z (2006) Chromatographic determination of plant saponins—an update (2002–2005). J Chromatogr A 1112:78–91

Park SY, Lee KY, Cho YJ, Park BY, Kim KJ, Lee NR, Kim DG, Kim YH, Hahn TW (2015) Efficacy of orally administered ginseng stem and leaf in chickens. Korean J Vet Res 55:1–7

Reichert CL, Salminen H, Weiss J (2019) Quillaja saponin characteristics and functional properties. Annu Rev Food Sci Technol 10(1):43–73

Saddoris-Clemons K, Osho S, Garcia M, Humphrey B (2024) Dietary inclusion of a proprietary combination of *Quillaja saponaria* and *Yucca schidigera* improves gut permeability and immune response under coccidia challenge in broilers. Animals 14(12):1737. https://doi.org/10.3390/ani14121737

Sánchez-Hernández C, Castañeda-Gómez del Campo JA, Trejo-Castro L, Mendoza-Martínez GD, Gloria-Trujillo A (2019) Evaluation of a feed plant additive for coccidiosis control in broilers herbals for coccidiosis control. Braz J Poult Sci 21(1):001–006

Siegler DS (1998) Saponins and cardenolides in plant secondary metabolism. Kluwer Academic Publishers, Boston/Dordrecht/London

Sparg SG, Light ME, van Staden J (2004) Biological activities and distribution of plant saponins. J Ethnopharmacol 94:219–243

Sun HX, Xie Y, Ye YP (2009) Advances in saponin-based adjuvants. Vaccine 27:1787–1796

Sun D, Jin X, Camerlink I, Tong M, Su J, Zhao F, Shi B (2019) Effects of Yucca schidigera extract on growth performance and antioxidative function of small intestine in broilers. J Anim Physiol Anim Nutr 103(2):529–536. https://doi.org/10.1111/jpn.13067

Suripta H, Astuti P, Nurtanti I (2023) Evaluation of semen quality of free-range chickens after addition of Putri Malu leaf herbal extract (Mimosa pudica Linn). Jurnal Penelitian Pendidikan IPA 9(9):7383–7389. https://doi.org/10.29303/jppipa.v9i9.4930

Thakur M, Singh K, Khedkar R (2020) Phytochemicals: extraction process, safety assessment, toxicological evaluations, and regulatory issues. In: Prakash B (ed) Functional and preservative properties of phytochemicals. Elsevier, pp 334–356. ISBN: 978-0-12-818593-3

Wina E, Pasaribu T, Rakhmani SIW, Tangendjaja B (2017) The role of saponin as feed additive for sustainable poultry production (Peran saponin sebagai pakan imbuhan dalam produksi unggas secara berkelanjutan). WARTAZOA 27(3):117–124. https://doi.org/10.14334/wartazoa.v27i3.1588

Yang Y, Laval S, Yu B (2014) Chemical synthesis of saponins. Adv Carbohydr Chem Biochem 71:137–226

Yoshiki Y, Kudou S, OKuBo K (1998) Relationship between chemical structures and biological activities of triterpenoid saponins from soybean. Biosci Biotechnol Biochem 62(12):2291–2299

Youssef IMI, Männer K, Zentek J (2020) Effect of essential oils or saponins alone or in combination on productive performance, intestinal morphology and digestive enzymes' activity of broiler chickens. J Anim Physiol Anim Nutr 104(4):1164–1176. https://doi.org/10.1111/jpn.13431

Zhang Y, Wang J, Qi B, Wu S, Zhang H, Qi G, Chen H, Luo H, Yin D, Lü F (2016) Evaluation of mango saponin in broilers: effects on growth performance, carcass quality, meat quality and plasma biochemical indices. Asian-Australas J Anim Sci 30(8):1143–1149

Chapter 5
From Field to Future: Analytical Assessment of Drying Methods for Carrot Sandge in Vidarbha Region, India

Simran Singh, Shweta Deotale, Sayantani Dutta, and Sheetal D. Deshmukh

Abstract This chapter deals with carrot *Sandge,* a traditional food adjunct of India, particularly from the Vidarbha region (Maharashtra). The study is focused on the impact of traditional and modern methods of drying on the nutritional quality and shelf life of carrot *Sandge*. The review provides an overview of worldwide production and nutritional profile of carrots, drying mechanism & physicochemical characteristics, as well as sensory evaluation for the carrot *Sandge*. It shows various drying layouts and the effect, for example, microwave drying can contribute to quality improvement without losing efficiency. It also discusses the sustainability of carrot *Sandge* with respect to reduction in food waste and energy used, as well as preservation of traditional knowledge. Problems such as quality control and food security are recognized, in addition to market expansion and product development. The study ends with a combination of knowledge-based traditional practices into modern technologies for exploiting sustainable ways to preserve cultural identity, simultaneously meeting current nutritional requirements and market demands.

Keywords *Sandge* · Carrots · Modern Drying Techniques · Sun Drying · Sustainability · Traditional Food

---

S. Singh · S. D. Deshmukh (✉)
Department of Food Technology, Laxminarayan Innovation Technological University, Nagpur, Maharashtra, India

S. Deotale
Visvesvaraya National Institute of Technology, Nagpur, Maharashtra, India

S. Dutta
Grupo de Investigación en Polifenoles,Química Analítica, Nutrición y Bromatología, Universidad de Salamanca, Salamanca, Spain

M. Thakur, T. Belwal (eds.), *Food Biodiversity and Consumer Engagements*, Sustainable Development and Biodiversity 46,
https://doi.org/10.1007/978-981-95-7777-4_5

5.1 Introduction

Traditional foods are a crucial component in the cultural heritage and dietary diversity of a region. They are culinary ambassadors signifying a place's history and features recipes, ingredients, or preparation methods. These foods provide nutrition, help keep family stories alive, promote connections within communities, and offer a means of transmitting culture from one generation to the next.

Indian foods are more traditional, with no other country can beat it by its variety of dishes and accompaniments; it has a flavor at different regions which offers widespread variety (Fig. 5.1). Indian culinary heritage includes main courses like *idli-dosa* from South Indian cuisine, *dalbatti-churma* in Rajasthani style, and *makki ki roti* with sarson ka saag as part of the Punjabi spread. Staple breads range from naan and roti in the North to appam in the South. Chutneys and pickles are typical food accompaniments, as well. Dairy-based sides such as raita and lassi are also common, while important spices like garam masala and panchphoron guide regional kitchens by differing from one another. Some of the other sweets served in different parts, such as rasgulla of Bengal, mysorepak of Karnataka, are eaten across India.

Sandge, these are traditional carrot fryums (Fig. 5.2) from the Vidarbha region of Maharashtra, India. Sesame seeds, grated carrot, and pumpkin are mixed with a

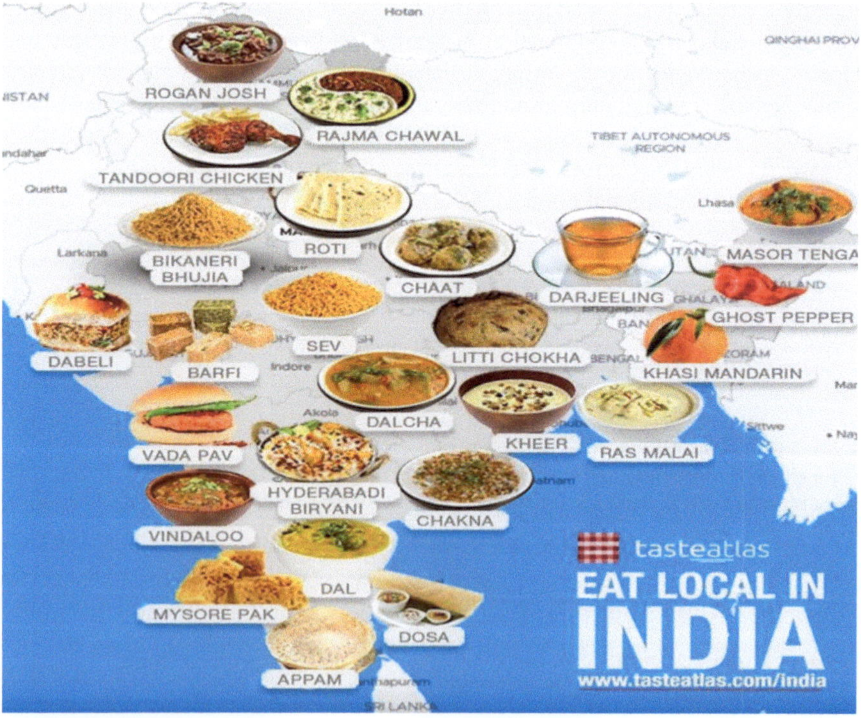

Fig. 5.1 Traditional foods of India

Fig. 5.2 Dried carrot sandge

variety of local flavorings, coriander leaves, and green chillies to form the *Sandge*. These balls are then rolled manually, after which the process of sun drying is carried out, normally taking 02–03 days (Khedkar et al. 2016).

Carrots, the important ingredient of *Sandge* are a storehouse of necessary nutrients. India is rich in carrot that produced 18.85 lakh metric tonnes annually (2020–21) (Government of India 2022). Nevertheless, with its seasonal availability and limited storage life, a huge chunk of the carrot harvest can be wasted. It is estimated that drying processes maintain the food security and shelf life of over 20% perishable crops around the world (Handbook of Postharvest Technology: Cereals, Fruits, Vegetables, Tea 2024). *Sandge* has the flavor of carrots, herbs, and spices. Coriander leaves in *Sandge* are are great sources of ascorbic acid, folic acid, riboflavin, fiber, and minerals like calcium, iron, and phosphorus. Moreover, sesame seeds enhance the nutty flavor and the crunchy feel of the *Sandge* while providing the many benefits of a great supply of essential nutrients, including fat, protein, dietary fiber, vitamins, and minerals (Chawla and Thakur 2013; Sesame Seeds Nutrition Facts and Analysis 2024).

Sandge is traditionally considered a side dish for various Indian foods like rice/khichadi. Although the traditional sun drying method is more economical, it is weather-dependent, time-consuming, and labor-intensive, and is also affected by contamination. In addition, there is a problem with the availability of market available *Sandge,* which are mostly not standardized and non-availability throughout the year. All of the available market *Sandge* have non-standardized formulas, which means various ingredient compositions, with a short shelf life. Usually, 3 months in the year, from March to May is the season when *Sandge* are available for sale (Khedkar et al. 2016).

Cultural, economic, and social changes in society caused by modernization and globalization have an impact on the food system. In the situation under consideration, the preservation of the cultural importance of *Sandge*, as well as the expansion of its scale and improvement of quality, becomes possible through the consideration of alternative methods of drying, like microwave drying.

In the broader context, it has been highlighted for various food products and has been proven to ensure the preservation of the value in food drying, such as nutrition, color, and taste of products, and to be efficient (Maskan 2000).

5.2 Carrots: Global Production, Shelf Life, and Nutritional Profile

5.2.1 Global Production

Carrots, scientific name—*Daucus carota* L., are some of the most popular root vegetables grown all over the world and used to prepare numerous delicious meals and desserts. These vegetables are also known for the large variety of nutrients they provide for the human body, as well as for their beneficial effects on the skin, improving its overall appearance (Sharma et al. 2012). They can be eaten raw, roasted, or fried, dried (Fig. 5.3), and are the ingredient of many dishes, soups, and salads. In 2020, the worldwide production was around 44.8 million tons. In these types of vegetables, China is the topmost producer, which accounts for almost 45% share in total production around the globe. The same holds true for those in Uzbekistan, the United States, and Russia. India also harvested a healthy production of 18.85 lakh metric tonnes in terms of the last season, and it is one crop that will mostly be rewarded by good prices, so the need for ample quantities even next year will remain strong (Government of India 2022).

5.2.2 Shelf Life

The shelf life of carrots is 5–7 days (without refrigeration), and they can be stored in a refrigerator for 3–4 weeks. Life depends on how it is stored at the right temperature and humidity. At 0 °C to −4 °C, at high relative humidity of 98–100%, fresh carrots can be kept for a shelf life ranging from 1 month to 3 weeks. However, the shelf life

Fig. 5.3 Dried carrots

cannot accurately reflect the taste and nutrition, and whether it can reach 3–4 weeks depends on the harvest time and storage conditions. Generally, carrots stored at room temperature last 3–5 days, and will deteriorate.

5.2.3 *High Nutritional Value*

Carrots contain a high amount of vitamins. It is rich in Vitamin A, which helps both with eye health as well as an immune booster in the form of beta-carotene, where carrot is 8285 μg/100 g. Fibers contribute 10% of the carrot composition, promoting digestion and gut health. Besides, carrots have dietary fiber, vitamin C, vitamin B6, vitamin K1, potassium, calcium, phosphorus, and magnesium. Carrots also contain antioxidants like lutein and alpha- carotene. Carrots have low protein content, contributing about 6%, and fat: 5% (Koca et al. 2007; Sharma et al. 2012).

Carrots can be considered a healthy food for many reasons. In humans, they support vision as well as immune function and are vital to cardiovascular health. Carrots are also a favorite among people, and their value is only supported by the fact that they make very tasty snacks and curd dishes. A second problem is that there are available only particular seasons in which fresh carrots can be used and cannot be consumed over the entire year. They also spoil quite fast if not kept in the refrigerator. That is why drying is the method of choice for preserving carrots.

5.3 Traditional Drying Techniques

Traditional drying methods are the oldest methods of preserving food. These techniques make foods last longer and better in flavor and nutrition. Food drying methods are not simply about ways to preserve food, but they also have an extensive engagement in Indian cooking.

Most dry items (Fig. 5.4) are used for taste-enhancing, like sun-dried tomatoes, dried mango (amchur), etc. (Adeyeye et al. 2022; Calín et al. 2020).

Table 5.1 shows various ancient drying techniques, their procedure, advantages, and examples of food products to be dried by these traditional drying techniques.

These techniques are crucial for ensuring food security, especially in the age before refrigerators. Although modern preservation methods exist, many of these traditional drying techniques are still employed today as they are very easy and affordable to middle class Indian family.

Fig. 5.4 Dried fruits and vegetables

Table 5.1 Various traditional drying techniques with associated advantages and food products to be dried

Traditional drying techniques	Procedure	Advantage	Example of food products
Sun drying	Laying food on clean cloths, mats, or trays and directly exposing it to sun	Requires less energy	Papads, vadams, chillies, kasurimethi, mangoes, onions, apricots
Shade drying	Storing food in unlit, well-ventilated locations	Used for heat-sensitive raw materials	Herbs and vegetables
Smoked drying	Food is held above a smoldering flame	Helps the food to dry out and adds its own flavor while preserving it	Meat or fish
Salt drying	Salting the food until enough moisture is drawn out	Creates an environment opposed to bacteria	Salt fish
Wind drying	Hanging food on lines or placing it on elevated platforms to allow the wind to dehydrate it naturally	Use of natural air currents to preserve food without additional energy input	Drying small fish and certain vegetables

5.4 Modern Drying Techniques

Over time, the process of drying food for preservation and production has been modernized to offer all types of advanced systems that drastically improve operational processing speed with quality control and guaranteed shelf life. These techniques are commonly used within the context of industrial food production and processing (Calín et al. 2020; Hasan et al. 2019). Table 5.2 shows various modern

Table 5.2 Various modern drying techniques with associated advantages and food products to be dried

Modern drying techniques	Procedure	Advantage	Food products to be dried
Freeze drying	Freezing the food and then reducing its ice content under vacuum using sublimation	Heat-sensitive food products retain color, flavor, and vitamins	Premium cooking ingredients or flavoring
Spray drying	Atomization of liquid food into fine droplets is sprayed continuously in a hot air current	Inexpensive, environmentally friendly, fast processing, high encapsulation efficiency, and product stability	Liquids and slurries, including powdered milk, instant coffee, and other food flavors-additives
Drum drying	Liquid or puree is spread as a layer of film on the hot drum. While the drum turns, the food dehydrates rapidly and easily comes away in a sheet	Making dried food products that are crushed into flakes or powder	Instant mashed potatoes, precooked cereals, some fruit, and vegetable powder
Vacuum drying	Pressure and temperature are lowered by reducing the boiling point of water	Heat-sensitive food products retain color, flavor, and vitamins	Fruits, vegetables, or herbs
Microwave drying	Water is removed by radiation, which is two or three centimeters above the food product	Shortens processing time	Fruits, vegetables, or herbs
Fluidized bed drying	Air is passed at high velocity through a porous material to cause fluid-like behavior and obtain dry particles	Particles moving in the hot air stream result in uniform drying and prevent clumping	Coffee beans, grains, and vegetables

drying techniques, their procedure, advantages and examples of food products to be dried by these traditional drying techniques.

The benefits of such modern drying techniques include improved control over the dry process, reduced processing times, and enhanced product quality compared to traditional approaches. This made possible year-round production, uniform quality, and the development of products that had not previously been feasible using traditional drying methods.

These methods need higher capital and energy consumption, which makes them expensive compared to traditional methods.

5.5 Mechanism of Drying

During the drying of carrot *Sandge*, all moisture has to be removed in order to inhibit fermentation and other similar processes, which lead to microbial growth (Drying Curve—Moisture Control n.d.). This drying process is done in two main stages, as shown in Fig. 5.5:

The surface of the food where moisture evaporates first and moisture migration (moves to the surface from inside). There is a decreasing drying rate due to the fact that the movement of moisture from inside starts becoming limiting. As moisture has to travel further and is in an environment where it can be slow, the drying rate slows down.

Carrot *Sandge* has high moisture content due to both the carrots and other herbs during processing. As heat is applied, moisture on the surface of the *Sandge* begins to evaporate. This creates a concentration gradient between the surface and the interior. Moisture from the interior of the *Sandge* moves towards the surface through two basic mechanisms: capillary action in the carrot tissue and diffusion through the protein matrix of the *Sandge*. Some water is bound to proteins and carbohydrates in the *Sandge*. This water requires more energy to remove than free water. As moisture is removed, the structure of *Sandge* may change, and proteins may denature and coagulate.

Also, carrot fibers may shrink and become more concentrated. The surface of the *Sandge* may dry faster than the interior, forming a harder outer layer which can slow down if further drying occurs if not properly managed. As water activity decreases, microbial growth is inhibited, contributing to preservation. The drying rate and final quality of the carrot *Sandge* will depend on factors such as temperature, relative humidity, air velocity, sausage size, and composition.

5.5.1 *The Drying Curve*

The drying curve of foods is a plot that shows the variation in moisture content and duration between different degrees of dehydration process (Fig. 5.6). Usually, this

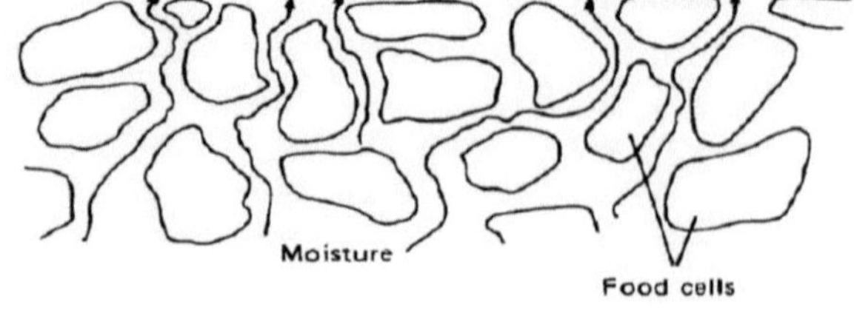

Fig. 5.5 Mechanism of drying in carrot *Sandge*

Fig. 5.6 Drying curve

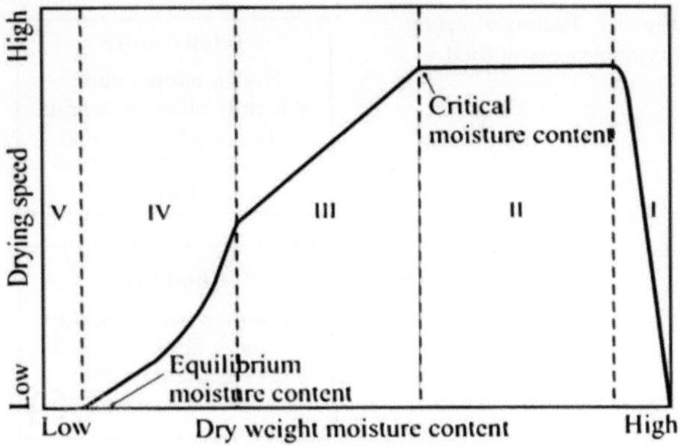

curve is divided into three specific phases, which are the initial transient period, constant rate phase, and falling rate stage. During the initial adjustment phase, the food temperature equilibrates with drying conditions, and surface moisture is evaporated. Only after this comes the constant rate period, which is a slow but linear drop in moisture content, while just free water close to the surface can evaporate away. In the second falling rate regime, moisture begins to evaporate from within the particle until a critical intermediate moisture content is reached and constant drying takes place (the final "drying period"). The final stage is sufficiently slow for moisture to move from the interior of the material through a network of moisture channels to reach an exposed surface (Drying Curve—Moisture Control n.d.).

The curve of this region usually displays an exponential fall, which corresponds to the increased complexity in migrating bound water from a product matrix. The shape and length of the different phases in this drying curve can differ depending on how much water is bound by a food, as well as factors like structure or other specific conditions (temperature, etc.).

This is critical for optimizing drying processes of some food items in preservation but also processing, where a proper selection of the best parameters and final quality prediction are necessary because they directly determine time and energy costs as well as product properties. Prakash et al. (2004) demonstrated this through a performance evaluation of blanched carrots dried by three different driers, highlighting that the choice of drying method significantly impacts both energy efficiency and the final quality attributes of the dried product.

The drying characteristics of food materials largely vary due to the porosity, homogeneity, and hygroscopic nature, etc. The food materials that are hygroscopic reach the falling rate sooner than non-hygroscopic foods (Drying Curve—Moisture Control n.d.).

The effectiveness of drying depends on several factors, as shown in Fig. 5.7. As the water is removed, several changes occur in the food shown in Fig. 5.8.

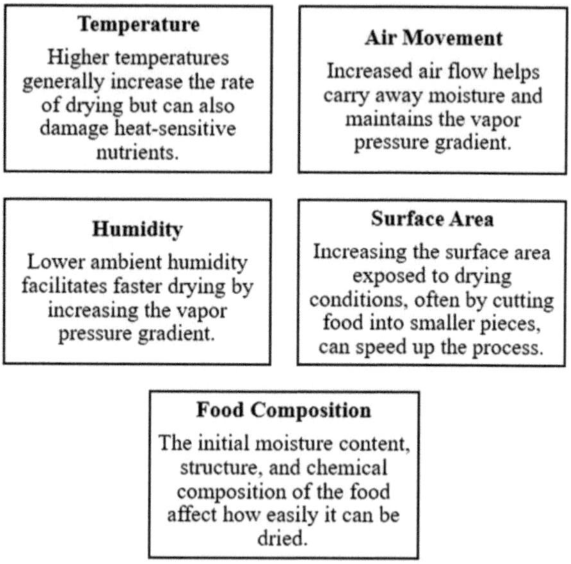

Fig. 5.7 Factors affecting drying process in food

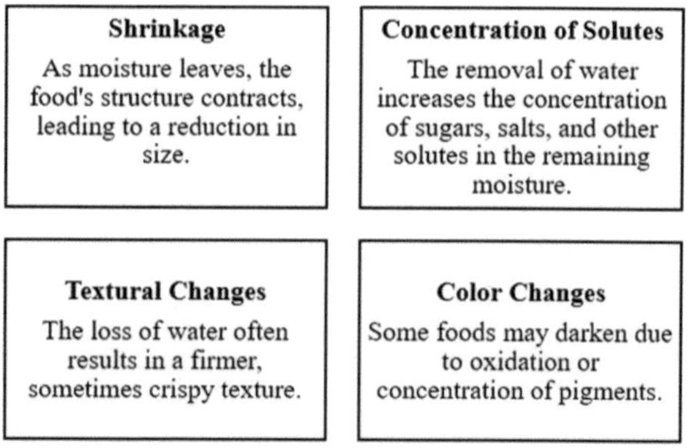

Fig. 5.8 Changes in food during drying process

5.6 Quality Parameters of Dried Carrot *Sandge*

5.6.1 Physical Properties

Carrot *Sandge* is a dried vegetable product made primarily from grated carrots mixed with other ingredients like pumpkin, tomato, spices, and seasonings. It typically has a crisp, brittle texture when dried.

The drying process decreases the moisture content and water activity of a product significantly. This leads to a very significant shrinkage and texture changes, compared to the fresh mixture. The *Sandge* puff up and become crunchy when fried. Its color varies from orange to brownish, depending on the ingredients and drying method used (Khedkar et al. 2016).

5.6.2 *Chemical Properties*

The drying process concentrates the nutrients in Carrot *Sandge*. It preserves most of the beta-carotene, fiber, and minerals from carrots and other ingredients. The method of drying has some effect on the vitamin C content. *Sandge* has relatively low moisture content and water activity levels, which helps in its stability. It is known to contain a range of antioxidants and polyphenols, in levels that depend on the ingredients used as well as the production process. The fat content goes up after frying. Freshly prepared *Sandge* has low free fatty acid levels, which indicates rancidity but it would increase during storage (Khedkar et al. 2016).

5.6.3 *Sensory Attributes*

The taste is intricate, with the sweetness of carrots and wonderful savory flavor from the other ingredients that she uses. It has a very nice and inviting scent. The exact sensory traits can differ from one recipe to another, depending on the methods employed in its production (Khedkar et al. 2016).

5.6.4 *Microbial Safety*

The low moisture content and water activity of dried carrot *Sandge* are unfavorable to the adequate growth, development, proliferation, and survival, including that of spoilage organisms so as help in prolonging shelf life. Sufficient drying to low moisture levels is necessary for ensuring microbiological safety. Nevertheless, establishing and implementing proper sanitary conditions during its preparation, as well as employing the right packaging and storage environment, is critical to avoid any potential contamination or spoilage. Moreover, the frying process for *Sandge*, when they are cooked to be eaten, provides an extra security against any potential microbial dangers (Khedkar et al. 2016).

5.7 Comparison of Drying Techniques

Increasing the use of new technologies from traditional to preserve food quality and nutritional value, especially in Carrots the microwave-vacuum-drying processing developed an alternative way instead of freeze drying, to be suggested as a new step for commercial process with the specification (Ignaczak et al. 2023).

The quality analysis based on moisture content did not differ much in comparison among solar cabinet, fluidized bed, and microwave oven drying methods, yet these

Table 5.3 Comparison of drying techniques: review

Methodology	Conclusion	References
Carrots were blanched for 1–3 min or pre-osmosed for 15 (OD), dried with different types of drying methods viz., conventional convection drying method (CD), microwave-convective drying (MW—CD), Microwave-vacuum-drying method (MVD)	The results indicate that microwave-vacuum drying is a fast, promising alternative to freeze-drying for the production of high-quality carrot snacks while maintaining bioactive compounds better than traditional methods	Ignaczak et al. (2023)
Carrots were dried using a solar cabinet drier, fluidized bed drier (at temperatures 50, 60, 70 °C), and microwave oven drier (at power levels 2, 3, 4)	The quality of the dried carrots from fluidized bed was superior with respect to color, rehydration properties, and β-carotene retention over conventional drying methods, i.e., microwave and solar, besides it offered better storage stability in thick polyethylene packaging	Prakash et al. (2004)
This review analyzed different drying techniques for fruits, vegetables, and herbs, evaluating their weaknesses and exploring improvements. It evaluated various quality parameters of dried products, including physical properties, chemical composition, and sensory attributes	Drying effectively preserves fruits, vegetables, and herbs; it also reduces volume and weight for cheaper transport and longer shelf life. However, it typically decreases various quality parameters. The optimal drying technique is different for each material, requiring specific evaluation. Novel or combined techniques should aim for high product quality while considering energy efficiency and environmental sustainability	Calín et al. (2020)

three treatments were far superior from conventional approach, while all we did was use this hybrid facility-infrastructure system aimed at saving energy (energy efficient carrots). These techniques were more effective at retaining the color, rehydration properties of the product, and nutrient content. However, a comprehensive survey of drying methods for different vegetables revealed that though the process is successful in food preservation and preventing large volume reduction during transportation time, it reduces quality characteristics (Table 5.3).

The best method differs per material, and it shows that we should utilize the right ways. There was a common theme across all of the studies: Novel or hybrid methods need to be developed that will maintain product integrity, conserve energy consumption, and be environmentally benign represents an ongoing requirement for innovations in food drying technologies (Calín et al. 2020).

5.8 Sustainable Aspects of Carrot *Sandge*

5.8.1 Food Waste Reduction

Carrot *Sandge*, a traditional dried carrot item, which has a great impact on food waste reduction. This process converts fresh carrots into a shelf-stable form, enabling the preservation of surplus carrots in peak harvest seasons. This is particularly critical when it comes to carrots, as they are highly perishable in their fresh form. By drying, it is possible for carrots to last from weeks up to several months, maybe even a year if preserved correctly. Now waste is reduced, as carrots that would spoil before anyone could eat them can be stored for up to 8 months and used a little at a time.

5.8.2 Energy Efficiency in Drying

The primary energy cost in producing Carrot *Sandge*, depends upon the method chosen for drying. They are, in fact, highly energy efficient since they were using free solar energy, albeit time-consuming. However, as mentioned in the research paper, solar cabinet drying has lower quality outputs. Newer techniques like fluidized bed drying and microwave drying can save energy up to a certain extent in the context of the time required for a certain size dry as compared to regular sun-drying, alongside keeping quality. According to the comparison done in the previous section, fluidized bed drying at moderate temperatures (~50 °C) was found to be a good optimization between energy consumption and product quality. Also, techniques such as heat recovery should be employed in mechanical drying processes to enhance the energy efficiency, or plant hybrid solar-assisted-mechanical-drying systems, which may deliver both advantages of a combined solar and mechanical drier.

5.8.3 Preservation of Traditional Knowledge and Cultural Heritage

Carrot *Sandge* is another side dish, a crucial part of traditional food preservation techniques. If we continue making and eating carrot *Sandge*, the traditional knowledge of food preservation that our ancestors have given as heritage still remains alive. This indigenous knowledge will typically cover not only the preservation process but also the choice of processing techniques, as well as its traditional uses in dishes once dried. Also, it promotes the intergenerational knowledge migration and the bond between the community.

5.8.4 Climate Resilience in Food Preservation

Sandge production has a number of advantages over current and other potential means, make them paradigm example where tradition meets technology.

- *Extended Shelf Life*: Production of *Sandge* will help build capacity in the processing community with the knowledge that they now have an additional market-ready product to be produced at any point during surplus and sold on demand. These dried carrot *Sandge* can be stored for a longer period of time, which will allow for year-round sale of this carrot product.
- *Lesser Usage of Cold Storage*: Carrot *Sandge*, unlike the fresh carrot, does not require to be stored in a refrigerator. This means that less energy is needed to preserve the food, making this a more resilient option with lower or no electricity requirement in places where there are frequent, if not regular, power outages (low-grid-access countries) and increased costs of living due to climate change.
- *Adaptability to Changing Production*: Climate change can cause varying crop yields. *Sandge* is made from carrots and can avoid wastage if carried over to next season. One year, when the harvest fails, these stored *Sandge* will give way to meager portions of fresh produce.
- *Transportation*: Dried carrots are lighter and more compact, so they might take less energy to transport, possibly reducing the carbon footprint overall.
- *Nutrient Retention*: Carrots can retain most of their nutrients, notably β-carotene when properly dried, leading to access year-round and helping communities become more adaptive with the nutrition depending on how climate conditions influence it.

To sum it all up, Carrot *Sandge* represents a holistic way of food preservation which addresses critical issues like reducing wastage of local produce and supporting the farmers around us while also adding to climate resilience as well as cultural heritage. But to realize the full benefits, energy-efficient green technologies must be used alongside traditional methods.

5.9 Challenges and Opportunities

5.9.1 Possible Challenges

The production of carrot *Sandge* includes:

- Quality control in terms of color, texture, and ensuring uniform nutritional properties, since the degree of drying might significantly affect the finished product.
- Food safety and sanitary issues, primarily associated with microbial resistance during drying, or contamination during the traditional sun drying process.

- Energy costs that might be substantial in the case of mechanical drying methods and affect the overall production overhead.
- Competition with fresh produce that might become available all year round in specific markets due to improved cold chains.
- Requirement for education and marketing efforts to create awareness among consumers about the existence of the product and its potential.

5.9.2 Possible Future Opportunities

Indicating the market potential for carrot *Sandge* includes:

- Its potential as a healthy product or part of the health food market.
- Position as a clean-label product due to minimal processing required.
- Potential exports in the countries where fresh carrots are not available all year round; creation of additional value-added products based on carrot *Sandge.*
- Possible inclination toward sustainable foods such as carrot *Sandge,* and its long product life, with no need for freezing, reducing refrigeration transportation overhead.

5.10 Conclusion

This study of a traditional food adjunct, carrot "*Sandge*" is situated at an interesting junction, where cultural heritage preservation meets modern scientific approach. Through this exploration of different methods for drying *Sandge* and its production possibilities have been identified to improve traditional foods without changing their cultural features. Traditional sun drying, compared to modern processes like microwave and air convective dryers, showed the potential of technological development toward quality parameters, nutrient preservation, and shelf-life enhancement. Among several drying techniques that proved to be able to fulfill the mentioned requisites, air convective drying seemed a very attractive alternative due to its energy efficiency and preservation of nutritional status and texture preference.

Moreover, the making of carrot *Sandge* also speaks about sustainability in food systems in many ways. This expands the life of carrots from fresh status, reduces food waste so as to contribute back into local agriculture and also saves traditional knowledge. Both tradition and innovation can be seen to have their place when the energy efficiency of different drying methods is evaluated, in addition to their impact on product quality. This also allowed key challenges to be identified, which included quality control and consumer awareness, as well as the identification of future areas for research and development.

In conclusion, the study of carrot *Sandge* not only contributes to the preservation of a traditional food product but also offers insights into sustainable food processing practices. It shows how traditional foods can be adapted to meet modern nutritional standards and market demands while maintaining their cultural heritage. Future research in this area could further optimize production methods, explore new markets, and investigate the potential health benefits of this traditional food adjunct.

References

Adeyeye SA, Ashaolu TJ, Babu AS (2022) Food drying: a review. Agric Rev 1(8):1–8

Calín-Sánchez Á, Lipan L, Cano-Lamadrid M, Kharaghani A, Masztalerz K, Carbonell-Barrachina ÁA, Figiel A (2020) Comparison of traditional and novel drying techniques and its effect on quality of fruits, vegetables and aromatic herbs. Foods 9(9):1261

Chawla S, Thakur M (2013) Coriandrum sativum: a promising functional and medicinal food. Medicinal Plants 5(2):59–65

Drying Curve—Moisture Control (n.d.) Retrieved August 9, 2024, from https://moisturecontrol.weebly.com/drying-curve.html

Food technology-I: lesson 11. Drying: principle, methods and applications

Government of India, M. of A. & F. W. D. of A. & F. W. E. & S. D. (2022) Agricultural-Statistics-at-a-Glance-2022

Handbook of postharvest technology: cereals, fruits, vegetables, tea. Retrieved April 30, 2024

Hasan MU, Malik AU, Ali S, Imtiaz A, Munir A, Amjad W, Anwar R (2019) Modern drying techniques in fruits and vegetables to overcome postharvest losses: a review. J Food Process Preserv 43(12):e14280

Ignaczak A, Salamon A, Kowalska J, Marzec A, Kowalska H (2023) Influence of Pre-Treatment and Drying Methods on the Quality of Dried Carrot Properties as Snacks. Molecules 28(17):6407

Khedkar R, Shastri P, Bawa AS (2016) Standardization, characterization and shelf life studies on sandge, a traditional food adjunct of Western India. Int J Environ Agric Biotechnol 1(2):238525

Koca N, Burdurlu HS, Karadeniz F (2007) Kinetics of colour changes in dehydrated carrots. J Food Eng 78(2):449–455

Maskan M (2000) Microwave/air and microwave finish drying of banana. J Food Eng 44(2):71–78

(PDF) Advances in production technology of carrot. Retrieved April 11, 2024

Prakash S, Jha SK, Datta N (2004) Performance evaluation of blanched carrots dried by three different driers. J Food Eng 62(3):305–313

Sesame seeds nutrition facts and analysis. Retrieved April 11, 2024

Sharma KD, Karki S, Thakur NS, Attri S (2012) Chemical composition, functional properties and processing of carrot—a review. J Food Sci Technol 49(1):22

Chapter 6
Osmotic Dehydration of Carrot Shreds for Instant Halwa Mix: Process and Storage Insights

Fasaha Ahmad, Shweta Deotale, Sayantani Dutta, and Sheetal D. Deshmukh

Abstract Carrot halwa is a popular traditional Indian sweet delicacy that is widely consumed, but the preparation process is time-consuming. In this study, an instant carrot halwa ready-mix (CHR) was developed from osmo dehydrated carrot and beetroot shreds. The carrot and beet root shreds in the ratio 1:0.1 were gradually roasted using 2% ghee in a stainless-steel kettle after the osmotic dehydration process. Grated khoa (25%) was then added gradually, with continuous heating and mixing. An antioxidant, tocopherol, was added to ghee along with blanched carrot shreds within legal limits. The heated carrot halwa mixture was spread in a thin layer on a stainless-steel plate and dried in a vacuum oven at 40–50 °C for 5–6 h. After drying and cooling to room temperature, the carrot halwa ready-mix was packed in multilayer laminated pouches, and a shelf-life study was conducted for 6 months to monitor quality changes. The dried carrot halwa mix was rehydrated with hot water (85 °C) in a 1:2 ratio of dried mix to hot water, with gentle mixing for sensory evaluation. It was observed that the instant carrot halwa mix remained stable and acceptable for 6 months under ambient conditions, but slight changes in nutritional parameters occurred. The moisture content increased from 2.16% to 2.62%, reducing sugars increased from 32.76% to 40.15%, and FFA (free fatty acids) increased from 0.121% to 0.156%. Water activity also increased from 0.368 to 0.380 during storage. In contrast, crude fiber decreased from 1.56% to 1.38%, total sugars decreased from 81.12% to 79.12%, protein decreased from 13.12% to 11.56%, β-carotene decreased from 8.14 mg/100 g to 7.56 mg/100 g, and fat decreased from 8.75% to 8.73%. However, the carrot halwa ready-mix remained organoleptically acceptable up to 6 months of storage. All the sensory

F. Ahmad · S. D. Deshmukh (✉)
Department of Food Technology, Laxminarayan Innovation Technological University, Nagpur, Maharashtra, India

S. Deotale
Visvesvaraya National Institute of Technology, Nagpur, MS, India

S. Dutta
Grupo de Investigación en Polifenoles, Química Analítica, Nutrición y Bromatología, Universidad de Salamanca, Salamanca, Spain

M. Thakur, T. Belwal (eds.), *Food Biodiversity and Consumer Engagements*, Sustainable Development and Biodiversity 46,
https://doi.org/10.1007/978-981-95-7777-4_6

parameters, including appearance, texture, taste, and overall acceptability, were comparable to the freshly prepared halwa, although there was a slight fading in the color of the carrot halwa ready-mix.

Keywords Carrot Halwa ready-mix · Osmo dehydrated carrot shreds · Beet shreds · Antioxidant · Storage stability · Nutritional parameter · Sensory quality

6.1 Introduction

The market of instant food mixes in the food industry is increasing in recent years, as there is growing demand for convenient and time-saving solutions among modern consumers, particularly working women (Dhiman et al. 2017). These instant mixes are formulated by combining a variety of ingredients, including vegetables, cereals, pulses, condiments, and spices, to create easy-to-prepare food products with enhanced quality attributes (Bala et al. 2015). The Indian food market has seen the emergence of various brands offering instant food mixes, such as Quick Bowl Dudhi Halwa, Surat, Jain Mix Vegetables, Chings Instant Veg Mix Soup, and Knorr Veg Mix Soup, which have gained popularity due to their convenience and time-saving benefits (Dhiman et al. 2017).

Carrot (*Daucus carota L.*), a widely cultivated root vegetable, is renowned for its exceptional nutritional profile, being rich in bioactive compounds such as carotenoids and dietary fibers, as well as essential vitamins like vitamin C, K, B_1, B_2, B_6, and folates (Arscott and Tanumihardjo 2010; Dias 2012; Thakur and Belwal 2023). These beneficial components contribute to the health-promoting properties of carrots, making them a valuable ingredient in the food industry (Raees-ul and Prasad 2015). The process of osmo dehydration, which involves the removal of moisture from carrot shreds using a concentrated sugar or salt solution, has been explored as a means of preserving the nutritional and sensory qualities of carrots for various food applications (Ali and Sharma 2020; Deshmukh et al. 2021).

Carrot halwa, a traditional and nutritious milk-based sweet preparation in India, is known by various regional names such as "Gajar-ka-Halwa," "Carrot Dessert," or "Carrot Pudding" (Raina et al. 2019; Vaghela et al. 2015). This dessert is considered highly acceptable and popular in the north Indian traditional dishes (Verma et al. 2018). The traditional recipe involves a heat desiccation process, where grated carrots are cooked with milk, ghee, sugar, khoa (reduced milk solids), dry fruits, and cardamom powder (Raina et al. 2019). This process of halwa preparation takes too much time depending on the quantity of raw materials and the whole process is lengthy and cumbersome.

The current study aims to develop an instant carrot halwa ready-mix (CHRM) using osmo dehydrated carrot shreds, khoa, and ghee, and to evaluate its physicochemical and sensory properties during 6 months of storage under ambient conditions. This approach seeks to leverage the convenience and popularity of instant food mixes while preserving the traditional flavors and health benefits associated with carrot-based delicacies.

6.2 Materials and Methods

Fresh, clean, sound orange carrots were procured from a local market in Nagpur, Maharashtra, India. Commercial sugar used for the preparation of the syrup was obtained from a local grocery shop. Khoa, a milk-based product, was obtained from Dive Bandhu Dairy shop, Nagpur. Commercial grade ghee from Dinshaw's company was used for the study.

6.2.1 Osmotic Dehydration

The carrots were peeled, de-cored, and shredded to a width of 3–4 mm, then blanched at 85 °C for 3–4 min. A sugar syrup with 50°B, 60°B, and 70°Brix was prepared. Two hundred grams of the carrot shreds were dipped into the sugar solution at room temperature (27 °C) for 2.5 h. During this osmosis process, water came out from the carrot shreds into the sugar syrup, and some solutes leached into the carrot shreds.

6.2.2 Preparation of Carrot Halwa Ready Mix

After the osmotic dehydration, the carrot and beet root shreds in the ratio 1:0.1 were gradually roasted using 2% ghee in a stainless-steel kettle. Grated khoa (25%) based on the total weight of carrot shreds was then added gradually, with continuous heating and mixing. On a stainless-steel plate, the heated carrot halwa mixture was spread thinly, and it was vacuum-dried for 5–6 h at 40–50 °C. After drying and cooling, the ready mix was packed in airtight multilayer laminated packaging materials,

For preparation of carrot halwa, the dried carrot halwa mix was rehydrated with hot water (85 °C) in a 1:2 ratio of dried mix to hot water, with gentle mixing and heating. Stepwise process for preparation of Carrot Halwa Ready Mix is given in Fig. 6.1.

6.2.3 Sensory Evaluation

A sensory evaluation of the Carrot Halwa Ready Mix (CHRM) was conducted using the Nine Point Hedonic scale method. The CHRM samples were assessed by a panel of ten judges based on a variety of sensory criteria, including colour, texture, flavour, and general acceptability. For the duration of the study, the same panel was utilized. The judges were shown the samples as they are typically consumed.

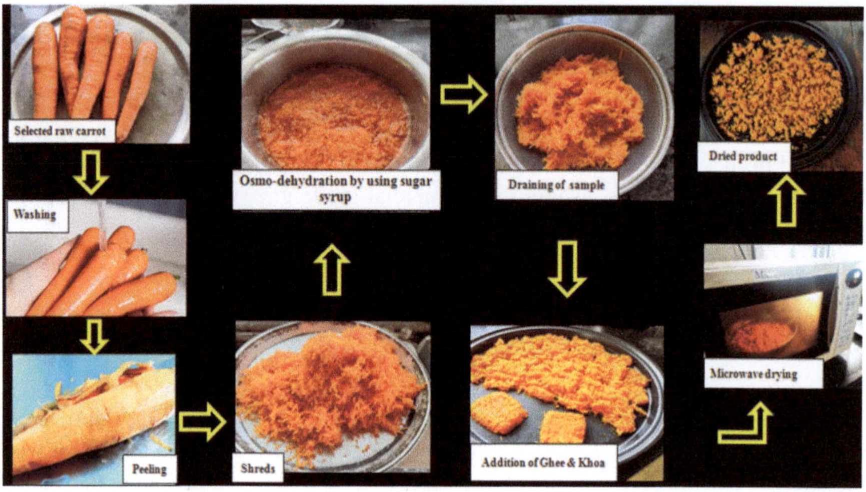

Fig. 6.1 Process for preparation of carrot Halwa Ready mix

6.2.4 *Physicochemical Analysis*

The following physicochemical analyses were performed on the CHRM samples:

Titratable acidity, pH, moisture, protein, free fatty acids (FFA), and total carotenoids were determined as per the methods given in Ranganna (2000). Water activity (a_w) of the CHRM samples, tempered at 25 °C, was measured using a water activity meter. Protein content was measured using the Kjeldahl method.

6.2.5 *Shelf-Life Study*

The best recipe of CHRM, based on nutritional and sensory evaluation, was selected for shelf-life studies. After being packaged in multilayer laminated packets, the chosen CHRM's shelf life was assessed monthly for 6 months in terms of its nutritional makeup and sensory qualities.

6.3 Results and Discussion

6.3.1 *Effect of Storage on Nutritional Quality and Sensory Attributes of Carrot Halwa Ready Mix*

The results in Table 6.1 show significant changes in various nutritional parameters during the 6-month storage period. The moisture content increased from 2.16% to 2.62%, reducing sugars increased from 32.76% to 40.15%, and fiber decreased from 1.56% to 1.38%. Water activity also increased from 0.368 to 0.380 during storage. In contrast, total sugars decreased from 81.12% to 79.12%, protein decreased from 13.12% to 11.56%, β-carotene decreased from 8.14 mg/100 g to 7.56 mg/100 g, and fat decreased from 8.75% to 8.73%. The increase in moisture content during storage was likely due to the absorption of small amounts of moisture from the atmosphere through the packaging material. An increase in reducing sugar may be due to the inversion of non-reducing sugar and conversion of polysaccharides to monosaccharides. The decrease in protein, total sugars, and β-carotene was associated with Maillard reactions and oxidation during storage. The decrease in fiber was attributed to the solubilization of polysaccharides. Despite these changes, the water activity remained below 0.50, indicating the product was still safe for consumption after 6 months of storage. The sensory attributes of the carrot halwa ready-mix were also found to be acceptable over the storage period.

Table 6.1 Effect of storage at ambient temperature on the nutritional quality of carrot halwa ready-mix from osmo dehydrated carrot shreds

Nutrients	Storage period		
	0 month	3 months	6 months
Moisture (%)	2.16	2.5	2.62
Titratable acidity (%)	0.0191	0.0201	0.0212
Total soluble solids (TSS)	83^0B	82^0B	81^0B
Ash (%)	1.64	1.62	1.63
Total sugars (%)	81.12	80.15	79.12
Reducing sugar (%)	32.76	35.16	40.15
Protein (%)	13.12	12.14	11.56
β-carotene (mg/100 gm)	8.14	7.98	7.56
Ascorbic acid (mg/100 gm)	24.18	22.03	21.12
Fat (%)	8.75	8.74	8.73
Free fatty acid (FFA) %	0.121	0.134	0.156
Fiber (%)	1.56	1.42	1.38
Water activity	0.368	0.37	0.38

Table 6.2 Sensory evaluation of carrot halwa ready-mix developed from osmo dehydrated carrot shreds

Sample	Storage period	Appearance	Color	Flavor	Taste	Texture	Overall acceptability
Freshly prepared	0 month	8.512	8.325	8.465	8.482	8.512	8.521
Halwa from ready-mix	0 month	8.511	8.318	8.431	8.426	8.485	8.452
Halwa from ready-mix	3 months	8.501	8.164	8.321	8.382	8.431	8.162
Halwa from ready-mix	6 months	8.489	8.162	8.289	8.245	8.321	8.012

Samples were evaluated using nine-point hedonic scales (1 dislike extremely, 9 like extremely)

6.3.2 Sensory Evaluation

The results showed that the sensory quality scores of the carrot halwa ready-mix decreased during the storage period. This trend was similar to the findings of previous studies on shelf-stable food products. However, the carrot halwa ready-mix remained organoleptically acceptable up to 6 months of storage (Table 6.2). All the sensory parameters, including appearance, color, texture, taste, and overall acceptability, were comparable to the freshly prepared halwa. While there was a slight fading in the color of the carrot halwa ready-mix, which was attributed to the oxidation of β-carotene during storage, it was still considered acceptable by the consumers. There was very little change in the texture and flavor of the carrot halwa ready-mix. It was expected that better packaging conditions would help retain the sensory characteristics of the carrot halwa ready mix, leading to improved acceptability over the storage period.

6.4 Conclusion

The study demonstrates that dehydrated, osmotically treated carrot shreds can be effectively utilized in the preparation of a β-carotene (8.14 mg/100 g) and protein-rich (13.12%) carrot halwa ready mix. This was achieved through appropriate processing and optimization of the carrot shreds along with other ingredients in the right proportions. The carrot halwa ready mix was found to be acceptable in terms of color, flavor, and texture even after 6 months of storage. The ready mix only requires 5–10 min for preparation, making it a convenient option, especially for working populations where the demand for ready-to-eat foods is increasing. Additionally, the formulated ready mix was found to be relatively low-cost, further enhancing its appeal and accessibility.

The development of sustainable ready mixes for traditional dishes like carrot halwa represents a combination of culinary heritage and environment. This innovative approach not only preserves cultural recipes but also addresses sustainability concerns in the food industry. In the production process of carrot halwa ready-mix, manufacturers can significantly reduce water usage, energy consumption, and waste generation. The use of locally grown, organic carrots for the preparation of traditional-based ready mixes not only supports farmers in getting income but also minimizes transportation-related problems. Advanced food processing techniques allow for the retention of nutrients and flavors while extending shelf life without artificial preservatives. Moreover, eco-friendly packaging solutions, such as compostable materials or reusable containers, further reduce the environmental impact. This sustainable ready-mix concept not only offers convenience to consumers but also educates them about the importance of sustainable food choices, potentially influencing broader consumption patterns towards more environmentally friendly options. Overall, this endeavor showcases the potential of using osmotically dehydrated carrot shreds to develop a shelf-stable, nutritious, and convenient carrot halwa ready mix.

References

Ali S, Sharma A (2020) Osmotic dehydration and assessment of quality attributes of seasonal vegetable crops: carrot and beetroot cubes. J Food Sci Technol 5(4):199–121. https://doi.org/10.25177/JFST.5.4.RA.10658

Arscott SA, Tanumihardjo SA (2010) Carrots of many colors provide basic nutrition and bioavailable phytochemicals acting as a functional food. Compr Rev Food Sci Food Saf 9(2):223–239. https://doi.org/10.1111/j.1541-4337.2009.00103.x

Bala A, Gul K, Riar CS (2015) Functional and sensory properties of cookies prepared from wheat flour supplemented with cassava and water chestnut flours. Cogent Food Agric 1(1):1019815. https://doi.org/10.1080/23311932.2015.1019815

Deshmukh SD, Gabhane S, Deshmukh DS (2021) Osmotic dehydration of carrot strips and modelling. J Phys Conf Ser 1913(1):012093. IOP Publishing. https://doi.org/10.1088/1742-6596/1913/1/012093

Dhiman AK, Vidiya N, Surekha A, Preethi R (2017) Studies on development and storage stability of dehydrated pumpkin based instant soup mix. J Appl Nat Sci 9(3):1815–1820. https://doi.org/10.31018/jans.v9i3.1444

Dias JS (2012) Nutritional quality and health benefits of vegetables: a review. Food Nutr Sci 3(10):1354–1374. https://doi.org/10.4236/fns.2012.310179

Raees-ul H, Prasad K (2015) Nutritional and processing aspects of carrot (Daucus carota)-a review. South Asian J Food Technol Environ 1:1–14. https://doi.org/10.46370/sajfte.2015.v01i01.01

Raina BL, Sharma S, Ahlawat VP (2019) Development and quality evaluation of instant carrot halwa mix. Int J Chem Stud 7(4):2864–2867

Ranganna S (2000) Handbook of analysis and quality control for fruit and vegetable products. Tata McGraw-Hill Education, New York

Thakur M, Belwal, T. (2023) Bioactive components. Springer, Singapore doi:https://doi.org/10.1007/978-981-19-2366-1_34

Vaghela NR, Modha HM, Gokhale AJ, Patel AM, Patel HG (2015) Process standardization for manufacture of ready-to-reconstitute carrot halwa mix. Indian J Dairy Sci 69:389–397

Verma A, Sharma HK, Gujral HS, Singh B (2018) Physicochemical and sensory properties of carrot-based dessert (gajrella). J Saudi Soc Agric Sci 17(2):152–159

Chapter 7
Exploring Panchgavya Ghrita as an Agri-Food Innovation for Human and Plant Health

Nikhil Raghuvanshi, Vikash Kumar, Nikhil Chandel, Alok Rai, Kshitij Parmar, Eid Sadun Alotaibi, and Ajay Tomar

Abstract Every living creature comprises five substances, i.e., water, air, earth, space, and fire, and those substances are collectively known as Panchabhootas. Health is maintained by the tridoshas, i.e., Pitha (fire), Vata (air), and Kapha (Phlegm). When the natural ratio of panchabhootas gets disturbed, it may cause the disease. The use of panchgavya has been stated by various scientists and researchers, and the Vedas have found it to be an amusing source of vital elements as well as hormones and minerals. *Bos indicus* (BI) or Kamdhenu is used for religious purposes as sensified animal by Hindus. There are multitudinous *Vedic* references. In the Ayurvedic treatment, every product has diverse potentials and uses in agriculture, health, and other fields. From ancient time cow-based products have been used for therapeutic processes, positive health, and pharmaceutical processes. The purpose of reverence for BI is the enormous therapeutic purpose of Kamdhenu-derived products like cow milk curd, cow urine, cow milk, cow ghee (butter), and cow dung. When it mixes in proper proportion, then this formulation is known as panchgavya and they have various therapeutic effects. Panchgavya has

---

Sanskrit Terms in English
Tridoshas, Pitha (fire), Vata (air), and Kapha (Phlegm), Panchbhoot, Panchgavya, Gavya, Charak, veerya, guna, prabhaava, and vipaka, rasapancaka, Apasmara chikitsa, Agni, Ghrita

---

N. Raghuvanshi
Faculty of Agricultural Science, GLA University, Mathura, Uttar Pradesh, India

V. Kumar · A. Rai
Department of Agriculture, Deen Dayal Upadhyaya Gorakhpur University, Gorakhpur, Uttar Pradesh, India

N. Chandel · K. Parmar · A. Tomar (✉)
Amity Institute of Organic Agriculture, Amity University Uttar Pradesh, Noida, Uttar Pradesh, India

E. S. Alotaibi
Department of Mathematics and Statistics, AlKhurmah University College, Taif University, Taif, Saudi Arabia

M. Thakur, T. Belwal (eds.), *Food Biodiversity and Consumer Engagements*, Sustainable Development and Biodiversity 46,
https://doi.org/10.1007/978-981-95-7777-4_7

immunostimulatory, hypoglycemic, anticonvulsant, and other beneficial effects on the human body. Soil fertility, earthworm development, preventing crop illnesses from bacteria and fungi, etc., are just a few of the ways Panchgavya can give back to Mother Nature. The purpose of this article is to provide a concise summary of the varied pharmaco-therapeutic value of the panchgavya for human and agricultural health.

Keywords *BOS indicus* · Soil · Disease · Agriculture

7.1 Introduction

In the history, it is well defined that Maharshi Vasishta (one of the most admired Vedic rishi) served the divine cow "*Kamdhenu*" and Maharshi Dhanvantari (the god of Ayurveda) offered to mankind a very beneficial medicine for various disease "*panchgavya*" (a combination of cow urine, cow dung, cow curd, cow milk and cow ghee (butter) (Revathi et al. 2021). All these five products are individually termed as "*gavya*" in Sanskrit and collectively known as "*panchgavya*". In the Samhita, with the help of different gavya, various therapeutic properties are mentioned. According to charak Samhita, panchgavya is used for the treatment of jaundice, fever, and epilepsy (Bhavana and Shreevathsa 2014; Chauhan 2018). A cow is treated as the backbone of the rural economy and Indian culture, and sustains life by the byproduct of the cow, so the cow in Indian culture is known as "*Gaumata*" and "*Kamdhenu*" because the gavya of the cow nourishes like a mother and serves as a store of medicines.

According to Ayurveda, the action of a drug is based on its veerya, guna, prabhaava, and vipaka. These, as combinations or as themselves, regulate the grade of drug action in the human body (Russo et al. 2012). Panchgavya, as a constituent as well as alone, has been shown to have anticonvulsant activities and neuroprotective in albino rats (Achliya et al. 2004). Providence of the drug continuously surmises on rasapancaka, and it goes in conjunction with current pharmacodynamics (Joshi et al. 2015). The action of a particular drug depends upon the action of the agni of a particular drug. Ayurveda is the science of the treasures of medicine of mineral origin, plants, and animals. In place of a single drug, they use a combination of different animals, plants, and minerals. Many of the drugs are mentioned in Ayurvedic psychiatry. In the Apasmara chikitsa, panchgavya ghrita is mentioned for various treatments (Bishnoi et al. 2021). (Radha and Rao 2014) observed the presence of naturally occurring beneficial microorganisms, predominantly bacteria, yeast, actinomycetes, photosynthetic bacteria, and certain fungi in organic liquid manures like panchgavya. The positive impact of Panchgavya alone or in combination with various organic nutrient management practices has been found in various crops. Panchgavya application is found to be more profitable than recommended fertilizer application and chemical spray (Sumit and Patel 2020). It contains a wide variety of helpful bacteria, such as *Azospirillum* spp., *Azatobacter* spp., *Pseudomonas* spp., and many more. It is a good source of antioxidants, minerals, lipids, carbs, and proteins. *Bos indicus* (BI) urine is useful in agriculture for the prevention and treatment of fungal and insect infestations (Jandaik et al. 2015). Cow urine has been shown to stimulate plant growth due to the availability of higher nutrient contents and other phytohormones (Kanaujia and Kumar 2018). Panchgavya's fungicidal and insecticidal effects are bolstered by its antimicrobial properties. The presence of plant growth hormones like IAA and GA in a mixture of naturally produced beneficial microorganisms including yeast (*Saccharomyces*), photosynthetic bacteria (*Rhodopseudomonas*), Actinomyces (*Streptomyces*), lactic acid bacteria (*Lactobacillus*), and fungi (*Aspergillus*) increases disease-free growth and yield in many crops (Naik et al. 2020). It can be sprayed on the leaves, used on the soil in conjunction with watering, or used as a seed treatment (Sangeetha and Thevanathan 2010). What sets botanicals apart is their simplicity of production by farmers and micro- and medium-sized firms (Tharmaraj et al. 2011). The Ayurvedacharyas prescribe Panchgavya as the apex of total health, both physiological as well as psychological. Based on the existing scientific evidence, this present review examines the ingredients included in each Panchgavya product as well as its potential health benefits, medical characteristics, agricultural uses, and role in nanotechnology (Table 7.1).

7.2 Historically Significant

The oldest Hindu writings, the Vedas, were written during the Vedic period. The duration is unknown. Philological and linguistic evidence suggests that the Rigveda, the oldest Veda, was written between 1700 and 1100 BCE, during the early Vedic period. The period ended around 500 BCE, and all Vedic Sanskrit literature has a terminus ante quem of 150 BCE (Praburaman et al. 2015). In Ayurveda, the mixture of five cow byproducts (urine, curd, milk, dung, ghee) is called panchgavya (Garg and Mudgal 2007; Sharma and Singh 2018). Panchgavya therapy (cow-therapy) is commonly used as an alternative therapeutic approach in India for the maintenance of human and animal health. It has attracted the attention of medical and veterinary professionals due to its antimicrobial and antifungal properties (Joseph and Sankraganseh 2011). Panchgavya is made up 5 ingredients.

1. *Gos'akr't* (Cow dung)
2. *Goksheera* (Milk)
3. *Godadhi* (Curd)
4. *Goghr'ta* (Ghee)
5. *Gomootra* (Cow's urine)

All the five ingredients are mixed in equal proportion and ghrita is prepared according to the proper preparation method regarding ghrita.

Table 7.1 Chemical and physical properties of Panchgavya

S. No	Parameters	Methods	Value	References
1.	Odor	Sensory evaluation	**Fruity**	Chakraborty and Sarkar (2019)
2.	Color	Visual evaluation	**Light brown**	Karthika et al. (2017)
3.	EC (dS m^{-2})	Conductivity meter method	**10.2**	Ali et al. (2011)
4.	pH	pH meter method	**5.5**	Chandrakala et al. (2007)
5.	Total Nitrogen (ppm)	Microkjeldhal method	**229**	Chakraborty and Sarkar (2019)
6.	Organic carbon (%)	Walkley and Black wet digestion	**17.4**	Kumar et al. (2017)
7.	Potassium (ppm)	Flame photometry and nitric-perchloric (9:4) digestion	**232**	Chakraborty and Sarkar (2019)
8.	Phosphorus (ppm)	Colorimetry using vanado-molybdo phosphoric yellow color method and nitric-perchloric (9:4) digestion	**209**	Chakraborty and Sarkar (2019)
9.	Calcium (ppm)	AAS and nitric-perchloric (9:4) digestion	**25**	Chakraborty and Sarkar (2019)
10.	Sulphur (ppm)	Turbidimetry and nitric-perchloric (9:4) digestion	**2.1**	Chakraborty and Sarkar (2019)
11.	Magnesium (ppm)	AAS and nitric-perchloric (9:4) digestion	**1.98**	Ali et al. (2011)
11.	Sodium (ppm)		**90**	Chandrakala et al. (2007)
12.	IAA (ppm)	Spectrophotometric method	**8.5**	Chakraborty and Sarkar (2019)
13.	GA (ppm)	Spectrophotometric method	**3.5**	Chakraborty and Sarkar (2019)
14.	Cytokinin (μg/ml)	Bioassay	**3.22**	Chakraborty and Sarkar (2019)
15.	Ascorbic acid (μg/ml)	Titrimetric method	**13.0**	Chakraborty and Sarkar (2019)
16.	Bacterial count (cfu ml^{-1})	viable plate method	$\mathbf{6.33 \times 10^8}$	Somdutt et al. (2021)
17.	Fungal count (cfu ml^{-1})	viable plate method	$\mathbf{5.1 \times 10^4}$	Somdutt et al. (2021)
18.		viable plate method	$\mathbf{3 \times 10^5}$	

(continued)

Table 7.1 (continued)

S. No	Parameters	Methods	Value	References
	Actinomycetes (cfu ml^{-1})			Somdutt et al. (2021)
19.	Dehydrogenase (μg ml^{-1})	Casida method	**2.77**	Somdutt et al. (2021)

7.2.1 *Goksheera (Cow milk)*

Guna: Guru, Snigdha
Rasa: Madhura (sweet)
Vipaaka: Madhura Dosha
Veerya: S'eeta (cold)
karma: Vaata, Kaphavardhaka, pitta samana

Cow milk is a balanced diet; it contains various active constituents, such as *Lactoperoxidase* enzyme, calcium, casein, vitamins A, B2, B12, potassium, iodine, proteins, and choline (Koksal et al. 2016). Cow milk is used as a healthy food because it contains low cholesterol, low calories, and high micronutrients, calcium, protein, and vitamins. It contains vitamins A, B complex group, and vitamin C, and carotenes (Haug et al. 2007). It helps to be effective in pain and curing fever, diabetes, tumors, and weaknesses. It also delays the aging processes (Lamb et al. 2015). Cow milk is also used as a fungicide for human health and animals (Marangoni et al. 2019). Immuno-globulins (IgA) are responsible for the antimicrobial effect, including lysozyme, lactoferrin, vitamin B12-binding protein, and lactoperoxidase (Funakoshi et al. 1982). It enhances skin fairness, and smoothing creams are made from cow milk (Kazimierska and Kalinowska-Lis 2021). Fat in cow milk is an active component for anti-cancer agents, which helps in skin cancer and breast cancer, and it controls the cell contact inhibition (Parodi 1997). It is also used for religious purposes for the preparation of "panchamrat" which is used as a prasaad (Krishna et al. 2011). Milk enhances nitrifying dhatus, ojus, rasaayana, sexual enhancer, and improves cognition. It is shown for incapacitating circumstances and can be used for recuperation. Cow's milk is a rich source of vital ingredients essential for the mental and physical health of human beings and hence viewed as a balanced diet (Hill et al. 2018).

7.2.2 *Godadhi (Curd)*

With the help of the coagulation of milk protein this milk this milk product is obtained.

Guna: Guru
Rasa: Amla (acidic)

Vipaaka: Amla (acidic)
Veerya: Ushna (hot)
Doshakarma: pitta, Kapha vardhaka, Vaatashamana

Cow curd is considered a blood purifier, "Vatanashak," "Tridoshnashak" used in malabsorption syndrome for therapeutic purposes, increases agni and bala. It is also used in dysuria and remittent fevers. In cow curd, various active constituents are present, such as vitamin B12, vitamin B5, probiotics bacteria, calcium and magnesium, probiotics bacteria, magnesium and magnesium, tryptophan, and folic acid (Raj Joshi and Adhikari 2019). Vitamin B12 prevents dementia and confusion, vitamin B5 relieves anxiety and stress. Folic acid enhances hemoglobin synthesis. Magnesium and calcium enhance the secretion of melatonin, reduce weakness and restlessness, and help induce sleep (Banks et al. 1986). Tryptophan is a 1° amino acid that produces melatonin and serotonin, which work as neurotransmitters involved in promoting relaxation and sleep. Probiotic bacteria increase immunoglobulins, enhance immunity, acts as natural cell killers (Wang et al. 2021). The cow curd is enriched with a huge amount of *lactobacillus* bacteria. The occurrence of *lactobacillus* bacteria also produces antifungal metabolites, e.g., phenyllactic acid, cyclic dipeptides, 3-hydroxylated fatty acid, and proteinaceous compounds, which were recently reviewed (Cortés-Zavaleta et al. 2014; Schnürer and Magnusson 2005).

7.2.3 Goghrita (Ghee)

Guna: Snigdha
Rasa: Madhura (sweet)
Vipaaka: Madhura (sweet)
Veerya: S'eeta (cold)
Doshakarma: pitta, Vaata saman

Cow ghee has various active constituents such as vitamin A, carotenoids, vitamin E, butyric acid, and conjugated linolenic acid (Kumar 2015). Carotenoids and vitamin E promote skin cell growth, reduce the risk of heart diseases and certain cancers, improve immune system health, and are antioxidants that fight against free radicals. Butyric acid mainly in the gastrointestinal tract reductions inflammation in the body and is now a suggested dietary addition for people with ulcerative colitis (Sharma and Chandola 2015). Cow ghee has advanced digestibility compared to any other vegetable and animal fat and an improved rate of absorption. Cow ghee is a very virtuous source of nutrition for the baby and mother. It is good for improving intelligence, improving vision, and immune enhancing. The mixture of some selected herbals and cow ghee is used to treat skin diseases, and when combined with honey, it heals wounds also (Udwadia 2011). Cow ghee in the formation of panchgavya has been specified by an increase in hemagglutination titer, neutrophil adhesion, and delayed type hypersensitivity responses in albino rats (Fulzele et al.

2003). Cow ghee, when mixed with some selected herbal plant than, is used to cure skin diseases, and when cow ghee is mixed with honey can ease the healing of wounds (Halder et al. 2021). The Ayurvedic formulation of panchgavya, containing *G. glabra*, *E. officinalis*, and cow's ghee, is tranquillizing in nature (Achliya et al. 2004).

7.2.4 *Gomaya (Cow Dung Juice)*

Guna: Laghu, rooksha
Rasa: Kashaaya, Tikta
Vipaaka: Kat'u (Ushna)
Veerya: S'eeta

It is rich in nitrogen and organic matter. The gobar gas liberated by its fermentation carries hydrogen, methane, carbon dioxide, etc. In cow dung, *Mycobacterium vaccae* bacterium is present, which stimulates a cluster of neurons in the brain that secrete serotonin hormone, which is a mood enhancer to feeling of good and happiness (Gupta et al. 2016). Cow dung is claimed to be beneficial for getting rid of dandruff, making skin soft. rid of bad odor and skin disease (Raja et al. 2021). Tooth powder made up from cow dung is used to fight germs or cavities, brings relief to ulcers, mouth boils, and removes bad breath, thus preventing problems related to the mouth. As compared to the cow dung cakes, fresh cow dung kills the microbes of TB and malaria. Smoke from ashes essentially causes our eyelids to open and close numerous times so that a good amount of water comes out from the eyes, which increases the vision power of an individual with age. Cow dung has prophylactic and antiseptic properties (Munshi et al. 2018). Cow dung also has antifungal properties. Fresh cow dung is pure until it has rested on the ground; after resting on the ground, it starts change (Akinde and Obire 2008). Cow dung is used for the bioremediation of several pesticides and is found to be productive because of the existence of a larger microbial population and a higher concentration of nutrients (Behera and Ray 2021; Randhawa and Singh Kullar 2011).

7.2.5 *Gomutra (Cow's Urine)*

Guna: Rooksha, Laghu, Teekshna
Rasa: Lavana, Kat'u
Veerya: Ushna
Vipaka: Kat'u
Dosha karma: Aggravates pitta, Vaata kapha shamana

Cow urine is one of the best as well as useful urine as compared to other seven urine present in the ayurvedic texts (Randhawa 2010). It is useful in microbial diseases and works as a germicide. It contains ammonia, and it is used externally as well as internally. It is effortlessly digestible. It increases the efficiency of antifungal, antimicrobial, and anticancer drugs (Patel et al. 2021). Cow urine helps them to survive and decreases apoptosis in lymphocytes, competently repairs the damaged DNA, and thus, it is useful in cancer therapy. Cow urine has been granted US patients (No. 6,410,059 and 6896907) for its therapeutic properties, specifically as an antibiotic, bio enhancer, and anticancer agent (Khanuja et al. 2002). Due to the ample number of constituents (copper, iron, nitrogen, manganese, sulfur, carbolic acid, chlorine, silicon, magnesium, calcium salts, citric, creatinine, enzymes, mineral salts, hormones, vitamins like A, B, C, D, E, gold acids, uric acid, etc.) present in the cow urine and which act as bio-reducing and/or stabilizing agents, there is no need to add external noxious stabilizing/capping additives. The chemical compounds identified in panchgavya and their biological activity in the human body are explained in Table 7.2. The detailed protocols of biogenic synthesis of NMs by cow urine are also described by the authors (Wang et al. 2006). Copper and iron help in the production of erythropoiesis and hemoglobin, control unwanted fat, and phosphorus supports the elimination of stones from the urinary tract (Camaschella et al. 2016). Calcium works as a germicidal; it helps in bone strengthening and the purification of blood. Ammonia, mucus, and the air of the body, stabilize bile, stabilizes blood. Nitrogen, NH2, N2 Removes toxins and blood abnormalities, activates the kidneys, natural restorative for the urinary tract.

7.3 Therapeutic Effects of Panchgavya on Human Health

Ayurvedic practitioners frequently turn to panchgavya substances to treat conditions like hyperlipidemia, leukoderma, arthritis, nutritional deficiencies, renal disorders, digestive tract problems, asthma, acid reflux, and more (Fig 7.1). These treatments show promise as effective cancer and HIV therapies (Patel et al. 2018).

7.3.1 Analgesic Effect

To assess analgesic activity, rats (150–200 g) of both sexes were given 2 2 ml of cow urine, 2 ml of its distillate orally diclofenac sodium (50 mg/kg, oral) as standard. The tail immersion test measured analgesia at 0, 30, 60, and 90 min after injection. Cow urine and distillate orally had analgesic efficacy comparable to diclofenac sodium. Steroids and volatile fatty acids generated the analgesic effect in other investigations utilizing advanced analytical methods. When the analgesic effects of cow urine and its distillate were tested in a rat tail immersion study, the cow urine group showed results comparable to the standard drug group, which had been given diclofenac sodium. Cow urine contains steroidal moieties and a few volatile fatty acids, the presence of which has been established through other studies using chemical and

Table 7.2 Chemical compounds identified in panchgavya and their biological activity in the human body

Biological activity	Chemical compound	References
Anti-HIV and anticonvulsant activity	Paraldehyde	Gocmen and Oktem (1994)
React with neurotransmitters and proteins	Acetaldehyde, tetramer	Gocmen and Oktem (1994)
Anti-HIV and anticonvulsant activity	Paraldehyde	Gocmen and Oktem (1994)
Anti-oxidant activity	2-Butanol, 3-chloro	Gocmen and Oktem (1994)
Antimicrobial activity	Methoxyacetic acid, butyl ester	Ganesh and Mohankumar (2017)
Antimicrobial, anti-inflammatory, antioxidant, Anticancer, antiarthritic, antihistaminic, antieczemic, and anticoronary activities	Butanoic acid, 3-hydroxy-, ethyl ester, (+/-)-	Ganesh and Mohankumar (2017)
Antibacterial, antifungal, antioxidant, nematicide, and pesticidal activities	Propanoic acid, 3-methoxy-, methyl ester	Ganesh and Mohankumar (2017)
Mosquitocidal efficacy and antimicrobial	2-Diphenyl-1-picrylhydrazyl, 2,20 -azino-bis (3-ethylbenzothiazoline-6-sulphonic acid)	Sathiyaraj et al. (2022)
Antimicrobial activity and bioactive compounds against Alzheimer's disease	Cholest-5-en-3-ol (3.beta.)-, acetate	Sathiyaraj et al. (2022)
Antioxidant	Di-n-propylmalonic acid	Sathiyaraj et al. (2022)
Anti-fungal and anti-mycotoxigenic activity	1-Methyl-1-(2-hydroxyethyl)-1-silacyclobutane	Sathiyaraj et al. (2022)
Anti-depression, anti-diabetic and anxiety	Pseduosarsasapogenin-5,20-dien methyl ether	Sathiyaraj et al. (2022)
No biological activity found	Butanamide, 3,n-dihydroxy-	Sathiyaraj et al. (2022)
Treatment of indigestion, cold, asthma, fever, nausea, vomiting, and diarrhea	2,3-o-benzal-d-mannosan	Sathiyaraj et al. (2022)
Antibacterial activity	Ethene, 1,10 -[oxybis (2,1-ethanediyloxy)]bis	Sathiyaraj et al. (2022)

instrumental analysis, and which are responsible for the urine's analgesic effects (Singh 2019).

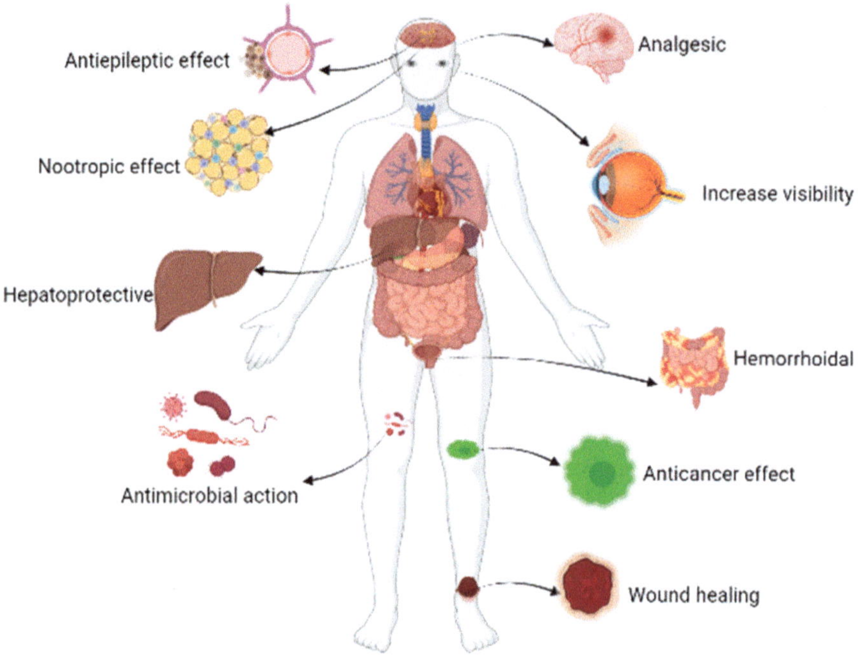

Fig. 7.1 Effects of panchgavya on human health

7.3.2 *Hepatoprotective Effect*

At a dose range of 150–300 mg/kg, p.o., PGG protected the livers of male Sprague Dawley albino rats from the toxicity of carbon tetrachloride (CCl4). Liver function was evaluated via estimation of serum marker enzymes using silymarin as a reference standard; these enzymes included alanine aminotransferase (ALT), aspartate aminotransferase (AST), alkaline phosphatase (ALP), and acid phosphatase (ACP) (Ahmad et al. 2019). PGG formulation was as effective as silymarin in preventing the CCl4-induced elevation of ALT, AST, ALP, and ACP. Thymidylate synthetase and thymidine kinase levels are increased by phospholipids during repair; these increases reach a maximum after 72 h, indicating liver regeneration (Devarshi et al. 1986).

7.3.3 *Anti-hemorrhoid activity*

Anal mucosa enlargement and proximal displacement are the telltale signs of hemorrhoids. Both male and female patients with hemorrhoids of grades I and II were treated with 250 mg of Gaumutra-ghana (extract of cow urine) twice daily for

30 days. The effects of drinking cow urine on the large intestine were beneficial. Clinical research has found that ingesting cow urine can help with pain, make bowel movements easier, and stop itching and bleeding (Lohsiriwat 2012; WTalokar et al. 2013).

7.3.4 Wound Healing Activity

Cows' ghee may have wound-healing properties. The high levels of saturated and unsaturated fatty acids are thought to have helped with the healing of wounds (Biyani et al. 2011). Researchers looked at how well cow ghee and an extract from Aegle marmelos leaves worked together to heal wounds by measuring their effects on a number of different parameters. These included wound size reduction, wound contraction, wound closure, and tissue regeneration in the affected area. In just 8 days, the combination accelerated the healing of wounds (Biyani et al. 2011; Shaikh et al. 2019). The topical application of 0.5 g of formulated gel containing both cow ghee and Aloe vera was studied for its effect on wound healing. Between 21 and 24 days, wound contraction ceased, leading to increased epithelization, tensile strength, and collagen formation. Similar therapeutic efficacy to that of 1% w/w framycetin sulphate cream (0.5 g) was observed in wound healing (Shaikh et al. 2019; Shivananjappa et al. 2020). The wound-healing properties of cow urine were tested using an excision wound model in male and female Wistar albino rats. The nitrofurazone standard was utilized. The results of the experiment demonstrated that cow urine has significant wound-healing ability. Cow urine was found to be more effective than nitrofurazone at wound healing after being applied externally to the affected area for 14 days, with 50% wound healing compared to 50.3% healing with nitrofurazone (Sanganal et al. 2011).

7.3.5 Effect on Eyes

Dry eyes, itching, burning sensations, and redness are all symptoms of computer vision syndrome (CVS), a widespread issue around the world. Treatment involves using lubricating eye drops, which can be harmful if used frequently because of the preservatives they contain. Since cow ghee is naturally lubricating, it can be safely used to treat CVS. Vitamin A, which is present in cow ghee, keeps the tear film from drying out and protects against sun damage and blindness (Mulik and Bhusari 2013). Vitamin A deficiency is a major cause of widespread blindness among children in India. A 250 ml serving of cow milk contains vitamin A and B_{12} sufficient to meet 75% daily vitamin A requirement of the pre-school child (Matte et al. 2012).

7.3.6 Anti-microbial Activity

Cow urine and cow urine distillate were evaluated for their antimicrobial activity against common pathogens like *Salmonella typhi, Bacillus subtilis, Pseudomonas aeruginosa,* and *Klebsiella pneumonia* (Ahuja 2012). Cow urine distillate showed maximum activity (zone of inhibition) against *P. aeruginosa* (7.06 0.05 mm), *S. typhi* (6.3 1.23 mm), and *E. coli* (8.0 0.17 mm) at 5, 10, and 15 ml (Shah et al. 2011). Additionally, both cow urine and distillate showed free radical scavenging activity. To a lesser extent than distillate, fresh cow urine was more effective. The antimicrobial activity of fresh cow urine was comparable to that of Ofloxacin, which was the gold standard (Jarald et al. 2008).

7.3.7 Anti-epileptic Effect

The anti-epileptic effects of PGG were studied in a rat model of convulsions induced by maximal electroshock (MES). Through boosting rat locomotor activity, PGG was found to have anti-convulsant properties. Also, the formulation effectively prevented the sleepiness brought on by the phenobarbitone. When treating epilepsy, the PGG is recommended as an adjunctive therapy (Devesh and John 2012).

7.3.8 Nootropic Effect

A diazepam-induced amnesia paradigm was compared nootropic effects of PG to piracetam (270 mg/kg). Transfer latency and scape transfer latency were assessed using the Elevated Plus Maze (EPM) and Morris Water Maze (MWM). Swiss albino mice (18–30 g) received 2.5, 5, and 10 g/kg PG, while Wistar rats (120 g) received 1.75, 3.5, and 7.0 g/kg (160–200 g). The 5 g/kg and 3.5 g/kg formulation groups had therapeutic intensity comparable to the gold standard in EPM and MWM testing (Shastra and Kalpana 2014). Similar studies found that 5 gm/kg of PG had a substantial effect on the EPM test, and that piracetam and PG at 3.5 gm/kg had a significant memory-enhancing action in the MWM test (Achliya et al. 2003).

7.3.9 Anti-cancer Effect

In a rat model of 7,12-dimethylbenz (a)-anthracene (DMBA)-induced mammary carcinogenesis, cow ghee upregulated cox-2 and PPAR-g in mammary glands, compared to soyabean oil. Cow ghee had lower tumor rates (26.6%), weight (1.67 g), and volume (1925 mm^3) than soy oil (65.4%, 6.18 g, 6285 mm^3) (Rani and Kansal 2011). (Malik et al. 2022) patented bioactive fraction from cow urine distillate (Go Mutra) as a bioenhancer of anti-infective, anti-cancer, and nutritional agents. These patents claim that cow urine and allopathic pharmaceuticals improve drug efficacy and recovery from infectious and malignant disorders Funde et al. (2022) reported that RCUD composition protects DNA from oxidative stress. This patent disclosed RCUD's composition and a stepwise process for accessing damaged DNA following treatment. Evidence suggests that panchgavya may also be effective as an anticancer agent (Jain et al. 2010; Rani and Kansal 2012).

7.4 Beneficial Effects of Panchgavya in Agriculture

Integrated pest management, integrated soil fertility management, and integrated disease management are just a few examples of how Panchgavya contributes to successful crop production. Here are some of Panchgavya's benefits, as described by an anonymous author (2018, p. Because of its high concentration of humid compounds and the fertilizing bio elements it contains, cow dung has been put to use for a variety of purposes since ancient times. To encourage plant growth, composted cow dung contains a high concentration of total nitrogen (0.74%), as well as a few hormones (Singh 2018) (Fig. 7.2).

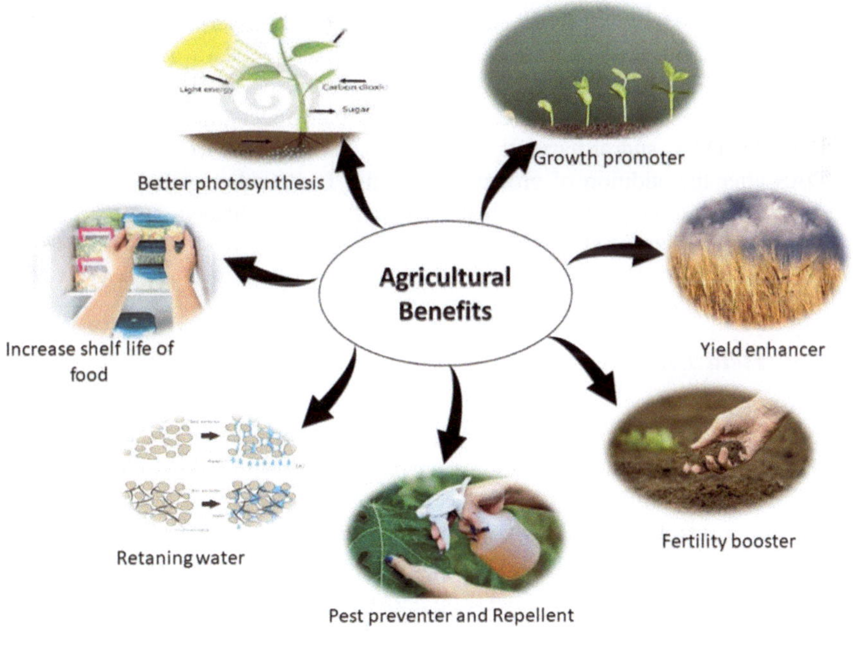

Fig. 7.2 Beneficial effects of Panchgavya in agriculture

7.4.1 Growth Promoter

In terms of fostering growth and providing immunity in the plant system, Panchgavya has the potential to play a crucial role. When they sprayed panchgavya on a chili crop, they saw new leaves emerge and turn a dark green color within 10 days (Ram et al. 2015). According to research by Vennila and Jayanthi (2008), Okra plant height and dry matter production were significantly boosted by using the full recommended dose of fertilizer in conjunction with a Panchgavya spray (2%) on the plants. The EMOs presented in Panchgavya have the potential to produce phytohormones that promote the growth of rice plants. These phytohormones may include proactive compounds that affect the receptivity of rice leaves' stomata. When leaves were dehydrated, the stomata of EMO-treated rice opened more slowly than those of water-treated control plants (i.e., they remained open longer). This indicated that EMO contained bioactive chemicals that boosted LAI by acting on the leaf stomata (Sah et al. 2018). Based on its overall performance on productivity, growth, quality of soil health, crops, and economics, biogas slurry with the Panchgavya combination was deemed the best organic nutrition practice for a sustainable maize-sunflower-green gramme system (Somasundaram et al. 2007). Turmeric, onion, paddy, ginger, banana, sugarcane, vegetables, and curry leaf were just some of the crops tested, and the results showed that Panchgavya improved the crops' growth, vitality, resistance to pests and diseases, and keeping quality (Naresh and Dhaliwal 2020). Spraying a panchgavya mixture on tomato, chilli, and cowpea seedlings accelerated their linear growth in both the shoot and root systems (Sarkar et al. 2014). Pinjari et al. (2014) conducted a field experiment in 2005–2006 to determine the impact of a 3% Panchgavya spray on sweet corn. He found that the spray had a significant impact on all of the growth characters studied, including plant height, the number of functional leaves, and dry matter accumulation in the crop at 60 DAS, 90 DAS, and at harvest. All growth characteristics were still unaffected 30 DAS after the addition of growth stimulants. The number of leaves on maize crops was found to be greater after being sprayed with Panchgavya (3%), according to research by Sachan et al. (2020).

7.4.2 Yield Enhancer

Increases in crop yield are the result of a number of interrelated physiological and biochemical activities that act on growing plants to alter their anatomy and morphology. Spraying with panchgavya has been shown to increase crop yields for a variety of reasons. It is possible that the smaller amounts of IAA and GA present in panchgavya, when foliar sprayed, would have stimulated the plant system, increasing the production of growth regulator in the cell system, and the action of growth regulators in the plant system, resulting in the necessary growth and development and a better yield (Kumar et al. 2018). This may be because panchgavya has a positive effect on the important yield attributes that have a significant positive correlation with seed and straw yield, such as the number of branches per plant and the number of pods and seeds per plant and pod (Kumar and Singh 2020). Grain production, in particular, was boosted by the use of Panchgavya because it boosted the production of growth-promoting substances (Dhanushkodi and Nageswari 2022). Foliar spray of Panchgavya at the branching and flowering stages produced the highest seed and stover yield of chickpeas (Panchal et al. 2017). Since an increase in growth and yield parameters at low concentrations of panchgavya was reported by (Rakesh et al. 2017), it is suggested that panchgavya can be used for spray after being diluted properly to 3%. The nutrients IAA and GA found in panchgavya are delivered to the plant through a foliar spray. Improvements in both growth and yield can be seen in a wheat crop treated with Panchgavya's seed-soaking and foliar spray tillering (Kumar et al. 2011).

7.4.3 Fertility Booster

By introducing more macronutrients, micronutrients, and beneficial microorganisms, panchgavya boosts soil fertility and health. It functions as an organic manure, enhancing soils' ability to retain water. Panchgavya stimulates the development and multiplication of beneficial soil microorganisms, leading to robust plant growth (Vijayakumari et al. 2012). The soil's microbial flora is vital to its overall health. Plant development and crop yield are both affected by the microorganisms that live in the rhizosphere environment around the roots. Improved agricultural sustainability was achieved through the introduction of beneficial microorganisms from Panchgavya and their subsequent establishment in the soil (Qu et al. 2020). Since the microorganisms presenting the rhizosphere environment around the roots influence the plant growth and crop yield, (Beaulah 2001) discovered that the beneficial micro-organisms from panchgavya and their establishment in the soil improved the sustainability of agriculture. Cattle manure contains a plant growth-promoting substance and other nutrients that support microbial development (Lupwayi et al. 2023). Panchgavya contains chemoautotrophs (Ammonifers and nitrifiers) that colonize on the leaves to boost ammonia uptake and total N supply (Micci et al. 2022). (Somasundaram et al. 2007) estimated that increasing the proportion of biogas slurry in the soil increased the total nutrient content and nitrogenase activity of sunflower, maize, and green gram (*Vigna radiata* L.). With the addition of biogas slurry with panchgavya, all crops were able to accumulate more nitrogen. The short-term plant growth test with *Cajanus cajan* seeds treated with Panchgavya showed enhanced length of root (19.4 cm) and shoot (16.9 cm), dry mass (147 mg), leaf area (14.57 cm^2), chlorophyll content (23 spad units), and photosynthetic activity (18.8 $\mu mol\ m^{-2}\ s^{-1}$) after 15 days of sowing (Leo Daniel Amalraj et al. 2013). Hence, these tests suggest that Panchgavya can be used as a low-cost preparation to support soil health.

7.4.4 Disease Preventers in Both Plants and Animals

They discovered that the application of liquid organics resulted in the acquisition of induced defence mechanisms by the tomato, chili, and cowpea, and they were able to quantify this effect using the polyphenol oxidase found in panchgavya. The results showed a 48.2% control of the disease in nursery beds for the tomato-damaging fungus *Pythium aphanidermatum* (Kumar et al. 1970). Panchgavya's potential as an antifungal was tested against *Rhizoctonia solani, Sclerotinia rolfsii, Fusarium solani,* and *Phytophthora colocasiae* (Rathore and Patil 2019). The antifungal activity of panchgavya was tested against five common soil-borne pathogens, and it was found to inhibit growth by 90% against *Fusarium solani* f. sp. pisi and *Fusarium oxysporum* f. sp. pisi and by 100% against *Fusarium rolfsii, Sclerotinia sclerotiorum*, and *Fusarium oxysporum* (Jandaik and Sharma 2016). Cow urine and dung were tested in vitro by Basak et al. (2002) for their ability to prevent cucumber wilt brought on by *Fusarium oxysporum* f. sp. cucumerinum and *Fusarium solani* f. sp. cucurbitae. Mucolytic bacteria were found in cow manure. Mycelial growth of *Fusarium oxysporum* f. sp. cucumerinum and *Helminthosporium sigmoideum* was inhibited by bacterial isolates that were detected in an extract of cow dung compost. (Ye et al. 2020). Bacillus strains inhibiting *R. solani, B. sorokiniaria, S. sclerotiorum, T. roseum, F. solani*, and *F. oxysporum* were isolated from cow dung by Solanki et al. (2015). *Bacillus subtilis* GB519, which is effective against *Fusarium oxysporum, Colletotrichum lagenarium, Sclerotinia sclerotiorum, Botrytis cinerea*, and *Rhizoctonia solani*, was isolated by Zhu et al. (2021) from compost. A number of important microorganisms, including *Aspergillus niger, Trichoderma harzianum, Bacillus cereus,* and *Bacillus subtilis*, were discovered in cow dung compost by (Muhammad and Amusa (2003). *Sclerotium rolfsii, Fusarium oxysporum, Pythium aphanidermatum, Hymenoptera maydis*, and *Rhizoctonia solani*, all of which can cause blight in seedlings, were sprayed with the microbes. *Bacillus subtilis, B. cereus. Pseudomonas aeruginosa* and *Rhizobium radiobacter* were found in cow dung manure by Charest et al. (2005). Tomato wilt is caused by *Fusarium oxysporum* f. sp. lycopersici successfully treated by animal dung and urine (Hajji-Hedfi et al. 2018). *Pythium ultimum, Rhizoctonia solani, Phytophthora, Fusarium oxysporum, Verticillium dahlia*, and *Sclerotinia* spp. were all reduced by 20–60% in compost-amended soils (Noble and Coventry 2005).

7.4.5 Water Retaining

Drought resistance water evaporation is decreased by a thin, oily film that forms on the leaves and stems. The plants' extensive and deep roots enable them to withstand prolonged dry spells. Both of the aforementioned elements help to ensure drought resistance and reduce the irrigation water requirement by 30%.

7.4.6 Panchgavya-Based Nanotechnology

Recent rapid advancements in nanotechnology can be attributed to its numerous and varied uses in fields as diverse as biomedicine, biosensors, agriculture, bio-imaging, biotechnology, pharmaceuticals, and water purification. In this perspective, panchgavya-based nanoparticles have become one of the functional areas of research in drug delivery (Prasad et al. 2020). Synthesizing copper nanoparticles (CuNPs) and panchgavya with biological properties is an important part of making a NPs that are beneficial in science. Multiple methods were used to determine the properties of copper nanoparticles, including Fourier transmission infrared spectroscopy, ultraviolet spectrophotometer, dynamic light scattering, X-ray diffraction analysis, High-resolution transmission electron microscope, scanning electron microscope, and X-ray fluorescent microscopy. In addition, their anti-inflammatory and cytotoxic effects have been measured (Arumugam et al. 2019). Synthesis of silver nanoparticles using 4 ml of Panchgavya filtrate in 96 ml of 1 mM $AgNO_3$ solution. Antibiotic-resistant bacteria, including Acinetobacter sp., Aeromonas sp., and Citrobacter sp., were tested, and the prepared nanoparticles were shown to have anti-microbial activity that increased with increasing concentration (Govarthanan et al. 2014). The antimicrobial activities of Ag NPs were investigated by Prabhu et al. (2014), who described a one-pot green synthesis of Ag NPs using aged cow urine as a reducing agent. Using the disc diffusion technique, it was found that the synthesized Ag NPs displayed potent antimicrobial activity against the following bacteria and fungi: *Streptococcus epidermis, Staphylococcus aureus, Bacillus subtilis, Salmonella typhi, Escherichia coli, Fusarium oxysporum,* and *Klebsiella pneumonia*. Synthesized Ag NPs from panchgavya showed significant antibacterial activity against *Pseudomonas* sp., according to research by N. Jain et al. (2019). The biosynthesis of spherical Ag NPs using Panchgavya and silver nitrate was described by Govarthanan et al. (2014). TEM, SEM, and XRD methods all confirmed the emergence of Ag NPs. Significant antibacterial activity was observed against *Aeromonas* sp., *Acinetobacter* sp., and *Citrobacter* sp. for the NPs in their as-synthesized form. Zinc oxide nanoparticles (ZnO) extracted from cow dung showed lower antimicrobial activity than bulk Zn and commercial ZnO nanoparticles when tested against *Escherichia coli* DH5-alpha. Based on the results of this research, ZnO may have a greater effect on seed germination and a lesser antibacterial capacity than previously thought (Javed et al. 2022). To investigate the antibacterial and catalytic properties of Ag_2ONPs. Vinay et al. (2019) described a simple method for biosynthesis using cow urine without the use of any noxious substances. Antibacterial activity was found to be quite high in these synthetic Ag_2O NPs against *Staphylococcus aureus, Pseudomonas desmolyticum, Escherichia coli,* and *Klebsiella aerogenes*. They also showed that methylene blue could be degraded photocatalytically by Ag_2ONPs for 4 h, with a degradation rate of up to 83.63%. Abdullah et al. (2015) described a concentration-dependent study in which Cd NPs were synthesized using cow urine, and their effect against a pathogenic bacterium, specifically *Klebsiella pneumonia, Staphylococcus aureus,* and *Pseudomonas aeruginosa,* was investigated. Cow urine encapsulated cellulose NPs were reported by Suk et al. (2018), who used the microemulsion technique to test the NPs' antimicrobial activity against *Bacillus subtilis* and *Aspergillus niger* via disc diffusion assay. The synthesis of spherical $CuFe_2O_4$ NPs using the sol-gel method and cow urine as a chelating agent was described by Satheeshkumar et al. (2019). Typically, $CuFe_2O_4$NPs range in size from 14.2 to 22.3 nm. Using urea and cow urine (Chamoli et al. 2020) synthesized graphene nanosheets (GNs). The electrical and optical performance of these synthetic GNs was significantly enhanced.

7.5 Conclusion and Future Perspectives

It follows that Panchgavya/Cowpathy, a modernized version of ancient science, is an approach with great potential moving forward. Panchgavya has shown promise as a treatment for a variety of human ailments, and it has the potential to make a positive contribution to society. Therefore, enlightening the populace about cows and panchgavya can help alleviate issues associated with a lack of food, fuel, shelter, health, nutrition, the elimination of poverty and unemployment, and a viable alternative energy source. Since only a small amount of literature is available, widespread promotion of the use of panchgavya in treating diseases in plants, humans, and animals is necessary. The scientific community should pay more attention to the panchgavya therapy because it has the potential to improve human health and could be widely adopted, promoted, and used if only it were backed by solid evidence. The use of chemical fertilizers and pesticides in farming has been cited as a potential threat to the surrounding ecosystem. To counteract this, organic manures like panchgavya can be used instead. In addition, panchgavya can function as fertilizer and pesticides, allowing for maximum productivity and financial returns. Invariably thermotolerant, microorganisms derived from the Panchgavya can be used in bio-based formulations of fertilizers, microbial enzymes, and growth regulators to compensate for the loss of crop productivity caused by global warming. Research into the chemistry, biology, microbiology, and pharmacology of the bioactive compounds found in Panchgavya, as well as clinical trials aimed at preventing and treating disease in humans and animals, is urgently needed. By working together, scientists, researchers, and clinicians can ensure that this alternative low-cost therapy with no side effects is as effective as it can be and win the public's trust. It's important that the panchgavya theory of Ayurveda spread not only among conventional households but also among the more enlightened members of today's scientific community. Having the backing of scientists, researchers, and clinicians is essential for bolstering the miraculous effects of cow urine therapy and the other panchgavya elements, which will encourage doctors and the general public to adopt and spread this wonderful therapy for the health benefits and help fight various diseases and disorders of both animals and humans.

References

Achliya GS, Kotagale NR, Wadodkar SG, Dorle AK (2003) Hepatoprotective activity of panchagavya ghrita against carbon tetrachloride induced hepatotoxicity in rats. Indian J Pharmacol 35(5):308–311

Achliya GS, Wadodkar SG, Dorle AK (2004) Evaluation of sedative and anticonvulsant activities of Unmadnashak Ghrita. J Ethnopharmacol 94:77–83

Ahmad N, Saleem M, Atta BM, Mahmood S (2019) Characterization of desi ghee extracted by different methods using fluorescence spectroscopy. J Fluoresc 29:1411–1421

Ahuja A (2012) Antimicrobial activities of cow urine against various bacterial strains. Int J Recent Adv Pharm Res 2:84–87

Akinde SB, Obire O (2008) Aerobic heterotrophic bacteria and petroleum-utilizing bacteria from cow dung and poultry manure. World J Microbiol Biotechnol 24:1999–2002

Ali MN, Ghatak S, Ragul T (2011) Biochemical analysis of Panchagavya and Sanjibani and their effect in crop yield and soil health. J Crop Weed 7:84–86

Arumugam DG, Sivaji S, Dhandapani KV, Nookala S, Ranganathan B (2019) Panchagavya mediated copper nanoparticles synthesis, characterization and evaluating cytotoxicity in brine shrimp. Biocatal Agric Biotechnol 19:101132

Banks JM, Clapperton JL, Muir DD, Girdler AK (1986) The influence of diet and breed of cow on the efficiency of conversion of milk constituents to curd in cheese manufacture. J Sci Food Agric 37:461–468

Basak A, Lee MW, Le TS (2002) In vitro inhibitory activity of cow urine and dung to *fusarium solani* f. sp. *cucurbitae*. Mycobiology 30:51–54

Beaulah A (2001) Growth and development of Moringa (*Moringa oleifera* Lam.) under organic and inorganic systems of culture. [Thesis]

Behera SS, Ray RC (2021) Bioprospecting of cowdung microflora for sustainable agricultural, biotechnological and environmental applications. Curr Res Microb Sci 2:100018

Bhavana KR, Shreevathsa (2014) Medical geography in Charaka Samhita. Ayu 35:371

Bishnoi S, Prajapat P, Kumar Mishra P, Mishra I, Radhakrishnan S, Radhakrishnan Rajasthan S (2021) A clinical study of Panchgavya Ghrita and Takradhara in the management of Mano-Avasada (depression). World J Pharm Res 10:1753–1770

Biyani DM, Verma PRP, Dorle AK, Boxey V (2011) Wound healing activity of cow ghee: a veterinary case report. Int J Ayur Med 2(3):115–118

Camaschella C, Pagani A, Nai A, Silvestri L (2016) The mutual control of iron and erythropoiesis. Int J Lab Hematol 38(Suppl 1):20–26

Chakraborty B, Sarkar I (2019) Quality analysis and characterization of Panchagavya, Jeevumrutha and Sasyamrutha. Int J Curr Microbiol Appl Sci 8:2018–2026

Chamoli P, Srivastava T, Tyagi A, Raina KK, Kar KK (2020) Urea and cow urine-based green approach to fabricate graphene-based transparent conductive films with high conductivity and transparency. Mater Chem Phys 242:122465

Chandrakala M, Bidari NS, Radder BIA (2007) Effect of FYM and fermented liquid manures on nutrients uptake by chilli (*Capsicum annuum* L.) and soil nutrient status at harvest. Asian J Hortic 4:19–24

Charest MH, Beauchamp CJ, Antoun H (2005) Effects of the humic substances of de-inking paper sludge on the antagonism between two compost bacteria and *Pythium ultimum*. FEMS Microbiol Ecol 52:219–227

Chauhan R (2018) Immunomodulatory properties of indigenous cow urine. MOJ Immunol 6(5): 302–303

Cortés-Zavaleta O, López-Malo A, Hernández-Mendoza A, García HS (2014) Antifungal activity of lactobacilli and its relationship with 3-phenyllactic acid production. Int J Food Microbiol 173: 30–35

Devarshi P, Kanase A, Kanase R, Mane S, Patil S, Varute AT (1986) Effect of mandur bhasma on lipolytic activities of liver, kidney and adipose tissue of albino rat during CCl4 induced hepatic injury. J Biosci 10(2):227–234

Devesh G, John PS (2012) Effect of Panchagavya Ghritra on some neurological parameters in albino rats. Asian J Pharm Clin Res 5(1):154–156

Dhanushkodi V, Nageswari R (2022) Influence of foliar application of Panchagavya and fish extracts and application of organic substances on yield and quality of ash gourd under organic farming. Int J Plant Soil Sci 34:1110–1114

Fulzele S, Satturwar P, Joshi SB, Dorle A (2003) Study of the immunomodulatory activity of Haridradi ghrita in rats. Indian J Pharmacol 35(1):51–54

Funakoshi S, Doi T, Nakajima T, Suyama T, Tokuda M (1982) Antimicrobial effect of human serum IgA. Microbiol Immunol 26(3):227–239

Funde S, Patil S, Palep H (2022) Anti-cancer effect of Panchagavya on human colon adenocarcinoma (HCT-116) cell line. Int J Pharm Sci Res 13(4):1771

Ganesh M, Mohankumar M (2017) Extraction and identification of bioactive components in *Sida cordata* (Burm.f.) using gas chromatography–mass spectrometry. J Food Sci Technol 54(10): 3082–3091

Garg AK, Mudgal V (2007) Organic and mineral composition of Gomeya (cow dung) from Desi and crossbred cows—a comparative study. Int J Cow Sci 3(1 and 2):17–19. https://www.indianjournals.com/ijor.aspx?target=ijor:ijcs&volume=3&issue=1and2&article=003

Gocmen B, Oktem N (1994) Paraldehyde Fuchsin staining and secretion of rumen ciliates of cattle. Turk J Zool 18(2):107–110. https://www.researchgate.net/publication/300178103_Paraldehyde_Fuchsin_Staining_and_Secretion_of_Rumen_Ciliates_of_Cattle

Govarthanan M, Selvankumar T, Manoharan K, Rathika R, Shanthi K, Lee KJ, Cho M, Kamala-Kannan S, Oh BT (2014) Biosynthesis and characterization of silver nanoparticles using panchakavya, an Indian traditional farming formulating agent. Int J Nanomedicine 9(1): 1593–1599. https://doi.org/10.2147/IJN.S58932

Gupta KK, Aneja KR, Rana D (2016) Current status of cow dung as a bioresource for sustainable development. Bioresour Bioprocess 3(1). https://doi.org/10.1186/S40643-016-0105-9

Hajji-Hedfi L, Regaieg H, Larayedh A, Chihani N, Horrigue-Raouani N (2018) Biological control of wilt disease complex on tomato crop caused by Meloidogyne javanica and Fusarium oxysporum f.sp. lycopersici by Verticillium leptobactrum. Environ Sci Pollut Res 25(19): 18297–18302. https://doi.org/10.1007/s11356-017-0233-6

Halder S, Anand U, Nandy S, Oleksak P, Qusti S, Alshammari EM, El-Saber Batiha G, Koshy EP, Dey A (2021) Herbal drugs and natural bioactive products as potential therapeutics: a review on pro-cognitives and brain boosters perspectives. Saudi Pharm J 29(8):879–907. https://doi.org/10.1016/J.JSPS.2021.07.003

Haug A, Høstmark AT, Harstad OM (2007) Bovine milk in human nutrition—a review. Lipids Health Dis 6:25. https://doi.org/10.1186/1476-511X-6-25

Hill JP, Rowan AM, Boland MJ, Landells VA (2018) Diabetes mellitus and consumption of Milk and dairy products. In: Reference module in food science. https://doi.org/10.1016/B978-0-08-100596-5.00968-9

Jain NK, Gupta VB, Garg R, Silawat N (2010) Efficacy of cow urine therapy on various cancer patients in Mandsaur District, India-A survey. Int J Green Pharm 4(1):29–35. https://doi.org/10.4103/0973-8258.62163

Jain N, Bhosale P, Tale V, Henry R, Pawar J (2019) Hydrothermal assisted biologicasynthesis of Silver nanoparticles by using honey and gomutra (Cow Urine) for qualitative determination of its antibacterial efficacy against pseudomonas sp. isolated from contact lenses. Eurasian. J Biosci 13(1):27–33

Jandaik S, Sharma V (2016) Antifungal potential of panchagavya against soil borne fungal pathogens associated with Capsicum nurseries. Int Invention J Agric Soil Sci 4(2):22–26

Jandaik S, Thakur P, Kumar V (2015) Efficacy of cow urine as plant growth enhancer and antifungal agent. Adv Agric 2015. https://doi.org/10.1155/2015/620368

Jarald E, Edwin S, Tiwari V, Garg R, Toppo E (2008) Antioxidant and antimicrobial activities of cow urine. Glob J Pharmacol 2(2):20–22

Javed Z, Tripathi GD, Mishra M, Gattupalli M, Dashora K (2022) Cow dung extract mediated green synthesis of zinc oxide nanoparticles for agricultural applications. Sci Rep 12(1). https://doi.org/10.1038/s41598-022-22099-y

Joseph B, Sankraganseh K (2011) Antifungal efficacy of Panchagavya. Int J PharmTech Res 3(1): 585–588. www.patft.uspto.gov

Joshi R, Kh R, Sk S, Tripathi M, Yk G (2015) Panchagavya Ghrita, an Ayurvedic formulation attenuates seizures, cognitive impairment and oxidative stress in pentylenetetrazole induced seizures in rats. Indian J Exp Biol 53:446–451

Kanaujia A, Kumar US (2018) Comparative assessment of laboratory produced and commercially available cow urine distillates. Int J Adv Res 6(8):404–406. https://doi.org/10.21474/IJAR01/7534

Karthika SR, Sajeena A, Girija VK, Jocob J, Heera G (2017) Antifungal activities of organic preparations, botanicals and nonhazardous chemicals against Rhizoctonia solani Kuhn causing sheath blight of rice. J Trop Agric 55(1):104–113

Kazimierska K, Kalinowska-Lis U (2021) Milk proteins—their biological activities and use in cosmetics and dermatology. Molecules 26(11). https://doi.org/10.3390/MOLECULES26113253

Khanuja SPS, Kumar S, Shasany AK, Arya JS, Darokar MP, Singh M, Sinha P, Awasthi S, Gupta SC, Gupta VK, Gupta MM (2002) Pharmaceutical composition containing cow urine distillate and an antibiotic. U.S. Patent 6,410,059

Koksal Z, Gulcin I, Ozdemir H, Koksal Z, Gulcin I, Ozdemir H (2016) An important Milk enzyme: Lactoperoxidase. In: Milk proteins – from structure to biological properties and health aspects. https://doi.org/10.5772/64416

Krishna M, Bandari S, Sangu PK, Goli PP, Ala N, Prasanna Kumar T (2011) A prospective study on Parpati Kalpana w.s.r to Panchamrut Parpati. Int J Ayurvedic Med 1(3). https://doi.org/10.47552/IJAM.V1I3.46

Kumar A (2015) Detection of vegetable oil and animal depot fat adulteration in anhydrous Milk fat (ghee) using fatty acid composition. MOJ Food Process Technol 1(3):1–5. https://doi.org/10.15406/MOJFPT.2015.01.00013

Kumar CS, Singh G (2020) Effect of Panchagavya on growth and yield: a review. Int J Curr Microbiol Appl Sci 9(12):617–624. https://doi.org/10.20546/ijcmas.2020.912.073

Kumar R, Hooda I, Karwasra S (1970) Efficacy of Mahapancha Gavya (MPG) in controlling damping-off in tomato caused by Pythium aphanidermatum. Bangladesh J Agric Res 35(1): 11–16. https://doi.org/10.3329/BJAR.V35I1.5862

Kumar S, Ganesh P, Tharmaraj K, Saranraj P (2011) Growth and development of blackgram (Vigna mungo) under foliar application of Panchagavya as organic source of nutrient. Curr Botony 2(3): 9–11. www.currentbotany.org

Kumar P, Shri P, Veer R (2017) Yield and soil organic matter dynamics as affected by the long-term use of organic and inorganic fertilizers under Rice—wheat cropping system in subtropical Mollisols. Agric Res 6(6):575–577. https://doi.org/10.1007/s40003-017-0282-6

Kumar P, Khan N, Deo Singh P, Singh A, Yogesh, Yadav V (2018) Panchagavya: a low cost organic input in organic farming-a review. Int J Chem Stud 6(6):575–577. http://agriculturaltips.com/prepare-

Lamb MM, Miller M, Seifert JA, Frederiksen B, Kroehl M, Rewers M, Norris JM (2015) The effect of childhood cow's milk intake and HLA-DR genotype on risk of islet autoimmunity and type 1 diabetes: the Diabetes Autoimmunity Study in the Young. Pediatr Diabetes 16(1):31–38. https://doi.org/10.1111/PEDI.12115

Leo Daniel Amalraj E, Praveen Kumar G, Mir Hassan Ahmed SK, Abdul R, Kishore N (2013) Microbiological analysis of Panchagavya, vermicompost, and FYM and their effect on plant growth promotion of pigeon pea (Cajanus cajan L.) in India. Org Agric 3(1):23–29. https://doi.org/10.1007/S13165-013-0042-2/METRICS

Lohsiriwat V (2012) Hemorrhoids: from basic pathophysiology to clinical management. World J Gastroenterol 18(17):2009–2017. https://doi.org/10.3748/wjg.v18.i17.2009

Lupwayi NZ, Hao X, Thomas BW, Stoeckli J, Mesina L, Polo RO (2023) Alteration of the soil microbiome and enzyme activities by forage-applied manure from cattle fed the methane inhibitor 3-nitrooxypropanol supplement. Appl Soil Ecol 183:104765. https://doi.org/10.1016/J.APSOIL.2022.104765

Malik YS, Kaur I, Sharma A (2022) Anticancer activity of cow urine distillate, an update. Int J Cow Sci 6(1):3–5. https://doi.org/10.18231/j.ijcs.2022.002

Marangoni F, Pellegrino L, Verduci E, Ghiselli A, Bernabei R, Calvani R, Cetin I, Giampietro M, Perticone F, Piretta L, Giacco R, la Vecchia C, Brandi ML, Ballardini D, Banderali G, Bellentani S, Canzone G, Cricelli C, Faggiano P et al (2019) Cow's milk consumption and health: a health professional's guide. J Am Coll Nutr 38(3):197–208. https://doi.org/10.1080/07315724.2018.1491016

Matte JJ, Guay F, Girard CL (2012) Bioavailability of vitamin B_{12} in cows' milk. Br J Nutr 107(1): 61–66. https://doi.org/10.1017/S0007114511002364

Micci A, Zhang Q, Chang X, Kingsley K, Park L, Chiaranunt P et al (2022) Histochemical evidence for nitrogen-transfer endosymbiosis in non-photosynthetic cells of leaves and inflorescence bracts of angiosperms. Biol 11(6). https://doi.org/10.3390/BIOLOGY11060876

Muhammad S, Amusa NA (2003) In-vitro inhibition of growth of some seedling blight inducing pathogens by compost-inhabiting microbes. Afr J Biotechnol 2(6):161–164. https://doi.org/10.5897/AJB2003.000-1033

Mulik SS, Bhusari DP (2013) Conceptual study of Goghrita Eye drops (Aschyotana) in computer vision syndrome. Asian J Multidiscip Stud 1(3):1–6. www.ajms.co.in

Munshi SK, Roy J, Noor R (2018) Microbiological investigation and determination of the antimicrobial potential of cow dung samples. Stamford J Microbiol 8(1):34–37. https://doi.org/10.3329/sjm.v8i1.42437

Naik K, Mishra S, Srichandan H, Singh PK, Choudhary A (2020) Microbial formulation and growth of cereals, pulses, oilseeds and vegetable crops. Sustain Environ Res 30(1):1–18. https://doi.org/10.1186/S42834-020-00051-X/TABLES/5

Naresh R, Dhaliwal S (2020) Effects of Kunapajala and Panchagavya on nutrients release, crop productivity and soil health. Asian Agri-History 24(2):147–161. https://www.researchgate.net/publication/349029073

Noble R, Coventry E (2005) Suppression of soil-borne plant diseases with composts: a review. Biocontrol Sci Tech 15(1):3–20. https://doi.org/10.1080/09583150400015904

Panchal P, Patel P, Patel A, Desai A (2017) Effect of Panchagavya on growth, yield and economics of chickpea (Cicer arietinum). Int J Chem Stud 5(2):265–267

Parodi PW (1997) Cows' milk fat components as potential anticarcinogenic agents. J Nutr 127(6): 1055–1060. https://doi.org/10.1093/JN/127.6.1055

Patel P, Joshi C, Funde S, Palep H, Kothari V (2018) Prophylactic potential of a Panchgavya formulation against certain pathogenic bacteria. F1000Res 7:1612. https://doi.org/10.12688/F1000RESEARCH.16485.1

Patel N, Kadu A, Yadav S (2021) Review on the importance of Mutra (urine) in Visha Chikitsa (treatment of poison). J Indian Syst Med 9(2):90. https://doi.org/10.4103/JISM.JISM_32_21

Pinjari SS, Talathi MS, Ranshur NJ, Bhondave TS (2014) Integrated nutrient management in maize. Indian J Tradit Knowl 13(2):398–403

Prabhu M, Mutnuri S, Dubey SK, Naik MM (2014) One-pot rapid synthesis of face-centered cubic silver nanoparticles using fermented cow urine, a nanoweapon against fungal and bacterial pathogens. J Bionanosci 8(4):265–273. https://doi.org/10.1166/JBNS.2014.1235

Praburaman L, Park JH, Govarthanan M, Selvankumar T, Oh SG, Jang JS et al (2015) Impact of an organic formulation (panchakavya) on the bioleaching of copper and lead in contaminated mine soil. Chemosphere 138:127–132. https://doi.org/10.1016/J.CHEMOSPHERE.2015.05.059

Prasad SR, Padvi MN, Suryawanshi SS, Shaikh YI, Chaudhary LS, Samant AP et al (2020) Bio-inspired synthesis of catalytically and biologically active palladium nanoparticles using

Bos taurus urine. SN Appl Sci 2(4):1–12. https://doi.org/10.1007/S42452-020-2382-3/FIGURES/9

Qu Q, Zhang Z, Peijnenburg WJGM, Liu W, Lu T, Hu B et al (2020) Rhizosphere microbiome assembly and its impact on plant growth. J Agric Food Chem 68(18):5024–5038. https://doi.org/10.1021/ACS.JAFC.0C00073

Radha TK, Rao DLN (2014) Plant growth promoting bacteria from cow dung based biodynamic preparations. Indian J Microbiol 54(4):413. https://doi.org/10.1007/S12088-014-0468-6

Raj Joshi D, Adhikari N (2019) Benefit of cow urine, milk, ghee, curd, and dung versus cow meat: mini review. Acta Sci Pharm Sci 3:2581–5423. https://doi.org/10.31080/ASPS.2019.03.0360

Raja MKMM, Manne R, Devarajan A (2021) Benefits of cow dung—a human ignored gift. J Nat Remedies 21(3):189–202. https://doi.org/10.18311/jnr/2021/26653

Rakesh S, Poonguzhali S, Saranya B, Suguna S, Jothibasu K (2017) Effect of Panchagavya on growth and yield of Abelmoschus esculentus cv. Arka Anamika. Int J Curr Microbiol Appl Sci 6(8):3090–3097. https://doi.org/10.20546/IJCMAS.2017.609.380

Ram M, Rao K, Kumar MS, Jha NK (2015) Comparative yield analysis of chilli (Capsicum annuum L.) by application of vermicompost and Panchagavya. J Chem Pharm Res 7(9):319–323. www.jocpr.com

Randhawa G (2010) Cow urine distillate as bioenhancer. J Ayur Integr Med 1(4):240. https://doi.org/10.4103/0975-9476.74089

Randhawa GK, Singh Kullar J (2011) Bioremediation of pharmaceuticals, pesticides, and petrochemicals with Gomeya/cow dung. ISRN Pharmacol 2011. https://doi.org/10.5402/2011/362459

Rani R, Kansal V (2011) Study on cow ghee versus soybean oil on 7,12-dimethylbenz(a)-anthracene induced mammary carcinogenesis & expression of cyclooxygenase-2 & peroxisome proliferators activated receptor-γ in rats. Indian J Med Res 133:497

Rani R, Kansal VK (2012) Effects of cow ghee (clarified butter oil) & soybean oil on carcinogen-metabolizing enzymes in rats. Indian J Med Res 136(3):460

Rathore MS, Patil AD (2019) Panchagavya, an organic amendment for inhibiting damping-off causing Fusarium solani and Sclerotium rolfsii under in-vitro conditions. Ind J Pure Appl Biosci 7(4):203–206. https://doi.org/10.18782/2320-7051.7693

Revathi N, Muthukumar MG, Govindarajan AA, Nandhini CA, Senthilkumar VT (2021) Review on Panchagavya: a boon to the societal need (agriculture and medicinal industries). Int J Chem Stud 9(2):113–117. www.chemijournal.com

Russo E, Scicchitano F, Citraro R, Aiello R, Camastra C, Mainardi P et al (2012) Protective activity of α-lactoalbumin (ALAC), a whey protein rich in tryptophan, in rodent models of epileptogenesis. Neuroscience 226:282–288. https://doi.org/10.1016/J.NEUROSCIENCE.2012.09.021

Sachan HK, Krishna D, Prasad A (2020) Influence of organic and inorganic sources of fertilizers on growth and yield of maize (Zea mays L.) in Fiji. Indian J Agric Res 55(6):769–772. https://doi.org/10.18805/IJARE.A-590

Sah AK, Singh M, Pradhan AK, Prasad S, Chatoopadhyaya N, Behera SK et al (2018) Effect of organic fertilizer and microbial inoculants on soil biological properties and yield of scented rice (Oryza sativa L.). J Appl Nat Sci 10(3):995–1002. https://doi.org/10.31018/JANS.V10I3.1831

Sanganal JS, Jayakumar K, Jayaramu GM, Tikare VP, Paniraj KL, Swetha R (2011) Effect of cow urine on wound healing property in wister albino rats. Vet World 4(7):317–321. https://doi.org/10.5455/vetworld.4.317

Sangeetha V, Thevanathan R (2010) Nitrate assimilation in Panchagavya treated plants. J Am Sci 6(2):80–86. http://www.americanscience.org

Sarkar S, Kundu S, Ghorai K (2014) Validation of ancient liquid organics—Panchagavya and Kunapajala as plant growth promoters. Indian J Tradit Knowl 13(2):398

Satheeshkumar MK, Ranjith Kumar E, Srinivas C, Prasad G, Meena SS, Pradeep I et al (2019) Structural and magnetic properties of $CuFe_2O_4$ ferrite nanoparticles synthesized by cow urine assisted combustion method. J Magn Magn Mater 484:120–125. https://doi.org/10.1016/J.JMMM.2019.03.128

Sathiyaraj S, Suriyakala G, Gandhi AD, Babujanarthanam R, Kaviyarasu K, Rajakrishnan R et al (2022) Chemical composition and mosquitocidal efficacy of Panchagavya against Anopheles stephensi, Aedes aegypti and Culex quinquefasciatus. J King Saud Univ Sci 34(4). https://doi.org/10.1016/j.jksus.2022.101960

Schnürer J, Magnusson J (2005) Antifungal lactic acid bacteria as biopreservatives. Trends Food Sci Technol 16(1–3):70–78. https://doi.org/10.1016/J.TIFS.2004.02.014

Shah CP, Patel DM, Dhami PD, Kakadia J, Bhavsar D, Vachhani UD, Trivedi MN, J JV (2011) In vitro screening of antibacterial activity of cow urine against pathogenic human bacterial strains. Int J Curr Pharm Res 3(2):91–92

Shaikh SS, Ukande MD, Murthy K, Shete RV, Solunke R, Shaikh MS (2019) Traditional remedies for wound healing: a review. J Drug Deliv Ther 9(4-s):761–764. https://doi.org/10.22270/JDDT.V9I4-S.3263

Sharma H, Chandola HM (2015) Ayurvedic approach to food and dietary supplements for the brain and neurologic health. In: Watson RR, Preedy VR (eds) Bioactive nutraceuticals and dietary supplements in neurological and brain disease: prevention and therapy. Academic, pp 173–177. https://doi.org/10.1016/B978-0-12-411462-3.00018-7

Sharma B, Singh M (2018) Isolation and characterization of bacteria from cow dung of Desi cow breed on different morpho-biochemical parameters in Dehradun. Int J Adv Pharm Biol Chem 4(2):1–6. https://www.researchgate.net/publication/287993616

Shastra R, Kalpana B (2014) Assessment of nootropic activity of Panchagavya Ghrita in animal models. Int J Sci Res Publ 5(8). www.ijsrp.org

Shivananjappa M, Prakruthi M, Manasa R, Naik RS (2020) Health benefits of ghee (clarified butter) – a review from ayurvedic perspective. IP J Nutr Metab Health Sci 3(3):64–72. https://doi.org/10.18231/J.IJNMHS.2020.014

Singh AK (2018) Exploitation of Panchagavya: a novel approach for the sustainable production of vegetable crops in Pindar valley of Uttarakhand. J Pharmacogn Phytochem 7(6):199–203. https://www.academia.edu/38933708

Singh S (2019) Pharmacotherapeutics of gomutra (cow urine). Pharma Innov J 8(6):707–712. www.thepharmajournal.com

Solanki MK, Singh RK, Srivastava S, Kumar S, Kashyap PL, Srivastava AK (2015) Characterization of antagonistic-potential of two *bacillus* strains and their biocontrol activity against *Rhizoctonia solani* in tomato. J Basic Microbiol 55(1):82–90. https://doi.org/10.1002/jobm.201300528

Somasundaram E, Amanullah M, Thirukkumaran K, Chandrasekaran R, Vaiyapuri K, Sathyamoorthi K (2007) Biochemical changes, nitrogen flux and yield of crops due to organic sources of nutrients under maize based cropping system. J Appl Sci Res 3(12):1724–1729

Somdutt, Bhadu K, Rathore RS, Shekhawat PS (2021) Jeevamrut and Panchagavya's consequences on growth, quality and productivity of organically grown crops: a review. Agric Rev 1:1–9. https://doi.org/10.18805/ag.r-2239

Suk KH, Gopinath SCB, Anbu P, Lakshmipriya T (2018) Cellulose nanoparticles encapsulated cow urine for effective inhibition of pathogens. Powder Technol 328:140–147. https://doi.org/10.1016/J.POWTEC.2018.01.010

Sumit P, Patel N (2020) Importance of soil microbes on Panchagavya-based fertilizer for sustainable agriculture: a review. J Pharmacogn Phytochem 9(3):2277–2284. www.phytojournal.com

Tharmaraj K, Ganesh P, Suresh K, Anandan A, Kolanjinathan K (2011) A critical review on Panchagavya—A boon plant growth. Int J Pharm Biol Arch 2(6):1611–1614. www.ijpba.info

Udwadia TE (2011) Ghee and honey dressing for infected wounds. Indian J Surg 73(4):278. https://doi.org/10.1007/S12262-011-0240-7

Vennila C, Jayanthi C (2008) Response of okra to integrated nutrient management. J Soils Crops 18:36–40

Vijayakumari B, Yadav RH, Gowri P, Kandari L (2012) Effect of Panchagavya, humic acid and micro herbal fertilizer on the yield and post-harvest soil of soybean (*Glycine max* L.). Asian J Plant Sci 11(2):83–85

Vinay SP, Udayabhanu NG, Chandrappa CP, Chandrasekhar N (2019) Novel Gomutra (cow urine) mediated synthesis of silver oxide nanoparticles and their enhanced photocatalytic, photoluminescence and antibacterial studies. J Sci Adv Mater Devices 4(3):392–399. https://doi.org/10.1016/J.JSAMD.2019.08.004

Wang X, Wang M, Song H, Ding B (2006) A simple sol–gel technique for preparing lanthanum oxide nanopowders. Mater Lett 60(17–18):2261–2265. https://doi.org/10.1016/J.MATLET.2005.12.142

Wang X, Zhang P, Zhang X (2021) Probiotics regulate gut microbiota: an effective method to improve immunity. Molecules 26(19). https://doi.org/10.3390/MOLECULES26196076

WTalokar O, Belge DR, Belge DS (2013) Clinical evaluation of cow-urine extract with special reference to Arsha (hemorrhoids). Int J Pharm Sci Invent 2:5–8. www.ijpsi.org

Ye X, Chen Y, Ma S, Yuan T, Wu Y, Li Y, Zhao Y, Chen S, Zhang Y, Li L, Li Z, Huang Y, Cao H, Cui Z (2020) Biocidal effects of volatile organic compounds produced by the myxobacterium *Corrallococcus* sp. EGB against fungal phytopathogens. Food Microbiol 91:103502. https://doi.org/10.1016/j.fm.2020.103502

Zhu F, Wang J, Jia Y, Tian C, Zhao D, Wu X, Liu Y, Wang D, Qi S, Liu X, Li L, Jiang Z, Li Y (2021) *Bacillus subtilis* GB519 promotes rice growth and reduces the damages caused by rice blast fungus *Magnaporthe oryzae*. PhytoFrontiers 1(4):330–338. https://doi.org/10.1094/PHYTOFR-12-20-0041-R

Chapter 8
Integrative Review of Lecithin Extraction and Quantification Techniques in Sustainable Food Systems

Harika Vasa, Shireesha Mukka, and Maheshwar Mekam

Abstract Lecithin, an essential phospholipid composed of glycerol esters, fatty acids, phosphoric acid, and choline, is chiefly sourced from soybeans, sunflower seeds, egg yolk, and milk for industrial applications. It upholds cell membrane stability, delivers antibacterial, anti-inflammatory, and antioxidant effects, and functions effectively as an emulsifier in foods, animal feeds, pharmaceuticals, and industrial products. Additionally, it helps safeguard against neurological, cardiovascular, and cerebrovascular disorders, advancing sustainable food practices. This chapter reviews the extraction methods, such as solvent processes using hexane or ethanol—achieving yields near 93% from soybeans and egg yolk—together with greener enzymatic and ultrasound-enhanced techniques. Quantification employs HPLC, TLC, and phosphorus assays to evaluate phospholipid content accurately.

Keywords Egg yolk · Emulsifier · Extraction methods · Lecithin · Phospholipids · Soya bean

8.1 Introduction

Lecithin is a naturally occurring compound found in various foods. It is described as a complex mixture of lipids, such as triglycerides, fatty acids, glycolipids, and phosphatidic acid (Burnett et al. 2017; Alhajj et al. 2020).

Lecithin exhibits amphiphilic characteristics, which contribute to its emulsification properties and make it a crucial functional ingredient in the food industry. In addition to its role in food production, lecithin is also used in the cosmetic industry for its skin-conditioning properties and in the pharmaceutical industry as lecithin can be used as an additive in medications. It also acts as an essential nutrient with important roles in the human body (Price et al. 2018). It plays a crucial function in the structure of cell membranes, helping to maintain their integrity and facilitating

H. Vasa · S. Mukka (✉) · M. Mekam
Extension and Training Division, ICMR-National Institute of Nutrition, Hyderabad, India

M. Thakur, T. Belwal (eds.), *Food Biodiversity and Consumer Engagements*, Sustainable Development and Biodiversity 46,
https://doi.org/10.1007/978-981-95-7777-4_8

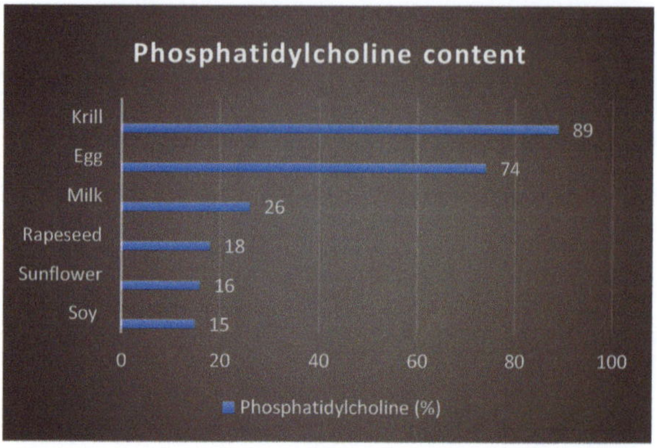

Fig. 8.1 Phosphatidylcholine content of food sources

proper communication between cells. From providing a smooth texture in chocolate to enhancing the effectiveness of skincare products, lecithins serve a multitude of purposes. Understanding the myriad sources and applications of lecithins underscores its importance in both nutrition and manufacturing. Overall, the versatility and benefits of lecithin make it a valuable and widely utilized ingredient in various aspects of our daily lives.

Lecithin can be extracted from both animal and plant sources. The most significant amount of lecithin is found in egg yolks and other oil seeds like soybean, sunflower, and rapeseed (Zhao et al. 2023). The phospholipid composition of the lecithin may differ significantly based on the source (vegetable or animal) and also on the method of obtention (Alhajj et al. 2020; Caparosa and Hartel 2020). Phosphatidylcholine, the functionally active component of lecithin, produces strong emulsifying properties and is found across all lecithin sources (Caparosa and Hartel 2020; Fig. 8.1).

This chapter provides a comprehensive and integrative analysis of Lecithin, which emphasizes its origin, extraction, and quantification methods, and also selected applications in food, pharmaceutical, and cosmetic industries. This work will help to understand the properties and functionality of lecithin.

8.2 Sources of Lecithins

Lecithin is widely available in nature. Sources of lecithins include a variety of plant and animal-based foods (Guiotto et al. 2015) and each botanical source has a different composition, which can be influenced by the specific source from which it is obtained, geographical region, storage conditions, and extraction and manufacturing processes (Bot et al. 2021).

Vegetable phospholipids are composed of phosphatidylcholine (PC), phosphatidylethanolamine (PE), phosphatidic acid (PA), phosphatidylinositol (PI),

Table 8.1 Phosphatidylcholine content in total phospholipids of different food sources

Source	Phosphatidylcholine (%)	References
Soy	15	Van Nieuwenhuyzen (2015)
Sunflower	16	Van Nieuwenhuyzen (2015)
Rapeseed	16.74–18.18	Mortensen et al. (2017)
Egg	74	Van Nieuwenhuyzen (2015)
Milk	26	Winther et al. (2011)
Krill	89	Van Nieuwenhuyzen (2015)

phosphatidylserine (PS), and small amounts of lysophosphatidylcholine (LPC), and other glycerol phospholipids. Soybeans are the primary source of commercial lecithin (Van Nieuwenhuyzen 2015), but the uses and benefits of soy lecithin are controversial, as it is claimed to pose health issues, which increased the demand for non-genetically modified (GM) soy-based lecithin (Li and Guo 2016). Despite the fact that research has demonstrated the similarity between lecithin derived from GMO soybeans and non-GMO varieties, the European market exhibits a preference for non-GMO lecithin (List 2015). As production and availability of non-genetically modified (GM) soybeans have decreased eventually (Jala and Prasad 2015), the demand for other commercially available oilseed-based lecithin, which is extracted from rapeseeds, cotton seeds, and sunflower has increased (List 2015). Rice bran lecithin is identified as an alternative to the GM-based lecithin due to its good nutritional profile and applications in food industries (Jala and Prasad 2015; Lehri et al. 2019).

As reported from literature, the animal-derived lecithins are primarily sourced from Egg yolk and followed by milk and from some marine species, including krill (Alhajj et al. 2020). Lecithin from eggs and egg yolks has been utilized as a nutritional food and an emulsifying ingredient for centuries (Van Nieuwenhuyzen 2015). The phospholipid content of egg yolk is more balanced than plant-derived phospholipids (Zhao et al. 2023). In milk, the phospholipids are present in fat globule membranes (Van Nieuwenhuyzen 2015). The utilization of milk lecithin is particularly intriguing because of its higher concentration of sphingomyelin (Price et al. 2018).

Lecithin can be derived from marine sources (various fish species and krill) as well (Rodríguez-González et al. 2019). Krill oil is extracted from small crustaceans and are important source of omega-3 fatty acids (Van Nieuwenhuyzen 2015). The main phospholipid class present is phosphatidylcholine, and they also contain omega-3 fatty acids like docosahexaenoic acid (DHA) and eicosapentaenoic acid (EPA) (Burri et al. 2012; Küllenberg et al. 2012).

Other than vegetable and animal origin lecithin, recent studies show interest in algae-derived phospholipids. Various phospholipids can be derived if proper extraction processes are established (Van Nieuwenhuyzen 2015).

The phosphatidylcholine content in total phospholipids of some important sources of Lecithin is listed below in Table 8.1.

8.3 Methods of Lecithin Extraction from Food Sources

8.3.1 Solvent Extraction Method

Lecithin is extracted from natural sources like soybeans, sunflower seeds, egg yolks, etc., by using organic solvents in a process known as solvent extraction. The solvent extraction method is an extremely effective technique employed to separate and extract lecithin by capitalizing on the varying selectivity of the solvent toward the different components of phospholipids. It is one of the traditional methods to separate Phospholipids (Zhao et al. 2023).

It utilizes a solvent to dissolve and extract its elements, producing a pure and concentrated extract. This approach is suitable for both large-scale industrial production and smaller laboratory settings. This is also widely utilized in the pharmaceutical, food industries, and cosmetic industries. Hexane, acetone, ethanol, and other organic solvents are commonly used for this purpose, depending on the extraction method. Lecithin can be obtained from fresh egg yolk through either single solvent extraction or mixed solvent extraction methods.

The Phospholipids were sequentially extracted from egg yolk using alcohols such as ethanol and butanol. A study was conducted to extract Lecithin from egg yolk using ethanol as a solvent. The purity of the resulting precipitate was enhanced by washing the precipitate with cold acetone (Dimethyl ketone) to a freezing temperature of −20 °C. The research findings indicated that by employing ethanol as the extracting agent and chilled acetone with freezing temperature, it is possible to achieve a purity level of 100% when extracting phospholipids from egg yolk (Gładkowski et al. 2012).

Phospholipids from egg yolk flakes can also be extracted by using two different alcohols, such as butanol and ethanol. A research investigation revealed that butanol was proven to be more efficient in extracting total lipid from egg yolk compared to ethanol. However, the concentration of phospholipids in the extract remained the same as that in the original yolk lipid. Interestingly, when aqueous ethanol solutions of 95% and 75% were used, the lipid extracts obtained during the initial stages of the sequential extraction had higher phospholipid concentrations. Despite that, it was found that a 75% ethanol concentration was the most effective for extracting phospholipids. This research successfully determined the ideal ethanol concentration for extracting phospholipids from dried egg yolk (Wang et al. 2017a).

The other richest source of lecithin is soybean; the phospholipids can also be extracted from soybean sources by solvent extraction. The study carried out by V.V. Patil et al. (2010) revealed that the purification of phosphatidylcholine from soya bean lecithin was achieved using extraction, precipitation, adsorption, and column chromatography. Among the solvents used, ethanol extraction resulted in a greater percentage of phosphatidylcholine. The optimal ratio for ethanol extraction was determined to be 1:7 (Patil et al. 2010). To increase the purity of phosphatidylcholine obtained from deoiled soya lecithin, a research investigation was carried out, which revealed that the extraction method utilized a blend of ethanol and isopropyl alcohol in different proportions to enhance both the purity and quantity of phosphatidylcholine. Furthermore, the purified phosphatidylcholine was subjected to additional purification via partitioning with varying ratios of acetonitrile–hexane and other solvent combinations (Jangle et al. 2013).

8.3.2 *Supercritical Fluid Extraction Technique*

Supercritical extraction is a specialized method utilized in diverse sectors such as pharmaceuticals, environmental assessments, and food techniques to extract high-purity compounds with minimal residue. By utilizing supercritical fluids such as carbon dioxide or water, which exhibit properties of liquids and gases, this technique can effectively dissolve target compounds without leaving behind any harmful residues. This innovative extraction process provides a secure, effective, and eco-friendly approach to isolating compounds with exceptional precision. This approach can be employed to extract lecithin with enhanced purity from marine sources such as scallops. To enhance the extraction selectivity, a two-step fractionation method was employed. Initially, the dried material underwent de-oiling using pure carbon dioxide, followed by the extraction of phospholipids using a mixture of carbon dioxide and ethanol. The phospholipid content in the extract was found to be highest at 58 g/100 g when ethanol was used as a cosolvent at a temperature of 45 ° C. Conversely, isopropanol demonstrated greater selectivity, resulting in an extract with a phospholipid purity of 90 g/100 g at 27 °C as co-solvent. In contrast to pure ethanol or conventional methanol-chloroform extractions, the carbon-dioxide-based mixtures demonstrated the ability to recover phospholipid content exceeding 35 gm per 100 gm of lipids was observed (Subra-Paternault et al. 2015).

8.3.3 *Enzymatic Method*

Enzymatic extraction is a highly efficient method for extracting lecithin from the cell membrane. It offers a more precise approach, resulting in improved purity and higher-quality products. By using specific enzymes, it is feasible to attain greater yields while minimizing processing time. A study was conducted to extract lecithin from chicken liver by enzymatic hydrolysis. This method included blending chicken liver in distilled water, deactivating enzymes with high heat, applying an ultrasonic wave for 10 min, then treating the mixture with neutral and protamex proteinase, followed by deactivation in boiling water, and introducing 95% ethanol with different solid–liquid ratios. The extraction procedure was conducted at 35 °C. The filtrate was evaporated, and the oil was gathered using petroleum ether (Huang et al. 2022). Enzymatic hydrolysis provides numerous benefits for extracting lecithin from chicken liver, especially in terms of selectivity, environmental impact, and cost-efficiency. Nevertheless, the success of the procedure relies on the conditions implemented, particular enzymes utilized, and the desired yield and purity of the lecithin.

8.3.4 Simultaneous Texturization and Extraction of Phospholipids (STEP) Technique

The STEP technique, which stands for simultaneous texturization and extraction of phospholipids, was developed to enhance the extraction of phospholipids like lecithin from food sources by integrating texturization of the food matrix with the extraction process. This innovative method is designed to improve the efficacy and efficiency of lecithin extraction. This simultaneous approach can shorten the total processing time when compared to sequential methods. The method was utilized to isolate phospholipids from liquid egg yolk through the application of various solvents, specifically 95% ethanol, 100% butanol, and 80% butanol, and the research indicates that every solvent has the capacity to modify the texture of liquid yolk and recover lipids with over 90% after five extractions. 80% butanol and 100% butanol extract total lipids more quickly than ethanol, while 95% ethanol exhibited a stronger preference for extracting phospholipid with high purity and yield of phospholipid. This approach eradicates the requirement for centrifugation, acetone, pre-drying, or cold therapies, which are usually needed in other extraction methods. A concentrated fraction of phospholipids can be achieved in a single step through the use of aqueous ethanol. Moreover, the solvent is capable of being reused after distillation. The novel feature of this research lies in its ability to both texture and strengthen the yolk protein network simultaneously and also enabling lipid extraction without generating protein fines that could hinder processing efficiency (Wang et al. 2017b).

There are a number of methods for the extraction of lecithin from various food sources; each of these methods has its own advantages and disadvantages. Among these methods, supercritical extraction is often the best technique for attaining top-notch purity levels and carrying out extensive manufacturing, in case of smaller operations or when combining texturization with extraction, simultaneous texturization and extraction of phospholipids technique (STEP) can be highly efficient, while enzymatic extraction is a highly efficient method for extracting lecithin from cell membranes. It is typically a good option for attaining specificity, affordability, and safeguarding delicate compounds. Ultimately, the most appropriate method to be chosen depends on specific requirements, which include scale, desired purity level, cost, and process intricacy.

8.4 Separation and Quantification Methods of Lecithin from Different Food Sources

Lecithin is a vital phospholipid that significantly contributes to various biological processes in the human body, including maintaining cell integrity and facilitating efficient nerve communication. By understanding the different amounts of lecithin found in various foods, individuals can make more informed choices regarding their nutrition and diet. Furthermore, analysing lecithin levels in diverse food items enables researchers to investigate its interactions with other nutrients and compounds, uncovering potential health benefits or risks linked to various eating patterns. This information is also essential for developing tailored dietary recommendations for specific health conditions or population groups. Additionally, assessing lecithin levels from different food sources can drive innovations in food and science technology resulting in the development of products that meet particular nutritional needs. Ultimately, examining lecithin concentrations in different foods is vital for understanding its impact on human health and enhancing dietary strategies for better overall well-being. Various techniques exist for measuring lecithin levels in a range of food sources, few of these separation & quantification methods are as follows in Fig. 8.2 and Table 8.2:

8.4.1 Column Chromatography

Column chromatography is a crucial method for isolating and purifying lecithin, which is widely used in biochemistry and analytical chemistry. This technique separates the phospholipid from a mixture by utilizing its unique properties, such as molecular weight and polarity. The sample containing lecithin is placed into a column filled with an adsorbent material, like silica gel or alumina, and as the sample progresses through the column. The various elements of the mixture engage with the adsorbent at different speeds. By choosing the right eluent, researchers can elute lecithin at a precise moment, allowing for the collection of purified lecithin free from contaminants. This method highlights the importance of chromatography in chemistry and its wide-ranging applications in various scientific fields (Zhao et al. 2023). The method used to extract phosphatidylcholine from soybean lecithin using silica gel as the adsorbent material showed a superior binding capacity when compared to alternative resins. The phosphatidylcholine with 84% purity and 65% yields was observed when silica gel was employed as the stationary phase in column chromatography (Patil et al. 2010). In the previous study it was found that Acetonitrile's selectivity in extracting Phosphatidylcholine from hexane fraction increased purity, while isopropanol enhanced the extraction rate. The synergistic effect between isopropanol and acetonitrile yielded superior results. The gradient elution method was used in a preparative silica column chromatographic technique, achieving phosphatidylcholine purity of up to 98% (Jangle et al. 2013).

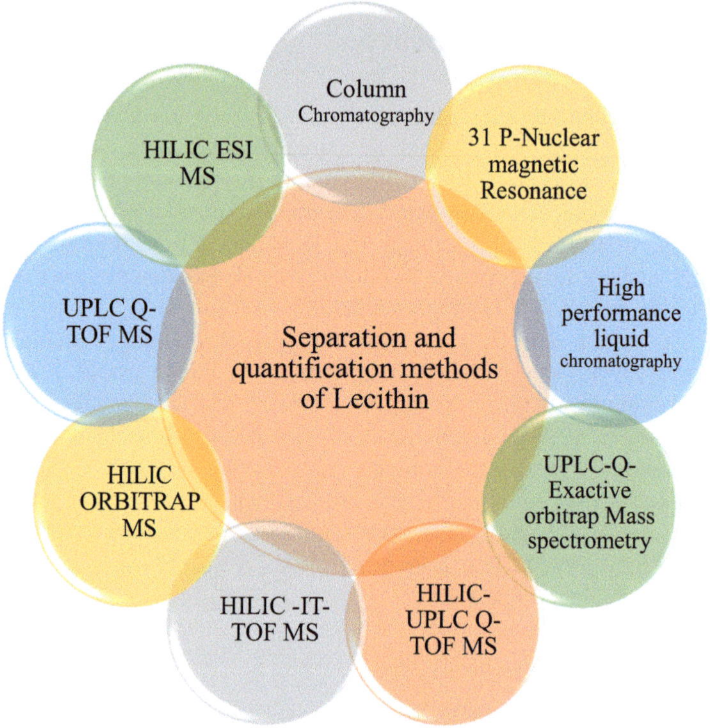

Fig. 8.2 Separation and quantification methods of lecithin

8.4.2 Liquid Chromatography

Liquid chromatography stands out as a highly effective analytical method used in laboratories to separate and identify components within mixtures. This method offers high resolution and sensitivity, making it useful in fields like pharmaceuticals, environmental analysis, and food testing. This adaptable technique relies on the concept of distributing solutes between a mobile phase and a stationary phase. The stationary phase can consist of either a solid or liquid substance contained within a column; in contrast, the mobile phase is generally composed of either a liquid or a gas that transports the sample through the column at a regulated speed. As the sample travels through the column, different components interact with the stationary phase to varying degrees, leading to their separation based on their and physical and chemical properties. Various kinds of liquid chromatography exist, where it can be combined with detection methods like ultraviolet or mass spectrometry for comprehensive information. Liquid chromatography serves as a crucial instrument in contemporary analytical chemistry, allowing researchers to effectively quantify, separate, and analyze compounds with accuracy and speed (Prado and Weber 2016; López-Porfiri et al. 2024).

Table 8.2 Extraction and quantification methods of lecithin from different food sources

Source	Extraction method	Quantification method	Inference	References
Soyabean lecithin (de-oiled)	Extraction, precipitation, adsorption	Column chromatography High-performance liquid Chromatography (HPLC)	The use of ethanol yields phosphatidylcholine with a purity of 84%	Patil et al. (2010)
Soyabean lecithin (de-oiled)	Solid–liquid extraction (ethanol-single solvent extraction) and liquid–liquid extraction	Column chromatography High-performance liquid Chromatography (HPLC)	High purity of phosphatidylcholine was obtained with acetonitrile and hexane system	Jangle et al. (2013)
Sunflower lecithin	Solvent extraction using absolute-ethanol and water mixtures and de-oiled using acetone	P-NMR (nuclear magnetic resonance) was used for determining the composition of phosphatidylcholine (PC) and phosphatidylinositol (PI)	The higher ethanol concentration in the extraction resulted in an enhanced yield of PC in the extract obtained	Guiotto et al. (2013)
Egg yolk	Extraction by organic solvents ethanol and chilled acetone	The purity of PL was determined by HPLC with CAD detector (CORONA TM charged aerosol detector)	Acetone chilled to −20 °C -yield 100% purity, the purity of PL determined by HPLC with CAD exhibits higher sensitivity and improved accuracy	Gładkowski et al. (2012)
Egg yolk modified to egg yolk flakes	Solvent extraction Varying concentrations of butanol and ethanol	Nuclear magnetic resonance (P-NMR)	Butanol proved to be more efficient in extracting total lipids, and 75% ethanol concentration was the most effective for extracting phospholipids	Wang et al. (2017a)
Egg yolk	Simultaneous texturization and extraction of phospholipids (STEP) Technique with 100% and 80% Butanol and 95% ethanol	31P nuclear magnetic resonance (NMR)	It was found that 95% ethanol was very effective in extracting phospholipids. This method eliminates the need for pre-drying, acetone, centrifugation, or chilling treatments	Wang et al. (2017b)
Scallops (Industry by products)	Supercritical fluid technology (First de-oiling, second phospholipid extraction by CO_2 and ethanol)	31P nuclear magnetic resonance (NMR)	The use of propanol-2 instead of ethanol allowed for considerably enhancing the purity of extracts	Subra-Paternault et al. (2015)
Donkey and cow milk	Solid-phase extraction (SPE) Hydrophilic interaction liquid	Ion trap-time of flight LCMS-IT-TOF mass spectrometer	HILIC column coupled to an hybrid ion trap-time of flight (IT-TOF) mass analyzer was developed for the	Donato et al. (2011)

(continued)

Table 8.2 (continued)

Source	Extraction method	Quantification method	Inference	References
	chromatography coupled to evaporative light scattering (HILIC-ELSD)		characterization of the unique phospholipid profile and fatty acid composition	
Milk from different species (Human, cow, camel, and mare)	Total milk lipids were measured by the modified Folch's method using chloroform and methanol Phospholipids are separated by a SPE (solid phase extraction) column	31P NMR (nuclear magnetic resonance) spectroscopy	31P NMR spectroscopy procedure is suitable to identify and quantify PL's in Milk samples, showing optimal sensitivity, high resolution, and simple sample preparation, and is ideal for accurately analysing the composition of bioactive phospholipids	Garcia et al. (2012)
Schizochytrium sp. S31	Bligh and dyer method Separated using an acetonitrile/ammonium formate gradient mobile phase	UPLC-Q-TOF-MS Hydrophilic interaction (HILIC) ultra performance liquid chromatography (UPLC) 13 coupled with quadrupole time of flight (Q-TOF) mass spectrometry (MS)	HILIC UPLC coupled with Q-TOF MS was utilized to examine the phospholipid composition through LC–MS/MS analysis in negative ion mode, the structures of 70 PL's were determined. The primary phospholipid found in Schizochytrium sp. S31 was phosphatidylcholine, representing 49.8% of the total phospholipids, with the predominant fatty acids being C16:0 and DHA	Li et al. (2016)
Bovine milk	Total milk lipids by modified Russo et al. method and dissolved in chloroform and methanol	PL analysis by LC–MS	The PLs were separated through (HILIC) and detected using an Orbitrap mass analyzer in positive mode. Tandem MS was used to confirm the structure of PLs in 32 bovine milk samples, revealing the relative abundance of different classes, significant variations in content, and correlations between major PL classes. This	Liu et al. (2015)

			method is highly sensitive and reproducible, providing valuable insights into PL variation and metabolism in various mammalian species	
Goat milk Soy milk Bovine milk	Folch and Bligh Method for isolation of total lipids using chloroform and methanol	Ultra-performance liquid chromatography (UPLC)-Q-Exactive Orbitrap Mass Spectrometry	(UPLC)-Q-Exactive Orbitrap Mass Spectrometry is a leading-edge technology, for the systematic identification of diverse compounds can be achieved	Li et al. (2017)
Chicken liver	Enzymatic hydrolysis Qualitative analysis by TLC	HPLC-ELSD	The study involved blending chicken liver, deactivating enzymes, treating with neutral and protomer proteinase, and introducing 95% ethanol with different solid–liquid ratios at 35 °C and the oil is collected using petroleum ether. Qualitative analysis by TLC and HPLC-ELSD quantification revealed 88.92% yield	Huang et al. (2022)

8.4.2.1 HPLC

HPLC, or high-performance liquid chromatography, is a prominent analytical method in chemistry utilized for the separation, identification, and quantification of components in mixtures. This technique involves the movement of a liquid solvent through a column filled with a stationary phase, enabling different interactions between the components and the column. The separation process is driven by the distinct retention times of each component. HPLC is highly regarded for its exceptional sensitivity, efficiency, and versatility in analyzing a wide range of compounds across various sample types. The detectors play a crucial role in reducing the loss of compounds in samples. Ultra-violet and Electrochemical detection techniques are favored for their straightforwardness and affordability. Innovations in stationary and mobile phases improve the analysis of complex samples. In essence, HPLC serves as a fundamental tool in analytical chemistry, providing researchers and scientists with a powerful approach for the thorough characterization and analysis of complex mixtures with remarkable accuracy (Fowler 2016; Hussain et al. 2019). This method is used to assess the purity of the phosphatidylcholine extracted from soybean source (Patil et al. 2010; Jangle et al. 2013).

HPLC can be coupled with detectors such as:

(a) *HPLC-ELSD*

The concentration and yield of phospholipids in chicken liver extracts after enzymatic hydrolysis were assessed using HPLC with an ELSD detector, which stands for evaporative light scattering detector (Huang et al. 2022). This method is also effective for determining the phospholipid content in infant formulas (Moloney et al. 2020).

(b) *HPLC-CAD*

This approach was utilized to evaluate the purity of phospholipids in the egg yolk extracts, which were the HPLC equipped with a CAD detector, where CAD is charged aerosol detector which has been demonstrated to exhibit higher sensitivity and improved accuracy. (Gładkowski et al. 2012). CAD is a more recent detector in comparison to the evaporative light scattering detectors (ELSD). Extensive evidence supports its superior sensitivity and precision when compared to the ELSD (Wang and Zhou 2017).

8.4.2.2 Ultra-Performance Liquid Chromatography (UPLC)

The core concept of ultra-performance liquid chromatography (UPLC) for separating components in a mixture is similar to that of high-performance liquid chromatography (HPLC), but the main difference is the particle size of the column's sorbent, which is less than 2 μm. UPLC is an advanced version of HPLC that offers faster, more efficient, and higher-resolution separations for various compounds. It achieves this by utilizing smaller particle sizes and higher pressures, and reduced elution time. This technique effectively separates a wide range of analytes, including both polar and non-polar compounds within complex sample mixtures (Batool and Menaa 2020; Ahmed et al. 2023), UPLC dramatically decreases the volume of mobile phase required by at least 80% in comparison to HPLC, while also providing much quicker run times. Furthermore, UPLC functions with a shorter column kept at higher temperatures, employing minimal injection volumes and reduced flow rates (Rahman 2018).

UPLC can be coupled with detectors such as:

(a) *UPLC Q-Exactive Orbitrap Mass Spectrometry*

The principle of UPLC-Q-EXACTIVE ORBITRAP mass spectrometry integrates ultra-high-performance liquid chromatography with high-resolution mass spectrometry, utilizing the Exactive Orbitrap analyzer. This method facilitates the precise identification and quantification of compounds within complex matrices, characterized by remarkable sensitivity and rapid analysis. This technique offers accurate mass measurement, interpretation of molecular structure, and also delivers outstanding analytical performance across various fields, such as proteomics, metabolomics, and environmental studies. The foundation of this innovative technology is its capability to separate and identify molecules according to their distinct mass-to-charge ratios, allowing researchers to investigate sample compositions with unparalleled detail and precision. This methodology has transformed the landscape of mass spectrometry, providing superior sensitivity, resolution, and speed for thorough analysis of complex sample mixtures, establishing it as an essential resource for contemporary research and development (Fu et al. 2020).

UPLC coupled with Q-Exactive Orbitrap Mass Spectrometry was employed to analyze the lipid content in soymilk, bovine milk, and goat milk. Quantification of lipid content was assessed and a comparison of the lipid profiles in three types of milk samples was also conducted. Different classes of lipids were quantified by this method. Soymilk was found to be abundant in phospholipids, while goat milk showed richness in medium chain triglycerides, ω-6 and ω-3 Fatty acids, particularly eicosapentaenoic acid and docosahexaenoic acid. Bovine milk, on the other hand, exhibited high levels of ceramide, triacylglycerol, and diacylglycerol. Furthermore, potential biomarkers among the lipids were identified for milk adulteration authentication purposes (Li et al. 2017).

(b) *HILIC-UPLC-Q-TOF MS*

HILIC-UPLC-Q-TOF MS is an advanced analytical method where it stands for hydrophilic interaction liquid chromatography–ultra-performance liquid chromatography combined with quadrupole time-of-flight mass spectrometry. This technique integrates the principles of hydrophilic interaction chromatography with the high efficiency of ultra-performance liquid chromatography, further augmented by the precise resolution and mass accuracy offered by the quadrupole time-of-flight mass spectrometry system. This approach effectively separates and identifies intricate mixtures of compounds with exceptional accuracy, rendering it an indispensable tool in the field of proteomics, metabolomics, and other disciplines that necessitate detailed chemical analysis. This technique offers rapid separation of polar substances and ionic substances and precise mass assessment. This method offers a comprehensive approach for examining a wide range of chemical substances in different samples. In conclusion, HILIC-UPLC-Q-TOF MS plays a crucial role in advancing scientific research and deepening our understanding of the chemical composition of intricate systems. This method offers a comprehensive approach for examining a wide range of chemical substances in different samples. In conclusion, HILIC-UPLC-Q-TOF MS plays a crucial role in advancing scientific research and deepening our understanding of the chemical composition of intricate systems (Velichkova et al. 2024).

The phospholipid profile of the marine microalgae *Schizochytrium* sp. S31 sample was analyzed using the hydrophilic interaction liquid chromatography (HILIC) coupled with ultra-performance liquid chromatography (UPLC) in conjunction with Q-TOF MS. This approach is highly effective for elucidating the structures of phospholipids. The analysis revealed that phosphatidylcholine was the predominant phospholipid in *Schizochytrium* sp. S31, accounting for 49.8% of the total phospholipid content. The combination of UPLC and mass spectrometry proves to be invaluable for biomarker discovery, as it fulfils stringent analytical criteria while providing exceptional sensitivity, mass accuracy, and resolution (Li et al. 2016).

8.4.2.3 Hydrophilic Interaction Liquid Chromatography (HILIC)

Hydrophilic interaction liquid chromatography (HILIC) is an important analytical method in separation science, designed to separate highly polar compounds according to their polarity and hydrophilicity. This technique employs a mobile phase that usually consists of a blend of water and organic solvent, which facilitates the elution and retention of analytes based on their hydrophilicity and polarity. HILIC provides excellent sensitivity and resolution, works well with a range of detection techniques, and is crucial for chromatographers working in areas such as pharmaceuticals, environmental science, and biochemistry. This method proves beneficial for analyzing compounds that are not effectively retained in conventional reversed-phase liquid chromatography, establishing its significance in the field of analytical chemistry (Cavazzini et al. 2023). The HILIC is equipped with IT-TOF MS or Tandem MS for the comprehensive analysis of a compound from a particular source.

HILIC can be coupled with detectors such as:

(a) *HILIC-IT-TOF MS*

This approach was used to determine the phospholipids in milk sources in which hydrophilic interaction liquid chromatography was utilized with a partially porous column to separate five phospholipids' classes extracted from milk samples by solid phase extraction then the ion trap-time of flight mass spectrometer enabled accurate detection, leading to a comprehensive characterization of the unique phospholipids profile and fatty acid composition of cow's and donkey's milk. Notably, cow's milk contained over 20 times more total PLs compared to donkey's milk (Donato et al. 2011).

(b) *HILIC-Orbitrap MS*

HILIC, coupled with an Orbitrap mass spectrometer, was employed for the detailed analysis and measurement of polar lipids in bovine milk samples. In this method Phospholipids were separated using hydrophilic interaction liquid chromatography and analyzed with an Orbitrap mass spectrometer in positive ionization mode. The structural identification of phospholipids was achieved through tandem mass spectrometry. When applied to 32 bovine milk samples, this method revealed the relative abundance of different classes, notable variations in phospholipid content across individual samples, and correlations between the major phospholipid classes. This approach is an essential tool for investigating the diversity and metabolism of critical phospholipid components in bovine and human milk, as well as across various mammalian species. It provides accuracy and reproducibility and also allows the simultaneous quantification of 70 different phospholipid species within a 45-min timeframe (Liu et al. 2015).

8.4.3 ^{31}P-NMR

^{31}P-Nuclear magnetic resonance spectroscopy is a rapid and accurate method for quantifying phospholipid classes, utilizing a single internal standard. NMR is inherently quantitative, non-destructive, and requires less sample preparation compared to other delicate methods. The magnitude of a specific signal's peak is straightforwardly proportional to the quantity of nuclei producing that signal. This establishes a linear correlation between the molar amount of the corresponding compound and the strength of the signal. The method can be effectively utilized for the analysis of phospholipid extracts without the need for prior separation (Wang and Zhou 2017). However, the ability to quantify individual phospholipid classes separately and accurately that greatly depends on experimental conditions.

The composition of phosphatidylinositol and phosphatidylcholine-enriched fractions obtained from sunflower extracts was analyzed using ^{31}P-NMR (Guiotto et al. 2013). The ^{31}P NMR spectroscopy technique is highly effective for the detection and measurement of phospholipids in milk samples. It provides a straightforward sample preparation, high resolution, and outstanding sensitivity, making it an excellent choice for accurately analysing the composition of bioactive phospholipids that may offer health benefits. Research was carried out to identify and measure phospholipids present in different types of milk, such as cow, camel, mare, and human, milk. The findings indicated that camel milk contained higher total phospholipid content than the other samples. This information could be valuable for selecting ingredients for infant formulas, functional foods, dietary supplements, and artificial nutrition (Garcia et al. 2012). This method is also effective for determining the phospholipid composition from egg yolk (Wang et al. 2017a, b) and also from the marine sources, such as scallops (Subra-Paternault et al. 2015).

8.4.4 *Other Methods*

Research investigations shown that phosphatidylcholine can be isolated and measured from various biological samples, including human milk, cell extracts, plasma, and tissues. The extraction techniques employed include liquid–liquid extraction or the Folch method, which is specifically utilized for obtaining phospholipids. Lecithin from human milk can then be separated and quantified using UPLC-Q-TOF-MS (Liu et al. 2023) and HILIC-ESI-MS system (Song et al. 2021), while the ultrahigh-performance liquid chromatography coupled to a triple-quadrupole mass spectrometer system is used to separate and assess the phospholipid levels in extracts from plasma, tissues, and cells. This method was developed to quantitatively analyze major classes of phospholipid in biological samples like tissue extracts, cell extracts, and plasma within few minutes. These approaches have the ability to quantify a diverse array of phospholipids with superior sensitivity and rapid processing capabilities (Huang et al. 2019).

There are a number of methods for lecithin separation and quantification; each of these methods has its own advantages and disadvantages. Among them, HPLC is a widely used method, powerful method, and reliable method for the separation of lecithin components. It is less sensitive compared to mass spectrometry while the ^{31}P NMR (nuclear magnetic resonance spectroscopy method is direct and non-destructive, specifically targeting phosphorus-containing compounds such as lecithin but this method has lower sensitivity than mass spectrometry techniques, necessitates a larger sample size, and has a restricted dynamic range while UPLC-Q Exactive orbitrap MS is ultra-performance liquid chromatography combined with high-resolution mass spectrometry is ideal for precise quantification and identification of lecithin in samples but it is expensive and requires skilled experts to perform operations. HILIC-UPLC-Q-TOF MS offers exceptional sensitivity and resolution make it particularly effective for polar compounds such as lecithin, and also offers outstanding mass accuracy. This is expensive and it requires meticulous method development. HILIC IT TOF MS is more effective for structural determination and quantification of polar lipids, but it offers lower mass accuracy and resolution than Q-TOF and Orbitrap systems. HILIC-Orbitrap MS method combines the robust capabilities of HILIC in isolating polar compounds with the high-resolution mass spectrometry of Orbitrap, which results in exceptional high sensitivity and accuracy for quantification. However, it is costly and demands specialized knowledge.

HILIC-Orbitrap mass spectrometry is possibly the most proficient method for quantifying lecithin in food products, attributing to its remarkable sensitivity, precision, and ability to isolate and quantify lecithin in intricate food matrices. The combination of HILIC for the separation of polar substances and Orbitrap mass spectrometry for precise mass measurement makes this approach particularly advantageous for lecithin analysis. However, UPLC-Q-Exactive Orbitrap mass spectrometry also offers similar performance and may be equally effective, which depends upon specific experimental procedures and the requirements (Table 8.2).

8.5 Applications of Lecithins

Lecithins have a broad range of applications across different sectors including Food, Pharmaceutical and Cosmetic sectors. Lecithins derived from plant and animal sources are commonly used in food manufacturing as emulsifiers. Moreover, lecithin can also be found in dietary supplements aimed at promoting brain health and maintaining proper nerve function. It plays a vital role in the body by aiding in the absorption of fats and fat-soluble vitamins. Due to its exceptional nutritional properties, they have a significant role in promoting health and overall well-being. The versatility of lecithins extends beyond the kitchen as they are utilized in various industries like pharmaceuticals and cosmetics for their emollient properties (Fig. 8.3). This chapter delves into some specific applications highlighted in various research articles.

8.5.1 Nutritional Significance of Lecithin in Promoting Overall Well-Being

The documented information regarding the nutritional properties in multiple research articles have highlighted the significance of fatty acid composition of lecithins and their use in preventing different health conditions and in promoting overall health. Lecithins sourced from plants and animals are known to contain essential fatty acids such as arachidonic acid, linoleic acid, and linolenic acid.

Fig. 8.3 Role of lecithin in various industries

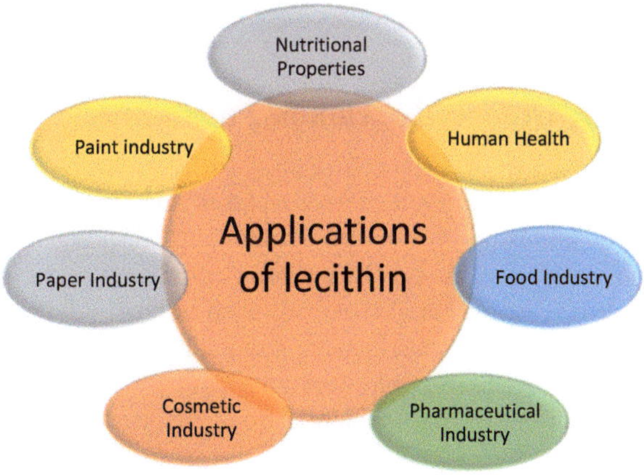

Notably, lecithins of marine origin are rich in polyunsaturated fatty acids, particularly eicosapentaenoic acid, which contributes to their health-promoting properties (Arab Tehrany et al. 2012; Küllenberg et al. 2012).

Lecithins can promote heart health due to its high levels of essential fatty acids and unsaturated fatty acids, and also help in increasing immunity (Li et al. 2017). Choline, a key component in lecithin synthesis, serves as a precursor to the neurotransmitter acetylcholine, essential for proper muscle functioning. Additionally, choline facilitates the transportation and metabolism of fats, thereby contributing to overall metabolic health (List 2015). Furthermore, the importance of choline in enhancing memory, cognitive performance, and brain function cannot be overstated (Zhao et al. 2023). The egg and soy phospholipids are the main source for choline (Van Nieuwenhuyzen 2015). Studies have revealed that egg lecithin has a role in delaying neurodegenerative diseases (Zhao et al. 2023).

It is widely recognized that alterations in the structure and function of neuronal membranes are linked to age-related physiological changes, including memory impairments and cognitive deficiencies. The decline in phospholipids, especially phosphatidylserine (PS), in neuronal membranes has been associated with these cognitive issues. Research has demonstrated that PS, predominantly derived from bovine brain, can enhance cognitive abilities in patients with cognitive decline. However, the utilization of PS sourced from vegetative origins is considered a safer alternative, offering similar cognitive benefits without the danger of developing encephalopathy (Moré et al. 2014).

It is noteworthy that lecithin plays a crucial role in enhancing cardiovascular health because it reduces excess LDL, or "bad cholesterol," and promotes the production of HDL, or "good cholesterol." Lecithin functions by lowering LDL levels and increasing HDL synthesis at the same time, strengthening our resistance to cardiovascular diseases. The anti-atherogenic reverse cholesterol transport (RCT) process depends critically on lecithin cholesterol acyltransferase (LCAT), according to documented literature. LCAT increases HDL's ability to transport cholesterol by converting unbound cholesterol in developing HDL molecules to cholesteryl esters and ultimately decreases the risk of cardiovascular diseases like atherosclerosis (Onaolapo et al. 2024).

8.5.2 Role of Lecithin in Food Industry

In the food industry, lecithin is an essential ingredient and it is multifunctional. It acts as an emulsifier to stabilize a wide range of products, including chocolate, margarine, and mayonnaise. This natural compound helps not only in enhancing texture and consistency but also extends shelf life by inhibiting oil separation. Furthermore, lecithin acts as a surfactant, facilitating the blending of ingredients that would otherwise be immiscible, thus contributing to the homogeneity of food formulations. Beyond this its diverse applications encompass improving the flow properties of powders, preventing sticking in baking, and even as a coating agent for food items (Jala and Prasad 2015; List 2015).

Lecithin, being a natural substance derived from a variety of food sources, minimizes the potential risks associated with its consumption and presents a minimal risk when compared to other emulsifiers that are produced through chemical methods. Additionally, the pure lecithin has a hydrophilic-lipophilic balance with an almost equivalent affinity to fats and water. This provides consumers with a wide range of options and better flexibility in application and makes lecithin as a superior emulsifier for culinary uses (Lecithin For Food n.d.).

In the baking industry, emulsifiers play a crucial role in enhancing dough strength and softening the crumb. Lecithin serves as a crumb softener and is regarded as a clean-label ingredient, providing a viable alternative to traditional additives (Tebben et al. 2022).

Lecithin is the preferred emulsifier in the chocolate industry as it provides yield stress and viscosity control, and also reduces production costs by decreasing the amount of cocoa butter required to achieve similar flow properties (Caparosa and Hartel 2020).

Lecithins act as efficient wetting and dispersing agents in both instant and baby food items. Initially, phospholipids (PLs) reduce the surface tension at the interface between fat and water, and then lecithin molecules form bonds with proteins. When lecithin is mixed with water or milk, it helps evenly distribute powdered ingredients. As a result, in the production of agglomerated whole milk powder (WMP) and fat-filled milk powder (FFMP), small quantities (from 0.05% to 0.2%) of lecithin are added. After the spray drying method is used, the warm powder is treated with lecithin during the agglomeration stage in a fluid bed unit. In infant milk formulations, egg lecithin, which includes docosahexaenoic acid (DHA) and arachidonic acid (ARA) fatty acids, is used as a replacement for breast milk. Phospholipids derived from egg have good absorption rates for ARA and DHA, are stable against oxidation, and offer protection to the intestinal lining by preventing the adhesion of harmful bacteria, which reduces the chance of severe diarrhea (Van Nieuwenhuyzen 2015).

An additional noteworthy use of lecithin in culinary applications is its role as a release agent for frying. Specifically refined liquid lecithins, when combined with waxes and vegetable oils, are employed in frying processes. Commercial pan-release sprays for home cooking are readily available in stores. A standard formulation for these pump sprays typically includes 65% vegetable oil, 25% medium-chain triglyceride (MCT) oil, and 10% hydrophilic lecithin (Van Nieuwenhuyzen 2015).

Lecithin is employed for its ability to preserve taste. It surrounds oil droplets within a phosphatidylcholine double layer, which not only protects the Flavors but also facilitates their transport, ensuring that the final product retains a robust taste, even in low-fat food formulations.

In the production of chewing gum, lecithin serves multiple roles, including softening, altering texture, as an ingredient in the gum base, and as an emulsifier or dispersing agent for Flavors (Lecithin For Food n.d.).

Lecithin also helps in extending the shelf life of food products. A research study has demonstrated that applying lecithin coatings, in combination with sodium alginate, gum, and agar, can boost the longevity of button mushrooms (Cavusoglu et al. 2021).

At present, vegetable lecithins are the predominant emulsifiers utilized in the food and cosmetic sectors, attributed to their low toxicity and significant nutritional benefits (Alhajj et al. 2020). With its wide range of benefits and functionalities, lecithin is high valued ingredient in the food industry.

8.5.3 Role of Lecithin in Pharmaceutical Industry

Lecithin serves multiple functions within the pharmaceutical sector. The following are some significant applications:

1. *Emulsifier and Solubilizer*: Lecithin functions as a potent emulsifier, aiding in the stabilization of oil-water emulsions. In pharmaceutical formulations, it enhances the solubility and bioavailability of drugs that are poorly soluble. By creating micelles or liposomes, lecithin encapsulates hydrophobic drugs, thereby facilitating their delivery and absorption within the body (Ameta et al. 2023).
2. *Excipient in Drug Formulations*: Lecithin is frequently used as an excipient in various pharmaceutical formulations, particularly in oral, topical, and parenteral products. It fulfils several roles, including improving drug dispersion, enhancing stability, and providing sustained-release characteristics (Van Nieuwenhuyzen 2015).
3. *Lipid Component in Liposomes and Nanoparticles*: Lecithin is a crucial ingredient in the formation of liposomes and lipid nanoparticles utilized for drug delivery. Liposomes are spherical vesicles made up of lipid bilayers, with lecithin being a primary lipid component. These structures are designed to encapsulate drugs, protect them from degradation, and target specific tissues or cells within the body (Arab Tehrany et al. 2012; Gutiérrez-Méndez et al. 2022).
4. *Surfactant in Inhalation Formulations*: Lecithin can act as a surfactant in inhalation formulations, enhancing the dispersibility of drugs intended for pulmonary delivery. It contributes to the formation of stable aerosols and improves drug deposition in the lungs, making it beneficial for respiratory therapies (Ghanshyam et al. 2011).
5. *Lecithin Microemulsions as Drug Carriers*: Lecithin microemulsions can be utilized for transdermal drug delivery and in the formulation of topical preparations that can be applied to the skin or mucous membranes (Murashova 2023).

   Overall, lecithin's versatility in enhancing drug delivery, stability, and solubility makes it a valuable ingredient in various pharmaceutical applications, contributing to improved therapeutic outcomes and patient compliance.

8.5.4 Role of Lecithin in Cosmetic Industry

The integration of lecithin into cream formulations has a history that spans many years. Lecithin plays a crucial role in the cosmetic industry due to its wide range of benefits and applications (Akcay et al. 2020). Lecithin extracted from soy and rice bran is widely used in the cosmetic industry (Jala and Prasad 2015; List 2015). This natural lipid is a versatile ingredient known for its emulsifying, moisturizing, and conditioning properties. Furthermore, this ingredient is valued for its ability to boost the absorption of other beneficial ingredients, allowing for a more effective delivery of active compounds into the skin. Given its versatility and compatibility with various formulations, lecithin is widely used in the formulation of cosmetics, hair care products, and even in lipsticks.

Lecithin offers numerous advantages for both skin and hair. Various products that utilize this ingredient include creams, cleansers, moisturizing liquid makeup, beauty lotions, and skin penetration enhancers (Lecithin (Antioxidant) n.d.). Lecithin serves multiple functions in cosmetics, such as facilitating application in lipsticks, contributing to the softness of creams, providing protection for hair, acting as a synergist with phenolic antioxidants in shampoos, and functioning as an emulsifier in prosthetic pastes. The typical inclusion rate of lecithin in these products ranges from 0.5% to 1% (Akcay et al. 2020).

Lecithin serves as an emollient, effectively softening the skin while enhancing its natural barrier to retain moisture. It is particularly beneficial for individuals with very dry skin and chapped lips. As an excellent emulsifying agent, lecithin facilitates the binding of various ingredients within a formulation, resulting in a product with improved texture. This ingredient significantly enhances the hydration of cosmetic products, imparting a luxurious feel (Lecithin (Antioxidant) n.d.).

The ingredient's natural origins and versatile characteristics render it highly desirable for formulators aiming to develop premium, performance-oriented products that align with the dynamic demands and preferences of consumers in the constantly evolving beauty sector.

8.5.5 Other Applications

Lecithin plays a vital role in both the paper and paint industries due to its versatile properties. In the paper industry, lecithin acts as an effective emulsifier and dispersing agent, helping to improve the smoothness and consistency of paper products. Additionally, it enhances the ink receptivity of paper surfaces, contributing to better print quality and appearance. In the paint industry, lecithin serves as a key ingredient in paint formulations, it functions as an interfacial agent, providing benefits such as improved flow, leveling, and adhesion properties. Its emulsifying properties help to stabilize paint emulsions and prevent separation, ensuring a homogenous mixture. Furthermore, lecithin acts as a natural and environmentally friendly alternative to synthetic additives in both industries, aligning with the growing demand for sustainable solutions (Van Nieuwenhuyzen 2015).

Lecithin plays a pivotal role in poultry and aquafeed as it helps to improve the overall quality of feeds by promoting better digestibility and absorption of nutrients in poultry and aquatic animals. Lecithin helps to improve the texture and palatability of feed, making it more appealing to animals and encouraging better feed intake. Deoiled lecithin is a common ingredient in poultry and aquafeed production. Soybean lecithin is particularly important in aquafeed due to its functions as a feed enhancer, growth promoter, immune system regulator, and antioxidant activator for various aquatic species (Van Nieuwenhuyzen 2015; Wee et al. 2023).

Furthermore, lecithin also plays a significant role in improving the health and growth of poultry and aquatic animals. It contains valuable nutrients such as choline, which is essential for proper liver function and development. In addition to its nutritional benefits, lecithin also acts as a natural antioxidant, helping to preserve the quality of feed and prevent oxidative damage. This can lead to improved shelf life of feed and better overall health of animals. Finally, the inclusion of lecithin in poultry and aquafeed is essential for promoting optimal growth, health, and performance in animals.

Overall, the versatile properties of lecithin make it a valuable ingredient in these diverse industries, contributing to the quality and performance of the final products.

The role of lecithin in different industrial sectors has been proven in various research studies and those are listed in Table 8.3:

8.6 Lecithin as a Sustainable Approach in Promoting a Healthier Food System

In recent years, the food industry has increasingly prioritized sustainability, focusing on environmentally responsible practices and ingredients. Lecithin has emerged as a noteworthy component for its potential in creating a sustainable approach to a healthier food system.

1. *Natural Emulsifier*: Lecithin, typically derived from soybeans, acts as a natural emulsifier in food products. Emulsifiers help blend ingredients that would otherwise separate, such as oil and water. By using lecithin, food manufacturers can create stable, uniform products without the need for synthetic emulsifiers thereby reducing reliance on chemicals in food production (Lecithin For Food n.d.)
2. *Reduction in Food Waste*: Lecithin offers a number of benefits when it comes to creating a more sustainable food system. Lecithin helps to prolong the duration of usability of many food products by enhancing their stability and texture. This reduces food waste by preventing spoilage and allows for longer storage times without compromising quality. As a result, there is less need to discard food due to deterioration, contributing to a more sustainable food supply chain. Lecithin is a byproduct of common food production processes, such as soybean oil

Table 8.3 Applications of lecithin in various sectors

Source	Nutritional property/for human health	Food industry	Pharmaceutical industry
Soy lecithin	High in polyunsaturated fatty acids (Arab Tehrany et al. 2012; Nguyen et al. 2014; Van Nieuwenhuyzen 2015)	Rheological property (Caparosa and Hartel 2020) Emulsifier (List 2015; Tebben et al. 2022)	Liposome encapsulation (Arab Tehrany et al. 2012; Adriana et al. 2014; De Assis et al. 2014; Van Nieuwenhuyzen 2015) Nutritional supplements Infant formulas (Drapala et al. 2015; List 2015)
Sunflower lecithin	High natural choline content, nutritionally beneficial unsaturated fatty acids (Nguyen et al. 2014; Sun et al. 2018; Grompone 2020)	Rheological properties in chocolate making (Caparosa and Hartel 2020) Emulsifying properties (Cabezas et al. 2012; Van Nieuwenhuyzen 2015)	–
Rice bran lecithin	Various health benefits like regulating the inflammatory reaction, lowering cholesterol, and cardiovascular risks (Liu et al. 2013)	Antioxidant (Jala and Prasad 2015) Emulsifier (Sun et al. 2018)	Liposome encapsulation (Adriana et al. 2014)
Rape seeds lecithin	Mostly contains monounsaturated fatty acids (Arab Tehrany et al. 2012; Nguyen et al. 2014)	Emulsifier (Nguyen et al. 2014)	Nanoliposomes (Arab Tehrany et al. 2012)
Egg yolk	Nutritional source of essential long-chain fatty acids, phosphatidylcholine, and choline (Van Nieuwenhuyzen 2015; Sun et al. 2018) Role in the regulation of lipid levels in bloods and also helps in the suppression of Alzheimer's disease. (Sun et al. 2018; Zhao et al. 2023)	Emulsifier Antioxidant (Food Safety and Quality: Detail n.d.) Wetting and dispersing agents in instant foods, used for infant food (Van Nieuwenhuyzen 2015)	Liposome(drug carrier) (Zhao et al. 2023) Used for parenteral nutrition and health food supplements (Van Nieuwenhuyzen 2015)
Dairy lecithin	Infant health improves memory (Contarini and Povolo 2013; Van Nieuwenhuyzen 2015; Sun et al. 2018; Alhajj et al. 2020;)	Rheological properties in chocolate making (Caparosa and Hartel 2020)	Used as a delivery system for liposoluble constituents (Contarini and Povolo 2013)

(continued)

Table 8.3 (continued)

Source	Nutritional property/for human health	Food industry	Pharmaceutical industry
Salmon	Essential fatty acids (Arab Tehrany et al. 2012)	–	Nanoliposomes (Arab Tehrany et al. 2012)
Krill oil	Rich in DHA & EPA (Van Nieuwenhuyzen 2015)	–	–

extraction. This means that by using lecithin in food production, we are able to reduce waste and make use of materials that would otherwise be discarded (Gupta and Gaur 2023).

3. *Healthier Formulations*: In food formulations, lecithin serves as a viable alternative to more heavily processed ingredients or additives that could pose adverse health effects. For instance, it can effectively replace partially hydrogenated oils, which contribute to the presence of harmful trans fats. The incorporation of lecithin in food products can be healthier option without sacrificing taste or texture (Puşcaş et al. 2020).
4. *Plant-Based Alternative*: Lecithin is naturally obtained from plant sources such as soybeans, sunflower, or rapeseed, making it a suitable alternative to animal-derived ingredients or synthetic additives. This aligns with the growing consumer preference for plant-based and sustainable food choices, contributing to a more ethical and environmentally friendly food system (The Eco-Friendly Potential of Soy Lecithin in Food and Beverage Manufacturing n.d.).
5. *Biodegradability and Environmental Impact*: Lecithin is biodegradable and non-toxic, which means it breaks down naturally in the environment without causing harm. This contrasts with many synthetic additives that may persist in ecosystems and contribute to environmental pollution. Using lecithin in food products thus supports a more environmentally sustainable approach to food production (Nyankson et al. 2015).

Lecithin can significantly contribute to creating a more sustainable approach in food production (Thakur 2024a, b). Its natural origins, functional properties, and health benefits make it a valuable ingredient for the food industry. By incorporating lecithin into food products, we can work toward building a healthier and more environmentally friendly food system for future generations.

8.7 Conclusion

The present chapter outlines the methodologies related to the sources, extraction, separation, and quantification of lecithin, along with its potential uses across various industrial sectors. The extraction techniques employed for lecithin derived from both plant and animal sources primarily include the organic solvent method, the simultaneous texturization and extraction of phospholipids (STEP) method, and the supercritical fluid extraction method. Research indicates that various separation and quantification techniques, such as liquid chromatography, column chromatography, HPLC, HILIC, and UPLC, can effectively isolate lecithin when paired with detectors like MS, tandem MS, IT-TOF, and ORBITRAP MS, facilitating a comprehensive analysis of phospholipids. Studies have demonstrated that HILIC-Orbitrap mass spectrometry and UPLC-Q-Exactive Orbitrap Mass Spectrometry are advanced methods that provide comparable performance, contingent upon specific experimental protocols and requirements. Furthermore, naturally sourced Lecithin has a broad spectrum of applications in diverse fields, contributing positively to human health, the pharmaceutical sector, and the cosmetic industry, with its emulsifying properties rendering it an ideal label-friendly ingredient. Nonetheless, further investigation is necessary to fully comprehend the applications, extraction, and quantification methods of Lecithin derived from marine sources.

References

Adriana RM, Leticia MDA, Maria IRM, Leonor ADSS (2014) Importance of lecithin for encapsulation processes. Afr J Food Sci 8(4):176–183. https://doi.org/10.5897/AJFS2013.1092

Ahmed F, Eticha T, Hymete A, Ashenef A (2023) Principles and applications of ultra-high-performance liquid chromatography. In: Núñez O, Sentellas S, Granados M, Saurina J (eds) High performance liquid chromatography—recent advances and applications. IntechOpen. https://doi.org/10.5772/intechopen.110540

Akcay TT, Karadeniz B, Sirin N, Aydin G, Goksu H (2020) Use of sunflower seed lecithin as an emulsifier in herbal cream preparation. Int J Traditi Complementary Med Res 1(3):125

Alhajj MJ, Montero N, Yarce CJ, Salamanca CH (2020) Lecithins from vegetable, land, and marine animal sources and their potential applications for cosmetic, food, and pharmaceutical sectors. Cosmetics 7(4):87. https://doi.org/10.3390/cosmetics7040087

Ameta RK, Soni K, Bhattarai A (2023) Recent advances in improving the bioavailability of hydrophobic/lipophilic drugs and their delivery via self-emulsifying formulations. Colloids Interfaces 7(1):16. https://doi.org/10.3390/colloids7010016

Arab Tehrany E, Kahn CJF, Baravian C, Maherani B, Belhaj N, Wang X, Linder M (2012) Elaboration and characterization of nanoliposome made of soya, rapeseed and salmon lecithins: application to cell culture. Colloids Surf B Biointerfaces 95:75–81. https://doi.org/10.1016/j.colsurfb.2012.02.024

Batool A, Menaa F (2020) Concentration and purification of seaweed components by chromatography methods. In: Sustainable seaweed technologies. Elsevier, pp 315–370. https://doi.org/10.1016/B978-0-12-817943-7.00013-5

Bot F, Cossuta D, O'Mahony JA (2021) Inter-relationships between composition, physicochemical properties and functionality of lecithin ingredients. Trends Food Sci Technol 111:261–270. https://doi.org/10.1016/j.tifs.2021.02.028

Burnett CL, Fiume MM, Bergfeld WF, Belsito DV, Hill RA, Klaassen CD, Liebler D, Marks JG, Shank RC, Slaga TJ, Snyder PW, Andersen FA (2017) Safety assessment of plant-derived fatty acid oils. Int J Toxicol 36(3_suppl):51S–129S. https://doi.org/10.1177/1091581817740569

Burri L, Hoem N, Banni S, Berge K (2012) Marine omega-3 phospholipids: metabolism and biological activities. Int J Mol Sci 13(11):15401–15419. https://doi.org/10.3390/ijms131115401

Cabezas DM, Madoery R, Diehl BWK, Tomás MC (2012) Emulsifying properties of different modified sunflower lecithins. J Am Oil Chem Soc 89(2):355–361. https://doi.org/10.1007/s11746-011-1915-8

Caparosa MH, Hartel RW (2020) Characterizing lecithin interactions in chocolate using interfacial properties and rheology. J Am Oil Chem Soc 97(12):1309–1317. https://doi.org/10.1002/aocs.12419

Cavazzini A, Catani M, Felinger A (2023) Hydrophilic interaction liquid chromatography. In: Liquid chromatography. Elsevier, pp 227–249. https://doi.org/10.1016/B978-0-323-99968-7.00030-8

Cavusoglu S, Uzun Y, Yilmaz N, Ercisli S, Eren E, Ekiert H, Elansary HO, Szopa A (2021) Maintaining the quality and storage life of button mushrooms (*Agaricus bisporus*) with gum, agar, sodium alginate, egg white protein, and lecithin coating. J Fungi 7(8):614. https://doi.org/10.3390/jof7080614

Contarini G, Povolo M (2013) Phospholipids in milk fat: composition, biological and technological significance, and analytical strategies. Int J Mol Sci 14(2):2808–2831. https://doi.org/10.3390/ijms14022808

De Assis LM, Machado AR, De Souza Da Motta A, Costa JAV, De Souza-Soares LA (2014) Development and characterization of nanovesicles containing phenolic compounds of microalgae *Spirulina* strain LEB-18 and *Chlorella pyrenoidosa*. Adv Mater Phys Chem 4(1):6–12. https://doi.org/10.4236/ampc.2014.41002

Donato P, Cacciola F, Cichello F, Russo M, Dugo P, Mondello L (2011) Determination of phospholipids in milk samples by means of hydrophilic interaction liquid chromatography coupled to evaporative light scattering and mass spectrometry detection. J Chromatogr A 1218(37):6476–6482. https://doi.org/10.1016/j.chroma.2011.07.036

Drapala KP, Auty MAE, Mulvihill DM, O'Mahony JA (2015) Influence of lecithin on the processing stability of model whey protein hydrolysate-based infant formula emulsions. Int J Dairy Technol 68(3):322–333. https://doi.org/10.1111/14710307.12256

Food safety and quality: Detail (n.d.) Retrieved August 12, 2024, from https://www.fao.org/food/food-safety-quality/scientific-advice/jecfa/jecfa-additives/detail/en/c/260/

Fowler BA (2016) Historical development of biomarkers. In: Molecular biological markers for toxicology and risk assessment. Elsevier, pp 17–37. https://doi.org/10.1016/B978-0-12-809589-8.00002-0

Fu C, Liu M, Li Y, Wang K, Yang B, Deng L, Tian J, Yang G, Zheng G (2020) UPLC-Q-Exactive Orbitrap MS analysis for identification of lipophilic components in *Citri Sarcodactylis Fructus* from different origins in China using supercritical CO_2 fluid extraction method. ACS Omega 5(19):11013–11023. https://doi.org/10.1021/acsomega.0c00854

Garcia C, Lutz NW, Confort-Gouny S, Cozzone PJ, Armand M, Bernard M (2012) Phospholipid fingerprints of milk from different mammalians determined by 31P NMR: towards specific interest in human health. Food Chem 135(3):1777–1783. https://doi.org/10.1016/j.foodchem.2012.05.111

Ghanshyam MU, Patel PR, Patel JK (2011) Formulation and characterization of solid lipid nanoparticles dry powder inhaler containing triamcinolone acetonide. Int J Res Pharm Chem 1(3):662–673

Gładkowski W, Chojnacka A, Kiełbowicz G, Trziszka T, Wawrzeńczyk C (2012) Isolation of pure phospholipid fraction from egg yolk. J Am Oil Chem Soc 89(1):179–182. https://doi.org/10.1007/s11746-011-1893-x

Grompone MA (2020) Sunflower and high-oleic sunflower oils. In: Shahidi F (ed) Bailey's industrial oil and fat products, 1st edn. Wiley, pp 1–54. https://doi.org/10.1002/047167849X.bio017.pub2

Guiotto EN, Cabezas DM, Diehl BWK, Tomás MC (2013) Characterization and emulsifying properties of different sunflower phosphatidylcholine enriched fractions. Eur J Lipid Sci Technol 115(8):865–873. https://doi.org/10.1002/ejlt.201200394

Guiotto EN, Tomás MC, Diehl BWK (2015) Sunflower lecithin. In: Polar lipids. Elsevier, pp 57–75. https://doi.org/10.1016/B978-1-63067-044-3.50007-8

Gupta R, Gaur S (2023) Lecithin as a functional ingredient in cereals. In: Valorization of biomass to bioproducts. Elsevier, pp 59–70. https://doi.org/10.1016/B978-0-12-822887-6.00014-0

Gutiérrez-Méndez N, Chavez-Garay DR, Leal-Ramos MY (2022) Lecithins: a comprehensive review of their properties and their use in formulating microemulsions. J Food Biochem 46(7):e14157. https://doi.org/10.1111/jfbc.14157

Huang Q, Lei H, Dong M, Tang H, Wang Y (2019) Quantitative analysis of 10 classes of phospholipids by ultrahigh-performance liquid chromatography tandem triple-quadrupole mass spectrometry. Analyst 144(13):3980–3987. https://doi.org/10.1039/C9AN00676A

Huang J, Lu F, Wu Y, Wang D, Xu W, Zou Y, Sun W (2022) Enzymatic extraction and functional properties of phosphatidylcholine from chicken liver. Poult Sci 101(6):101689. https://doi.org/10.1016/j.psj.2021.101689

Hussain N, Sun DW, Pu H (2019) Classical and emerging non-destructive technologies for safety and quality evaluation of cereals: a review of recent applications. Trends Food Sci Technol 91: 598–608. https://doi.org/10.1016/j.tifs.2019.07.018

Jala RCR, Prasad RBN (2015) Rice bran lecithin: compositional, nutritional, and functional characteristics. In: Polar lipids. Elsevier, pp 35–55. https://doi.org/10.1016/B978-1-63067-044-3.50006-6

Jangle RD, Magar VP, Thorat BN (2013) Phosphatidylcholine and its purification from raw de-oiled soya lecithin. Sep Purif Technol 102:187–195. https://doi.org/10.1016/j.seppur.2012.10.002

Küllenberg D, Taylor LA, Schneider M, Massing U (2012) Health effects of dietary phospholipids. Lipids Health Dis 11(1):3. https://doi.org/10.1186/1476-511X-11-3

Lecithin (Antioxidant): Cosmetic Ingredient INCI (n.d.) Retrieved August 11, 2024, from https://cosmetics.specialchem.com/inci-ingredients/lecithin

Lecithin For Food (n.d.) Giiava. Retrieved August 11, 2024, from https://giiava.com/lecithin-for-food/

Lehri D, Kumari N, Singh RP, Sharma V (2019) Composition, production, physicochemical properties and applications of lecithin obtained from rice (Oryza sativa L.)—a review. Plant Sci Today 6(sp1):613–622. https://doi.org/10.14719/pst.2019.6.sp1.682

Li J, Guo Z (2016) Identification and quantification of phenolic compounds in rapeseed originated lecithin and antioxidant activity evaluation. LWT 73:397–405. https://doi.org/10.1016/j.lwt.2016.06.039

Li L, Chang M, Tao G, Wang X, Liu Y, Liu R, Jin Q, Wang X (2016) Analysis of phospholipids in Schizochytrium sp. S31 by using UPLC-Q-TOF-MS. Anal Methods 8(4):763–770. https://doi.org/10.1039/C5AY02795K

Li Q, Zhao Y, Zhu D, Pang X, Liu Y, Frew R, Chen G (2017) Lipidomics profiling of goat milk, soymilk and bovine milk by UPLC-Q-Exactive Orbitrap Mass Spectrometry. Food Chem 224: 302–309. https://doi.org/10.1016/j.foodchem.2016.12.083

List GR (2015) Soybean lecithin: food, industrial uses, and other applications. In: Polar lipids. Elsevier, pp 1–33. https://doi.org/10.1016/B978-1-63067-044-3.50005-4

Liu L, Waters DLE, Rose TJ, Bao J, King GJ (2013) Phospholipids in rice: significance in grain quality and health benefits: a review. Food Chem 139(1–4):1133–1145. https://doi.org/10.1016/j.foodchem.2012.12.046

Liu Z, Moate P, Cocks B, Rochfort S (2015) Comprehensive polar lipid identification and quantification in milk by liquid chromatography–mass spectrometry. J Chromatogr B 978–979:95–102. https://doi.org/10.1016/j.jchromb.2014.11.036

Liu Y, Qiao W, Liu Y, Zhao J, Liu Q, Yang K, Zhang M, Wang Y, Liu Y, Chen L (2023) Quantification of phospholipids and glycerides in human milk using ultra-performance liquid chromatography with quadrupole-time-of-flight mass spectrometry. Front Chem 10:1101557. https://doi.org/10.3389/fchem.2022.1101557

López-Porfiri P, Gorgojo P, González-Miquel M (2024) Recovery of bio-based organic acids using green technologies. In: Reference module in chemistry, molecular sciences and chemical engineering. Elsevier. https://doi.org/10.1016/B978-0-443-15742-4.00010-7

Moloney C, O'Connor D, O'Regan J (2020) Polar lipid, ganglioside and cholesterol contents of infant formulae and growing up milks produced with an alpha lactalbumin-enriched whey protein concentrate. Int Dairy J 107:104716. https://doi.org/10.1016/j.idairyj.2020.104716

Moré MI, Freitas U, Rutenberg D (2014) Positive effects of soy lecithin-derived phosphatidylserine plus phosphatidic acid on memory, cognition, daily functioning, and mood in elderly patients with Alzheimer's disease and dementia. Adv Ther 31(12):1247–1262. https://doi.org/10.1007/s12325-014-0165-1

Mortensen A, Aguilar F, Crebelli R, Di Domenico A, Frutos MJ, Galtier P, Gott D, Gundert-Remy U, Lambré C, Leblanc J, Lindtner O, Moldeus P, Mosesso P, Oskarsson A, Parent-Massin D, Stankovic I, Waalkens-Berendsen I, Woutersen RA, Wright M, Dusemund B (2017) Re-evaluation of lecithins (E 322) as a food additive. EFSA J 15(4):4742. https://doi.org/10.2903/j.efsa.2017.4742

Murashova NM (2023) Lecithin microemulsions as drug carriers. Colloid J 85(5):746–756. https://doi.org/10.1134/S1061933X23600689

Nguyen MT, Van De Walle D, Petit C, Beheydt B, Depypere F, Dewettinck K (2014) Mapping the chemical variability of vegetable lecithins. J Am Oil Chem Soc 91(7):1093–1101. https://doi.org/10.1007/s11746-014-2455-9

Nyankson E, DeCuir MJ, Gupta RB (2015) Soybean lecithin as a dispersant for crude oil spills. ACS Sustain Chem Eng 3(5):920–931. https://doi.org/10.1021/acssuschemeng.5b00027

Onaolapo MC, Alabi OD, Akano OP, Olateju BS, Okeleji LO, Adeyemi WJ, Ajayi AF (2024) Lecithin and cardiovascular health: a comprehensive review. Egypt Heart J 76(1):92. https://doi.org/10.1186/s43044-024-00523-0

Patil VV, Galge RV, Thorat BN (2010) Extraction and purification of phosphatidylcholine from soyabean lecithin. Sep Purif Technol 75(2):138–144. https://doi.org/10.1016/j.seppur.2010.08.006

Prado R, Weber CC (2016) Applications of ionic liquids. In: Application, purification, and recovery of ionic liquids. Elsevier, pp 1–58. https://doi.org/10.1016/B978-0-444-63713-0.00001-8

Price N, Fei T, Clark S, Wang T (2018) Extraction of phospholipids from a dairy by-product (whey protein phospholipid concentrate) using ethanol. J Dairy Sci 101(10):8778–8787. https://doi.org/10.3168/jds.2018-14950

Puşcaş A, Mureşan V, Socaciu C, Muste S (2020) Oleogels in food: a review of current and potential applications. Foods 9(1):70. https://doi.org/10.3390/foods9010070

Rahman M (2018) Application of computational methods in isolation of plant secondary metabolites. In: Computational phytochemistry. Elsevier, pp 107–139. https://doi.org/10.1016/B978-0-12-812364-5.00004-3

Rodríguez-González T, Cerezo Valverde J, García B (2019) Performance of marine lecithin supplemented feeds for the common octopus (Octopus vulgaris) ongrowing: changes in proximate composition and lipid classes' profile. Fishes 4(3):47. https://doi.org/10.3390/fishes4030047

Song S, Liu TT, Liang X, Liu ZY, Yishake D, Lu XT, Yang MT, Man QQ, Zhang J, Zhu HL (2021) Profiling of phospholipid molecular species in human breast milk of Chinese mothers and comprehensive analysis of phospholipidomic characteristics at different lactation stages. Food Chem 348:129091. https://doi.org/10.1016/j.foodchem.2021.129091

Subra-Paternault P, ThongDeng H, Grélard A, Cansell M (2015) Extraction of phospholipids from scallop by-product using supercritical CO2/alcohol mixtures. LWT-Food Sci Technol 60(2):990–998. https://doi.org/10.1016/j.lwt.2014.09.057

Sun N, Chen J, Wang D, Lin S (2018) Advance in food-derived phospholipids: sources, molecular species and structure as well as their biological activities. Trends Food Sci Technol 80:199–211. https://doi.org/10.1016/j.tifs.2018.08.010

Tebben L, Chen G, Tilley M, Li Y (2022) Improvement of whole wheat dough and bread properties by emulsifiers. Grain Oil Sci Technol 5(2):59–69. https://doi.org/10.1016/j.gaost.2022.05.001

Thakur M (2024a) Sustainable food systems (Volume I), World sustainability series. Springer, Cham. https://doi.org/10.1007/978-3-031-47122-3_1

Thakur M (2024b) Sustainable food systems (Volume II), World sustainability series. Springer, Cham. https://doi.org/10.1007/978-3-031-46046-3_1

The Eco-Friendly Potential of Soy Lecithin in Food and Beverage Manufacturing (n.d.) Retrieved August 13, 2024, from https://www.lecitein.com/index.php/blog/the-eco-friendly-potential-of-soy-lecithin-in-food-and-beverage-manufacturing

Van Nieuwenhuyzen W (2015) Production and utilization of natural phospholipids. In: Polar lipids. Elsevier, pp 245–276. https://doi.org/10.1016/B978-1-63067-044-3.50013-3

Velichkova S, Foubert K, Theunis M, Pieters L (2024) HILIC UPLC/QTof MS method development for the quantification of AGEs inhibitors—troubleshooting protocol. Comb Chem High Throughput Screen 27(4):584–598. https://doi.org/10.2174/1386207326666230706120451

Wang T, Zhou D (2017) Advances in phospholipid quantification methods. Curr Opin Food Sci 16: 15–20. https://doi.org/10.1016/j.cofs.2017.06.007

Wang H, Yao L, Lee S, Wang T (2017a) Extraction of phospholipids from egg yolk flakes using aqueous alcohols. J Am Oil Chem Soc 94(2):309–314. https://doi.org/10.1007/s11746-016-2939-x

Wang H, Yao L, Lee S, Wang T (2017b) Simultaneous texturization and extraction of phospholipids from liquid egg yolk using renewable solvents. Eur J Lipid Sci Technol 119(2):1500523. https://doi.org/10.1002/ejlt.201500523

Wee W, Téllez-Isaías G, Abdul Kari Z, Cheadoloh R, Kabir MA, Mat K, Mohamad Sukri SA, Rahman MM, Rusli ND, Wei LS (2023) The roles of soybean lecithin in aquafeed: a crucial need and update. Front Vet Sci 10:1188659. https://doi.org/10.3389/fvets.2023.1188659

Winther B, Hoem N, Berge K, Reubsaet L (2011) Elucidation of phosphatidylcholine composition in krill oil extracted from *Euphausia superba*. Lipids 46(1):25–36. https://doi.org/10.1007/s11745-010-3472-6

Zhao F, Li R, Liu Y, Chen H (2023) Perspectives on lecithin from egg yolk: extraction, physicochemical properties, modification, and applications. Front Nutr 9:1082671. https://doi.org/10.3389/fnut.2022.1082671

[illegible]

Chapter 9
Advances in Artificial Intelligence in Smart Farming: Catalyzing Innovation and Sustainability

Swarnima Dey, Chaiyavat Chaiyasut, and Yogesh Kumar

Abstract The evolution of agriculture has been shaped by the transition from traditional farming practices to modern, technology-driven methods. This shift, now represented by Agriculture 4.0 or smart farming, incorporates advanced technologies such as the Internet of Things (IoT), artificial intelligence (AI), cloud computing, and automation to enhance productivity and efficiency. These innovations enable real-time monitoring of critical agricultural parameters like crop health, soil conditions, and weather patterns. AI plays a pivotal role in smart farming by enabling precision agriculture, yield prediction, pest and weed control, livestock management, and water optimization. AI algorithms process vast amounts of data from sensors, drones, and satellite imagery to offer insights that help farmers optimize resources, reduce costs, and minimize environmental impact. Additionally, AI supports automated systems for tasks such as harvesting, market analysis, and risk management, improving decision-making and resource allocation. However, the adoption of these technologies also introduces challenges, particularly regarding data privacy and cybersecurity. In rural areas, where internet infrastructure may be limited, ensuring the protection of sensitive agricultural data becomes crucial. As smart farming technologies continue to evolve, they promise to revolutionize agriculture by providing farmers with the tools to increase efficiency, sustainability, and profitability while addressing global food security concerns. This chapter explores

S. Dey
Department of Food Technology, Faculty of Science and Humanities, SRM University Delhi-NCR, Sonepat, Haryana, India

Amity Institute of Food Technology, Amity University, Uttar Pradesh, Noida, Uttar Pradesh, India

C. Chaiyasut
Innovation Center for Holistic Health, Nutraceuticals and Cosmeceuticals, Chiang Mai, Thailand

Y. Kumar (✉)
Department of Food Technology, Faculty of Science and Humanities, SRM University Delhi-NCR, Sonepat, Haryana, India

M. Thakur, T. Belwal (eds.), *Food Biodiversity and Consumer Engagements*, Sustainable Development and Biodiversity 46,
https://doi.org/10.1007/978-981-95-7777-4_9

the transformative role of AI and other smart farming technologies in reshaping the future of agriculture.

Keywords Smart farming · Artificial intelligence · IoT sensors · Cloud computer · Data analysis

9.1 Introduction

The evolution of agriculture has been deeply influenced by rural farming practices, which have transitioned from traditional manual labor and animal power to modern, technology-driven methods. This progression, marked by the use of basic tools in earlier eras, has now advanced to the deployment of automated equipment that significantly enhances productivity and operational efficiency (Chen and Hengjinda 2019). While traditional farming techniques, rooted in sensitivity to environmental conditions, have long been the foundation of agriculture, they often face challenges related to efficiency, scalability, and sustainability. In recent years, the integration of technological advancements has sought to address these challenges, setting the stage for the emergence of smart farming or Agriculture 4.0 (Rahaman et al. 2024).

Smart farming represents a revolutionary approach to agriculture, leveraging cutting-edge technologies such as the Internet of Things (IoT), artificial intelligence (AI), cloud computing, and sophisticated hardware and software systems (Shaikh et al. 2022). These technologies enable real-time monitoring and analysis of critical agricultural parameters, such as crop health, soil conditions, and weather patterns. Through the use of IoT-connected sensors, smart farming systems continuously collect vast amounts of data, which are processed either locally or on cloud servers using advanced AI algorithms. This real-time data analysis empowers farmers with unprecedented insights, enabling optimized decision-making and improved farm productivity (Patel and Bhatia 2024).

Key components of smart farming include hardware such as sensors, unmanned aerial vehicles (UAVs), global positioning systems (GPS), and automated machinery, as well as software platforms for data analysis and decision support (Boursianis et al. 2022). Innovations in this field have focused on enhancing sensor accuracy, improving UAV efficiency, and refining software capabilities for better data processing (Boursianis et al. 2022). However, despite these advancements, there remain critical challenges in ensuring data privacy and security, particularly in rural farm monitoring. The deployment of IoT and AI in rural areas introduces vulnerabilities due to limited cybersecurity infrastructure, raising concerns about protecting sensitive agricultural data, including farmers' personal and financial information and farm productivity statistics (Demestichas et al. 2020).

Moreover, the growing reliance on cloud-based infrastructure for data storage and processing amplifies privacy risks. Farmers, often unfamiliar with digital security measures, face significant challenges in safeguarding their data. This underscores the need for privacy-centric solutions that can address these vulnerabilities while maintaining the benefits of smart farming technologies (Hazrati et al. 2022).

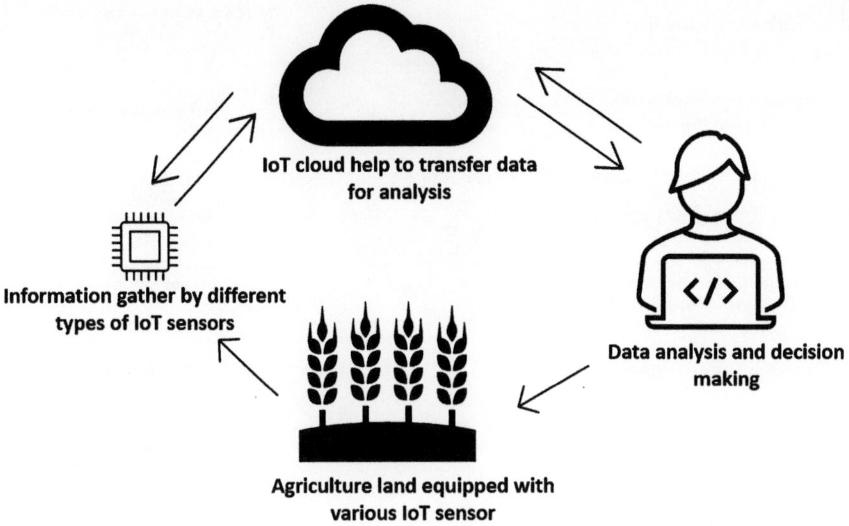

Fig. 9.1 Smart farming functions with three main components (IoT sensors, IoT cloud, and data analysis)

Typically, smart farming systems consist of three primary components: IoT sensors, cloud servers, and human experts (Lytos et al. 2020) (Fig. 9.1). IoT sensors, strategically deployed across farms, serve as the primary data collection points, gathering precise information on crops and environmental conditions. Cloud servers act as the central hub for data processing and storage, utilizing AI algorithms to generate actionable insights. Human experts, such as farmers or farm managers, play a crucial role in interpreting these insights and making informed decisions, often adjusting IoT devices manually to optimize farm operations (Lytos et al. 2020).

This chapter explores the transformative potential of smart farming technologies, delving into their components, benefits, and challenges. Special attention is given to addressing data privacy and security concerns, which are critical for ensuring the sustainable adoption of these technologies in rural agricultural settings. Through this exploration, the chapter aims to provide a comprehensive understanding of the role of smart farming in shaping the future of agriculture.

9.2 Role of Artificial Intelligence (AI) in Agriculture

Artificial intelligence (AI) is making significant strides in transforming agriculture. It is playing a key role in enhancing productivity, efficiency, and sustainability in farming practices (Shaikh et al. 2022). The breakdown of how AI contributes to various aspects of agriculture (Fig. 9.2):

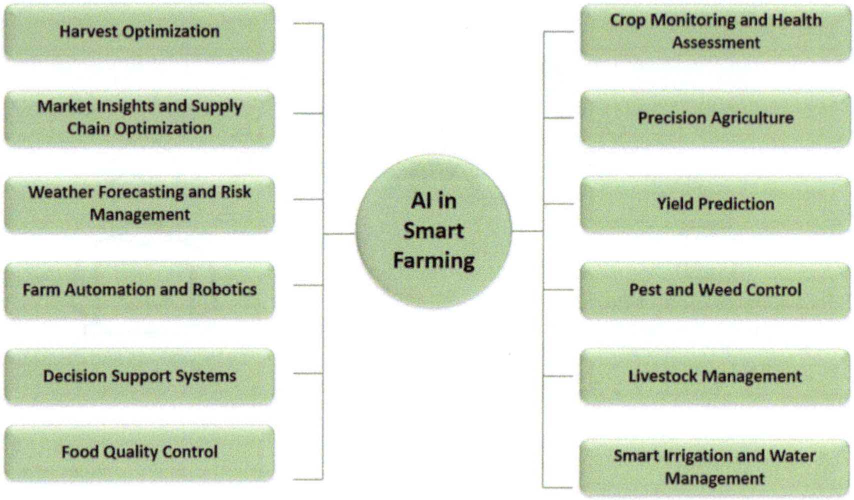

Fig. 9.2 AI contributes to various aspects of smart farming

9.2.1 Crop Monitoring and Health Assessment

AI-powered sensors and drones equipped with machine learning models are helping farmers monitor crop health in real-time. By analyzing images and data from sensors, AI can identify early signs of diseases, pests, or nutrient deficiencies. This allows for quicker intervention, minimizing crop loss and reducing the need for excessive pesticide use (Kaushik et al. 2021; Thakur and Belwal 2024).

9.2.2 Precision Agriculture

AI enables precision farming by processing large volumes of data from various sources, including IoT sensors, weather data, and satellite imagery. This helps farmers apply resources (like water, fertilizers, and pesticides) more efficiently and in a targeted manner. The result is optimized yields, lower costs, and reduced environmental impact (Shafi et al. 2019).

9.2.3 Yield Prediction

AI algorithms analyze historical crop data, weather patterns, and soil conditions to predict crop yields. This information is crucial for farmers and agricultural businesses to make informed decisions on resource allocation, harvest timing, and market strategies. Yield prediction also helps in forecasting food supply and addressing food security concerns (Akkem et al. 2023).

9.2.4 Pest and Weed Control

AI systems can detect and identify pest infestations and weed growth through image recognition and data analysis. These AI-driven tools can differentiate between crops, pests, and weeds, enabling farmers to apply targeted treatments, reducing the overall use of harmful pesticides and herbicides. This leads to better environmental outcomes and healthier crops (Mohamed et al. 2021).

9.2.5 Livestock Management

AI technologies, including wearable devices and tracking systems, help monitor livestock health and behavior. These devices collect data on animals' movement, feeding patterns, and health conditions, which is then analyzed by AI models. This enables farmers to detect early signs of illness or distress, improving animal welfare and productivity (Singh et al. 2022).

9.2.6 Smart Irrigation and Water Management

AI is instrumental in optimizing water usage in agriculture through smart irrigation systems. By analyzing weather patterns, soil moisture levels, and crop water requirements, AI systems can adjust irrigation schedules to ensure crops receive the right amount of water. This helps conserve water, reduce costs, and increase crop resilience in areas facing water scarcity (Alves et al. 2023).

9.2.7 Harvest Optimization

AI-powered systems can assess the optimal time for harvesting based on various factors such as crop ripeness, weather conditions, and market demand. Automated systems can also be used for harvesting, reducing labor costs, and increasing the speed and efficiency of the process (Kiani et al. 2022).

9.2.8 Market Insights and Supply Chain Optimization

AI tools can analyze market trends, consumer preferences, and supply chain data to provide farmers and agribusinesses with insights into demand fluctuations, pricing strategies, and inventory management. This helps farmers make informed decisions about what to grow, when to harvest, and how to transport goods efficiently, ultimately reducing waste and maximizing profitability (Lioutas and Charatsari 2020).

9.2.9 Weather Forecasting and Risk Management

AI is improving weather forecasting models, providing farmers with accurate predictions of temperature, precipitation, and other climatic conditions. With better weather forecasts, farmers can make proactive decisions regarding planting, irrigation, and harvest timing. AI can also help manage risks associated with climate change by predicting potential adverse weather events and guiding farmers on risk mitigation strategies (Pujahari et al. 2022).

9.2.10 Farm Automation and Robotics

AI is at the core of farm automation, with robots and autonomous machines handling tasks like planting, weeding, harvesting, and soil analysis. These robots are equipped with advanced sensors and AI algorithms to navigate fields, perform tasks accurately, and work in a way that reduces the need for manual labor. Automation increases productivity, reduces human error, and optimizes resource use (Javaid et al. 2022).

9.2.11 Decision Support Systems

AI-based decision support systems integrate data from various sources, such as IoT devices, satellite imagery, weather forecasts, and historical trends. These systems assist farmers in making data-driven decisions on planting, irrigation, pest control, and harvesting. By considering a range of factors simultaneously, AI helps farmers make more accurate, informed choices to improve farm productivity and sustainability (Javaid et al. 2022).

9.2.12 Food Quality Control

AI applications are helping ensure food quality by monitoring the conditions in which crops are grown and harvested. Through AI-powered sensors, farmers can assess factors like crop maturity, moisture content, and temperature, ensuring the best quality product reaches the market. AI can also be used in sorting and grading produce, improving consistency and reducing human error (Agrimonti et al. 2021).

9.3 Importance of AI in Addressing Modern Agricultural Challenges

Artificial intelligence (AI) plays a crucial role in addressing modern agricultural challenges by providing innovative solutions to improve productivity, efficiency, and sustainability. With the growing global population, climate change, and resource limitations, AI is becoming essential in modern farming. AI helps optimize resource use, such as water, fertilizers, and pesticides, through precision agriculture. This reduces waste, lowers costs, and ensures that resources are applied more effectively, leading to higher yields with fewer inputs (Chen et al. 2022). By analyzing vast amounts of data from sensors, satellites, and weather forecasts, AI enables farmers to make informed decisions. This includes predicting crop yields, identifying pests or diseases early, and determining the optimal time for harvesting, all of which lead to more efficient farm management (Saadane et al. 2022). Climate change, unpredictable weather patterns, and pest outbreaks pose significant risks to crops. AI's ability to forecast weather, monitor crop health, and predict potential threats helps farmers mitigate these risks and protect their crops and livestock (Shaikh et al. 2022). AI-powered automation, including robotics and autonomous machinery, reduces the reliance on manual labor, which is often in short supply. Automation increases operational efficiency and ensures that tasks such as planting, weeding, and harvesting are done accurately and efficiently (Charania and Li 2020). AI helps optimize farming practices in a way that minimizes environmental impact. For example, smart irrigation systems powered by AI reduce water usage, while precision pest control reduces the need for harmful chemicals, leading to more sustainable farming practices (Akkem et al. 2023). AI also helps farmers understand market trends, consumer demand, and supply chain dynamics. This enables better planning and decision-making, ensuring farmers can meet market demands and maximize profits (Akkem et al. 2023).

9.4 Applications of AI in Smart Farming

Artificial intelligence (AI) has a profound impact on smart farming, offering innovative solutions to increase productivity, reduce costs, and promote sustainability. The key applications of AI in smart farming (Table 9.1):

9.4.1 *Crop Monitoring and Management*

AI-powered systems utilize sensors, drones, and satellite imagery to continuously monitor the health of crops. AI algorithms analyze this data to detect diseases, pests, nutrient deficiencies, or environmental stress. This enables timely intervention, improving crop health and reducing losses (Kaushik et al. 2021).

9.4.2 *Yield Prediction Models*

AI uses historical data, weather patterns, and real-time field conditions to predict crop yields. These models allow farmers to forecast production more accurately, optimize resource allocation, and plan better for harvest and market demand (Akkem et al. 2023).

9.4.3 *Weed and Pest Control*

AI-powered computer vision and machine learning algorithms can identify weeds and pests in real-time. This enables the use of targeted, automated spraying systems that apply herbicides and pesticides only where needed, reducing chemical use and increasing efficiency (Mohamed et al. 2021).

9.4.4 *Smart Irrigation Systems*

AI-based irrigation systems monitor soil moisture levels, weather forecasts, and crop water requirements to optimize irrigation schedules. This helps conserve water, reduce costs, and improve crop growth, particularly in regions facing water scarcity (Alves et al. 2023).

Table 9.1 Applications, data sources, and collection methods of artificial intelligence (AI) in smart farming

Applications	Data sources	Collection methods	References
Crop monitoring and management	AI algorithms	Action occurs across the crop, the cycle begins and ends at the crop level, and specialized sensors then record the crop's reaction	Saiz-Rubio and Rovira-Más (2020)
Yield prediction models	Farm management information system (FMIS)	The challenge can be met with the aid of advanced sensing technologies in agriculture, which offer precise applications of phytosanitary products based on detailed information on soil, crop status, and environmental conditions. This reduces the need for herbicides and pesticides, improves water use efficiency, and increases crop yield and quality	Saiz-Rubio and Rovira-Más (2020)
Weed and pest control	Proximal sensing	Monitoring platforms operate through the ground level with help of robots that acquire information for plant status, disease incidence, and infestations	Saiz-Rubio and Rovira-Más (2020)
Smart irrigation systems	Fuzzy-based decision support system (DSS)	DSS taking as input variables soil moisture, rain forecast to provide irrigation management decisions	Saiz-Rubio and Rovira-Más (2020)
Autonomous Machinery	Agricultural robots integrating AI	Robots can reduce the farm operating costs	Saiz-Rubio and Rovira-Más (2020)
Livestock health monitoring	Sensors	Wearable sensors that are attached to or within cattle can track a variety of activities, including eating, rumination, body temperature, pH, laying behavior that helps livestock early disease detection, management, and operations	Džermeikaitė et al. (2023)
Soil analysis and management	Proximal sensing	Monitoring platforms operate through the ground level with help of robots that acquire information for plant status, disease incidence, and infestations	Saiz-Rubio and Rovira-Más (2020)
Farm management platforms	Farm management information system (FMIS)	It is a management information system made to help farmers with a number of activities, from field work	Saiz-Rubio and Rovira-Más (2020)

(continued)

Table 9.1 (continued)

Applications	Data sources	Collection methods	References
		evaluation to operational planning, execution, and documentation	
Drone-based applications	Internet of Things (IoT) with Unmanned Aerial Vehicles (UAVs)	Requirements for automated, real-time environmental parameter monitoring with both aboveground and subsurface sensors	Almalki et al. (2021)
Supply chain and logistics optimization	Precision agricultural technologies (variable rate technologies (VRTs), electronic maps, yield monitors, and guidance farming systems)	The automation of numerous tasks like planting, seeding, harvesting, and soil sample is self-sufficient and intelligent	Karunathilake et al. (2023)
Food quality assessment	Internet of things (IoT), big data, and Artificial Intelligence (AI)	Monitor a number of factors, including temperature, humidity, lighting, water and air quality, and other observable growth aspects, in order to increase future output by keeping a traceable record	Moghayedi et al. (2022)
Climate resilience planning	Internet of things (IoT)	Utilizing information and communication technology (ICT) to enhance agricultural output, quality, yield, and profitability while lowering environmental impact (e.g., precise pesticide application, effective irrigation, etc.)	Adamides et al. (2020)
Biodiversity preservation	Remote sensing	Data analyse for atmosphere, soil and plant data	Adamides et al. (2020)
Agri-Fintech solutions	IoT and Fintech	Fintech, IoT, and big data are critical technologies that help agricultural practitioners identify farming methods and make particular judgments by providing useful facts	More and Aslekar (2022)
Blockchain integration	IoT-AWS cloud, ESP32, the smart farm security monitoring framework, and the Ethereum Rinkeby smart contract mechanism	The distributed ledger architecture used by the framework to store and manage data obtained from IoT devices deployed in smart farms offers safe, tamper-proof data storage and guarantees the accuracy and legitimacy of the data	Aliyu and Liu (2023)

9.4.5 Autonomous Machinery

AI enables the development of autonomous machinery, such as tractors and harvesters, that can operate with precision in the fields. These machines perform tasks like planting, cultivating, and harvesting without human intervention, reducing labor costs and increasing operational efficiency (Mim et al. 2025).

9.4.6 Livestock Health Monitoring

AI applications in livestock monitoring use sensors and wearables to track animals' health and behavior. AI analyzes the data to detect early signs of illness, stress, or abnormal behavior, allowing farmers to provide timely care and improve overall animal welfare (Singh et al. 2022).

9.4.7 Soil Analysis and Management

AI algorithms process data collected from sensors and drones to analyze soil health, including moisture levels, nutrient content, and pH. This information guides decisions regarding fertilization and irrigation, helping farmers optimize soil conditions for better crop growth (Chen et al. 2022).

9.4.8 Farm Management Platforms

AI-driven farm management platforms integrate data from various sources (sensors, satellite imagery, weather forecasts) to provide farmers with actionable insights. These platforms assist in decision-making, resource optimization, and efficient farm management across multiple areas (Mohamed et al. 2021).

9.4.9 Drone-based Applications

AI-powered drones are used for crop monitoring, pest control, soil analysis, and even planting seeds. Drones equipped with high-resolution cameras and sensors collect real-time data that AI processes to provide valuable insights into crop health, growth patterns, and other agricultural parameters (Inoue 2020).

9.4.10 Supply Chain and Logistics Optimization

AI optimizes the agricultural supply chain by predicting demand, managing inventory, and optimizing transportation routes. This improves the efficiency of food distribution, reduces waste, and ensures that fresh produce reaches the market in optimal condition (Rajabzadeh and Fatorachian 2023; Thakur and Belwal 2024).

9.4.11 Food Quality Assessment

AI applications in food quality assessment involve the use of computer vision and machine learning to grade and sort harvested produce based on size, color, and ripeness. These systems ensure that only high-quality products reach consumers, maintaining consistency in the food supply chain (Akkem et al. 2023).

9.4.12 Climate Resilience Planning

AI-powered climate prediction models analyze weather patterns and simulate future climate scenarios to help farmers adapt to changing conditions. These models allow farmers to anticipate risks such as droughts, floods, or temperature extremes and plan for climate resilience (Islam et al. 2022).

9.4.13 Biodiversity Preservation

AI tools help monitor and protect biodiversity by analyzing ecosystems, tracking wildlife, and assessing the health of natural habitats. AI applications in agriculture can identify areas that need conservation efforts and ensure sustainable farming practices that protect local biodiversity (Garske et al. 2021).

9.4.14 Agri-Fintech Solutions

AI-powered agri-fintech platforms use data to provide farmers with financial insights, such as credit scoring, insurance recommendations, and investment opportunities. These platforms help farmers access financial services that were previously unavailable, improving their financial stability and growth prospects (Rayhan et al. 2024).

9.4.15 *Blockchain Integration*

AI can be integrated with blockchain technology to enhance traceability and transparency in the agricultural supply chain. By recording every transaction and product movement on an immutable blockchain, farmers and consumers can ensure the authenticity, quality, and ethical sourcing of agricultural products (Chen et al. 2023).

9.5 Conclusion

The integration of artificial intelligence (AI) into agriculture is reshaping the landscape of modern farming by providing innovative solutions to many of the challenges that farmers face today. From crop monitoring and management to livestock health, AI is enabling farmers to optimize every aspect of their operations. AI-powered systems help improve crop health by identifying early signs of diseases and pests, reducing the need for excessive chemical use. Precision farming technologies, such as smart irrigation and autonomous machinery, enhance resource efficiency, decrease labor costs, and increase overall productivity. AI also plays a pivotal role in predicting crop yields, assessing soil health, and improving decision-making through advanced data analytics. By leveraging machine learning algorithms, farmers can make more accurate forecasts, optimize planting schedules, and allocate resources more effectively. Additionally, AI-driven solutions help manage supply chains by ensuring the timely delivery of high-quality produce, minimizing waste, and maximizing profitability.

Furthermore, AI's potential extends to environmental sustainability by helping farmers adapt to climate change through predictive weather models and risk management strategies. With its ability to enhance food quality assessment, preserve biodiversity, and facilitate agri-fintech solutions, AI is transforming agriculture into a more resilient, efficient, and sustainable industry. Ultimately, the adoption of AI in agriculture is crucial for addressing the growing demand for food amidst global challenges such as climate change, resource scarcity, and population growth. By embracing these advanced technologies, farmers can navigate these challenges effectively, ensuring a secure and sustainable food future.

References

Adamides G, Kalatzis N, Stylianou A, Marianos N, Chatzipapadopoulos F, Giannakopoulou M, Neocleous D (2020) Smart farming techniques for climate change adaptation in Cyprus. Atmos 11(6):557

Agrimonti C, Lauro M, Visioli G (2021) Smart agriculture for food quality: facing climate change in the 21st century. Crit Rev Food Sci Nutr 61(6):971–981

Akkem Y, Biswas SK, Varanasi A (2023) Smart farming using artificial intelligence: a review. Eng Appl Artif Intell 120:105899
Aliyu AA, Liu J (2023) Blockchain-based smart farm security framework for the internet of things. Sensors 23(18):7992
Almalki FA, Soufiene BO, Alsamhi SH, Sakli H (2021) A low-cost platform for environmental smart farming monitoring system based on IoT and UAVs. Sustainability 13(11):5908
Alves RG, Maia RF, Lima F (2023) Development of a Digital Twin for smart farming: irrigation management system for water saving. J Clean Prod 388:135920
Boursianis AD, Papadopoulou MS, Diamantoulakis P, Liopa-Tsakalidi A, Barouchas P, Salahas G, Goudos SK (2022) Internet of things (IoT) and agricultural unmanned aerial vehicles (UAVs) in smart farming: a comprehensive review. Internet of Things 18:100187
Charania I, Li X (2020) Smart farming: agriculture's shift from a labor intensive to technology native industry. Internet of Things 9:100142
Chen JIZ, Hengjinda P (2019) Applying AI technology to the operation of smart farm robot. Sens Mater 31:1777
Chen Q, Li L, Chong C, Wang X (2022) AI-enhanced soil management and smart farming. Soil Use Manag 38(1):7–13
Chen HY, Sharma K, Sharma C, Sharma S (2023) Integrating explainable artificial intelligence and blockchain to smart agriculture: research prospects for decision making and improved security. Smart Agricultural Technology 6:100350
Demestichas K, Peppes N, Alexakis T (2020) Survey on security threats in agricultural IoT and smart farming. Sensors 20(22):6458
Džermeikaitė K, Bačėninaitė D, Antanaitis R (2023) Innovations in cattle farming: application of innovative technologies and sensors in the diagnosis of diseases. Animals 13(5):780
Garske B, Bau A, Ekardt F (2021) Digitalization and AI in European agriculture: a strategy for achieving climate and biodiversity targets? Sustainability 13(9):4652
Hazrati M, Dara R, Kaur J (2022) On-farm data security: practical recommendations for securing farm data. Front Sustain Food Syst 6:884187
Inoue Y (2020) Satellite-and drone-based remote sensing of crops and soils for smart farming–a review. Soil Sci Plant Nutr 66(6):798–810
Islam Z, Sabiha NE, Salim R (2022) Integrated environment-smart agricultural practices: a strategy towards climate-resilient agriculture. Econ Anal Policy 76:59–72
Javaid M, Haleem A, Singh RP, Suman R (2022) Enhancing smart farming through the applications of Agriculture 4.0 technologies. Int J Intell Netw 3:150–164
Karunathilake EMBM, Le AT, Heo S, Chung YS, Mansoor S (2023) The path to smart farming: innovations and opportunities in precision agriculture. Agriculture 13(8):1593
Kaushik SK, Mishra VN, Punia M, Diwate P, Sivasankar T, Soni AK (2021) Crop health assessment using Sentinel-1 SAR time series data in a part of Central India. Remote Sens Earth Syst Sci 4(4):217–234
Kiani F, Randazzo G, Yelmen I, Seyyedabbasi A, Nematzadeh S, Anka FA, Muzirafuti A (2022) A smart and mechanized agricultural application: from cultivation to harvest. Appl Sci 12(12): 6021
Lioutas ED, Charatsari C (2020) Smart farming and short food supply chains: are they compatible? Land Use Policy 94:104541
Lytos A, Lagkas T, Sarigiannidis P, Zervakis M, Livanos G (2020) Towards smart farming: systems, frameworks and exploitation of multiple sources. Comput Netw 172:107147
Mim MI, Sultana F, Hasan MR (2025) AI-Powered autonomous farming: the future of sustainable agriculture. Eur J Theor Appl Sci 3(1):11–31
Moghayedi A, Richter I, Owoade FM, Kapanji-Kakoma KK, Kaliyadasa E, Francis S, Ekpo C (2022) Effects of urban smart farming on local economy and food production in urban areas in African cities. Sustainability 14(17):10836
Mohamed ES, Belal AA, Abd-Elmabod SK, El-Shirbeny MA, Gad A, Zahran MB (2021) Smart farming for improving agricultural management. Egypt J Remote Sens Space Sci 24(3):971–981

More A, Aslekar A (2022) Role of ICT & Fintech in Indian agriculture. In: 2022 International Conference on Decision Aid Sciences and Applications (DASA). IEEE, pp 900–904

Patel B, Bhatia J (2024) A comprehensive review of internet of things and cutting-edge technologies empowering smart farming. Curr Sci 126(2):137

Pujahari RM, Yadav SP, Khan R (2022) Intelligent farming system through weather forecast support and crop production. In: Application of machine learning in agriculture. Academic, pp 113–130

Rahaman M, Lin CY, Pappachan P, Gupta BB, Hsu CH (2024) Privacy-centric AI and IoT solutions for smart rural farm monitoring and control. Sensors 24(13):4157

Rajabzadeh M, Fatorachian H (2023) Modelling factors influencing IoT adoption: with a focus on agricultural logistics operations. Smart Cities 6(6):3266–3296

Rayhan MJ, Rahman SM, Mamun AA, Saif ANM, Islam KA, Alom MM, Hafiz N (2024) FinTech solutions for sustainable agricultural value chains: a perspective from smallholder farmers. Bus Strategy Dev 7(2):e358

Saadane R, Chehri A, Jeon S (2022) AI-based modeling and data-driven evaluation for smart farming-oriented big data architecture using IoT with energy harvesting capabilities. Sustain Energy Technol Assess 52:102093

Saiz-Rubio V, Rovira-Más F (2020) From smart farming towards agriculture 5.0: a review on crop data management. Agronomy 10(2):207

Shafi U, Mumtaz R, García-Nieto J, Hassan SA, Zaidi SAR, Iqbal N (2019) Precision agriculture techniques and practices: from considerations to applications. Sensors 19(17):3796

Shaikh TA, Rasool T, Lone FR (2022) Towards leveraging the role of machine learning and artificial intelligence in precision agriculture and smart farming. Comput Electron Agric 198: 107119

Singh A, Jadoun YS, Brar PS, Kour G (2022) Smart technologies in livestock farming. In: Smart and sustainable food technologies. Springer Nature Singapore, pp 25–57

Thakur M, Belwal T (2024) Advances in postharvest and analytical technology of horticulture crops. Springer, Singapore. https://doi.org/10.1007/978-981-97-7247-6_1

Chapter 10
Indigenous Wild Foods of the South Pacific: Nutrition, Ethnobotanical Studies and Ecosystem Function

Ankit Paliwal, Asaeli Naika, and Arti Pillay

Abstract The traditional ecological knowledge of wild foods is vital for improving South Pacific island nations' food security and sustainability. This knowledge holds the insight and practices related to the use of wild food plants and Indigenous food systems, which have been an integral part of these communities' resilience and nutritional well-being for centuries. TEK has assisted not only in securing food security but also in promoting biodiversity, sustainable land management and the preservation of traditional diets. The erosion of TEK due to globalisation and modernisation poses a significant threat to preserving traditional food systems. This loss could lead to decreased biodiversity and increased vulnerability to food insecurity in South Pacific island nations. We can maintain the region's sustainability and food and economic security by fostering a relationship between TEK and modern scientific knowledge.

Keywords Traditional ecological knowledge · Wild foods · Food security · Sustainability · South Pacific

10.1 Introduction

The South Pacific region is diverse in ecology, climate, population and economy. The region comprises 22 island countries, which vary significantly in land mass, island topography, development and culture (Ferguson et al. 2022). These island nations face particular economic and sustainable development challenges, principally attributed to their small size, remoteness, population rise, limited resources, high vulnerability to climatic and natural disasters and international market fluctuations (Gough et al. 2010; Shah et al. 2018), Based upon common challenges, global organisations mostly ignore the significant socioeconomic and ecological

A. Paliwal (✉) · A. Naika · A. Pillay
Department of Applied Sciences, CETVET, Fiji National University, Suva, Fiji

M. Thakur, T. Belwal (eds.), *Food Biodiversity and Consumer Engagements*, Sustainable Development and Biodiversity 46,
https://doi.org/10.1007/978-981-95-7777-4_10

differences that influence food security across these islands and often inaccurately frame them as a homogenous region (Allen 2015).

The diversity of the South Pacific region can be broadly divided into geographical, cultural and socioeconomic differences. The region's geographic diversity includes high islands with volcanic soils, like Fiji and Vanuatu, low-lying atolls with limited land and freshwater resources, like Kiribati and Marshal islands, and vast Exclusive Economic Zones (EEZs). Most of these island nations are highly vulnerable to natural hazards such as cyclones, earthquakes, tsunamis and rising sea levels, which can severely disrupt food production and distribution systems. With climate change, the occurrence and magnitude of these natural hazards have also increased, impacting their agricultural yields, marine wildlife and fish stocks and the availability of freshwater resources (Gough et al. 2010; Allen 2015; Lyons et al. 2020; Ferguson et al. 2022). On the cultural side, the South Pacific is home to a rich embroidery of Indigenous cultures, each with unique traditions, knowledge systems and relationships with nature. Traditional ecological knowledge (TEK) refers to the cumulative body of knowledge, beliefs and practices developed by Indigenous communities over generations through their environmental interactions. It is fundamental for interpreting the relationship among food, culture and the environment in the Pacific Islands. This knowledge has been accumulated over a long period and has been a valuable tool for food production, resource management and environmental stewardship (Gough et al. 2010; IFAD 2016; Shah et al. 2018). For these nations, traditional ecological knowledge has played a crucial role in shaping sustainable food practices and resource management for centuries. However, globalisation and modernisation have led to cultural shifts, influencing dietary habits and impacting traditional food systems. The traditional calendar, intricately linked to the availability of food sources and the cultivation cycle of staple crops like yam, has become less relevant with changing agricultural practices (Magiri et al. 2022). Traditional food storage and preservation methods, crucial for ensuring food security during emergencies, have been overshadowed by modern preservation techniques (Vogliano et al. 2021).

The increasing integration of Pacific Island economies into the global market has led to a noticeable shift from traditional diets rich in fibre and nutrients to modern diets dominated by imported processed foods. (Allen 2015; Shah et al. 2018) The easy availability and affordability of a wider variety of processed foods have overshadowed traditional foods' cultural significance and nutritional value. (Vogliano et al. 2021) The transition from conventional polyculture farming to commercially oriented monoculture farming systems has significantly impacted food security in Fiji. Colonial influences and the desire for cash-based economies led to the establishment of large-scale plantations for crops like sugarcane and coconut, displacing traditional crops like giant swamp taro. This shift was further fuelled by policies prioritising export-oriented cash crops over conventional food crops (Shah et al. 2018). Population growth and urbanisation, further added pressure on food systems and resources, increasing dependence on imported food. The situation becomes more complicated, particularly in remote areas and among vulnerable populations with inequitable access to food and resources (Allen 2015; Mabhaudhi et al. 2019).

The recent overreliance on imported food and resources has put an economic and social burden on these island nations. Food security is a significant concern, and a sustainable food system is essential to ensuring that all people have access to sufficient, safe and nutritious food (Ferguson et al. 2022). For Pacific Island societies, the traditional food systems are deeply embedded in their cultural fabric, where food is not merely sustenance but plays a vital role in social customs, ceremonies and traditions (Sievert et al. 2019). The transition away from traditional diets and agricultural practices towards imported and processed foods has led to a rise in non-communicable diseases (NCDs) like obesity, diabetes and micronutrient deficiencies (Lyons et al. 2020; Shah et al. 2018). This transition has also resulted in cultural deterioration, as traditional food practices and knowledge are gradually being lost, historically ensuring food security and nutritional well-being in these island nations (Shah et al. 2018). The shift towards modern monoculture farming practices also results in a reduction in agrobiodiversity and soil degradation.

The main objective of this chapter is to combine TEK practices related to wild food systems, covering diverse wild foods and the South Pacific, as well as their utilisation and cultural significance. The chapter will also discuss the key challenges and opportunities for preserving TEK.

10.2 Traditional Ecological Knowledge (TEK)

Traditional ecological knowledge is the core of understanding the intertwining relationship among food, culture and the environment of Pacific Island nations. It includes many practices and knowledge that have sustained Pacific Island communities for centuries, including agricultural practices, fishing practices, food preservation and sharing (Shah et al. 2018). With centuries of collective knowledge, Indigenous communities have developed a deep understanding of their local environments, enabling them to cultivate various crops adapted to specific conditions. This is visible from the traditional Fijian calendar, which manages resource use based on natural phenomena and the cultivation cycle of staple crops like yam. Additionally, polyculture farming systems, which involve growing multiple crops together, have acted as natural germplasm conservatories, safeguarding valuable landraces and promoting agrobiodiversity in the region (Shah et al. 2018). This traditional environmental knowledge also plays a significant role in managing marine wildlife resources, including sustainable fishing practices based on knowledge of fish behaviour and migration patterns. These traditional management systems have proven helpful in maintaining healthy fish stocks for generations (Ferguson et al. 2022). Since these island nations are particularly vulnerable to cyclones and other climatic extremes, food storage and preservation have always been a key part of the traditional system, ensuring food availability during times of scarcity or natural disasters.

TEK fundamentally differs from Modern Scientific Knowledge (MSK) in its foundations and goals. TEK, rooted in Indigenous South Pacific cultures, is holistic, place-based and intergenerational, emphasising sustainability, spiritual reciprocity and communal resource governance (e.g. Fijian vanua systems or Māori kaitiakitanga). In contrast, MSK is reductionist, universalising and driven by empirical analysis, prioritising innovation, scalability and economic efficiency. While TEK relies on oral traditions and adaptive observation (e.g. lunar fishing calendars), MSK depends on controlled experiments and quantitative modelling.

Despite these contrasts, their synergies offer transformative potential. TEK's nuanced understanding of local ecosystems complements MSK's technological tools—for instance, combining Indigenous rotational farming with soil science to combat land degradation or merging traditional marine management (qoliqoli in Fiji) with satellite data for conservation. Challenges like power imbalances and intellectual property rights persist, but equitable collaboration can bridge these systems. By valuing TEK as a parallel and equally valid knowledge framework, consolidative approaches can address global problems like biodiversity loss and climate change while honouring cultural sovereignty and fostering resilient, culturally grounded solutions.

10.3 Role of Wild Foods in the South Pacific

TEK of wild foods in the South Pacific region is vital to food security, nutrition and cultural heritage. This knowledge is based upon a deep understanding of local ecology and the environment, including plant and animal species and their sustainable uses (Naika 2020; Lugo-Morin 2022; Vogliano et al. 2021). Wild food plants (WFPs) have historically played a crucial role in food and nutrition security, particularly for rural and indigenous communities. Still, they are increasingly threatened by land use changes, climate impacts, overexploitation and urbanisation, leading to a decline in their availability and associated traditional knowledge (Borelli et al. 2020). Wild food sources, such as wild yams, Tahitian chestnuts and edible ferns, have played an important role in food security in the past. In addition, traditional wild foods such as fish, crabs, shellfish and opossums continue to be consumed in the Pacific Islands (Shah et al. 2018; Vogliano et al. 2021). Wild foods are often considered as nutrient-rich with essential vitamins and minerals. For instance, wild green leafy vegetables such as edible ferns or *ota* and mangrove ferns such as *boreti* that grow well in the Pacific are rich in protein, minerals, vitamins (such as A, B, C and K), antioxidants and fibre (Lyons et al. 2020). Wild foods have been a critical component of food security for many centuries, particularly during crop failure. For example, in the past, wild food sources were a major contributor to food security after cyclone or other natural calamity (Shah et al. 2018). Apart from their role in food and nutritional security, these wild food sources are highly influenced by the culture and history of the region, with specific dishes associated with events such as weddings, funerals and religious ceremonies (Marrero and Mattei 2022).

Based on their origin, these wild foods can be divided into two broad categories: plant and animal. Historically, a wide variety of wild plants are consumed in the Pacific region, including green leafy vegetables, fruits and nuts, root crops and other wild plants. Edible ferns, kangkong, amaranth, taro leaves, sweet potato and cassava are commonly green leafy. Tahitian chestnut, cut nut, tropical almond, Tahitian vi apple, Malay apple, oceanic lychee and soursop are among the traditional fruits and nuts found in the region. Yams, sweet potatoes, cassava and taro (dalo) are common root crops, often referred to as "hard food" or "ground provisions," that are used to be grown in home gardens (Guell et al. 2021). Wild yam, giant taro, sago palm and mangrove ferns are other wild plants that are consumed in the region (Lyons et al. 2020; Shah et al. 2018). Wild animals mainly include seafood like fish, crabs and shellfish, which are essential sources of protein and are part of the diet in modern times too. Apart from these opossums, wild birds and their eggs are also consumed in a few South Pacific nations (Nunn et al. 2024).

Wild plants (together with fruits) in diets have provided basic nutritional requirements for mankind throughout history (Umerah and Nnam 2019). Their low consumption is indicative of unbalanced and unhealthy diets significantly limited in vitamins, minerals and certain bioactive compounds, which is linked to a high risk of several non-communicable diseases (NCDs) that can lead to a high morbidity and/or mortality rate (Ezzati et al. 2002). It is for this reason that a daily consumption of at least 400g of vegetables and fruits has been recommended by WHO in order to prevent micronutrient deficiencies and chronic illnesses such as diabetes, heart disease and obesity, especially in underdeveloped countries (WHO 2003; Lock et al. 2005). For a healthy and balanced diet, green leaves are an important constituent of food, although their intake (together with fruits) is generally considered low globally (Imathiu 2021).

Wild green leafy plants are an integral constituent of traditional diets (Mibei et al. 2011) for many Indigenous People worldwide. Wild plant species play a dynamic part in food systems. Wild plants have been reported to be essential in enhancing quality diets in developing countries, particularly amongst the rural population. (Xikombiso and Mbhenyane 2017; Mungofa et al. 2020). Several studies have reported that Wild plants address gaps in nutrition by providing healthful, less expensive and nutrient-dense food substitutes (Ranum et al. 2014; Maseko et al. 2018) due to their high concentration of essential nutrients (Bua and Onang 2017; Meldrum et al. 2018) and phytochemical content (Gupta et al. 2017; Maseko et al. 2018; Otang-Mbeng and Mashabela 2020) with protective properties against diseases. Consequently, including such underutilised indigenous and traditional crops in food systems could improve the nutritional status of many impoverished individuals due to their high nutritional value (Mabhaudhi et al. 2019).

While these wild plants can be grown comparatively at lower management costs and on poor marginal soil, they have also remained underutilised due to a lack of awareness of their health-protective properties over their exotic alternatives (Nnamani et al. 2009). These species or crops remain neglected and underutilised because of their suppressed potential and limited attractiveness compared to commodity food crops. They are mainly cultivated by local farmers in their areas of origin using their traditional knowledge with limited or no agricultural extension services or research inputs. Information on the chemical composition of wild plants tends to be either not reported or limited (Imathiu 2021) or anecdotal and abundant in scientific literature (Okeno et al. 2003; Mibei et al. 2011). Today, despite their potential role in addressing food and nutrition insecurities, wild food plants remain neglected and underutilised (Nnamani et al. 2009; Njume et al. 2014). For example, Solomon Islands alone have over 120 edible and nutrient-rich indigenous plant species. Wild foods such as ferns, opossums, fish, crabs and shellfish are still consumed, although their consumption is declining (Vogliano et al. 2021) (Table 10.1).

10.4 TEK-Based Practices for Wild Food Utilisation

Traditional farming methods often employ polyculture systems such as mixed cropping and act as natural germplasm conservatories for plants of ethnobotanical significance and valuable landraces. Traditional agroforestry systems have been credited with enabling the people of the Pacific Islands to be self-sufficient and healthy (Shah et al. 2018). A neo-traditional approach incorporating scientific and local indigenous knowledge can improve farming systems and be more adaptive for local communities. This involves preserving locally adapted landraces, promoting local food crops and utilising organic farming practices. It also means utilising the contributions of nature in various, low-cost external input agroecological production and harvesting methods (Nunn et al. 2024; Shah et al. 2018). Traditional knowledge systems (TKS) can also enhance the resilience of food systems by utilising local resources and eliminating external inputs. This is particularly important in the event of climate change and natural disasters. TEK emphasises the importance of biodiversity, which is essential for food system resilience (Kuhnlein 2015). Understanding the cultural acceptance and values of food sources and production practices is vital. Local communities should be informed of their desires, preferences, challenges and constraints when designing food system solutions (Guell et al. 2021). Traditional food preservation techniques in the South Pacific are deeply rooted in the region's environment, cultural practices and resourcefulness.

The traditional Pacific method of food staple preservation by fermentation in pits was most common for starch-rich food. Poi Pits (Hawai'i and Polynesia) or Davuke (Fiji) involves underground fermentation, where a hole in the ground is used for preserving foodstuffs through a process of fermentation. The depth of the pit deepens with the amount of food to be stored; usually it is a foot deep. It is then covered with multiple layers of plantain leaves. Once it dried, peeled staple foods such as breadfruit, plantain, taro or cassava are added. The staple food heap is then again covered with layers of plantain leaves to prevent the soil from being in contact with the food. In the end, the pit is covered with rocks and soil. The process is a lactic acid

Table 10.1 Wild food sources of South Pacific

Category	Food item	Description	Regions/uses
Starchy staples	Breadfruit (*Artocarpus altilis*)	Roasted, fermented, or pounded into flour	Polynesia, Micronesia, Melanesia
	Wild yams (*Dioscorea spp.*)	Tubers like *ufi* (Samoa/Tonga) and *kao* (Vanuatu) were foraged from forests	Samoa, Tonga, Vanuatu, Papua New Guinea
	Taro (*Colocasia esculenta*)	Wild varieties are used in dishes like *palusami* (wrapped in leaves with coconut)	Fiji, Samoa, Hawai'i
	Sago palm (*Metroxylon spp.*)	Starch extracted from pith	Papua New Guinea, Solomon Islands
Protein sources	Wild fish (Tuna, Mahi-mahi, Reef Fish)	Caught with nets, spears, or hooks	Coastal regions across the South Pacific
	Sea urchins (*Tripneustes gratilla*)	Harvested from reefs; eaten raw or roasted	Fiji, Tonga, Samoa
	Wild pigs (*Sus scrofa*)	Hunted in forests; cooked in earth ovens (*umu/lovo*)	Polynesia, Melanesia
	Fruit bats (*Pteropus spp.*)	Roasted or stewed	Samoa, Fiji, Vanuatu
	Edible insects (grubs, coconut grabs)	Sago grubs in PNG; coconut crabs in atolls	Papua New Guinea, Kiribati, Tuvalu
Fruits & Nuts	Pandanus (*Pandanus tectorius*)	Fruits/seeds eaten raw or cooked	Micronesia, Polynesia
	Wild bananas (*Musa spp.*)	Small, starchy varieties foraged in forests	Melanesia, Polynesia
	Tropical almond (*Terminalia catappa*)	Nuts eaten raw or roasted	Coastal regions, lowland forests
	Noni (*Morinda citrifolia*)	Fermented for medicinal use or famine food	Widely distributed
Leafy Greens & Vegetables	Ferns (*Diplazium spp.*)	Young fronds (*ota* in Fiji) are eaten raw or cooked	Fiji, Samoa, Vanuatu
	Bele (*Abelmoschus manihot*)	Wild hibiscus leaves are used in soups	Melanesia, Polynesia
	Seaweed (*Limu*)	Harvested from reefs (e.g. *limu kohu* in Hawai'i)	Hawai'i, Cook Islands
Coastal & Marine Foods	Mangrove crabs (*Scylla serrata*)	Delicacy caught in mangrove swamps	Fiji, Papua New Guinea
	Clams and shellfish	Giant clams and cockles gathered from tidal flats	Atolls, coastal Melanesia
	Sea cucumbers (*Holothuria spp.*)	Dried and traded as *bêche-de-mer*	Melanesia, Micronesia

(continued)

Table 10.1 (continued)

Category	Food item	Description	Regions/uses
Forest Foods	Wild Taro (*Cyrtosperma merkusii*)	Grows in swampy forests; famine food	Kiribati, Tuvalu, Marshall Islands
	Bush potatoes (*Pueraria spp.*)	Tuberous roots dug from forest floors	Papua New Guinea, Solomon Islands
	Candlenut (*Aleurites moluccanus*)	Nuts pressed for oil or used in relishes	Hawai'i, Samoa
	Tree fern pith	Scraped from trunks as emergency starch	New Zealand (Māori), Fiji
	Mangrove fruit	Boiled or fermented during scarcity	Coastal atolls, mangrove regions

fermentation and the nutritional quality of the preserved food is similar to that of the fresh one. This method preserves food for months at a time with the occasional change of the plantain leaves(Aalbersberg et al. 1988).

Earth oven, primarily a cooking method, also known as the hangi (Māori), lovo (Fiji), or umu (Samoa), involves slow-cooking food in an underground oven lined with hot stones. The slow steaming process partially preserves meats and vegetables for short-term storage. Vakalavalava involves grated cassava mixed together with the yabia, grated coconuts, sugar wrapped in leaves and then boiled or put in an earth oven.

Bila or madrai vaViti is a fermented cassava dessert. Cassava is peeled, washed properly, cut into even pieces and submerged in water for four to five days. Fermented cassava pieces are then dried in the sun. Afterwards, it is then pounded into smaller pieces, removing the fibers often present on it and then neatly wrapped in plantain leaves before adding to boiling water.

There are various ways through which seafood can be preserved. Kokoda (Fiji) involves marination of raw fish in citrus (like lime or lemon juice) and coconut milk, then fermented briefly. While Miti Hue (Tonga) is a traditional sauce in Polynesian cuisine made from fermented coconut cream, seafood or meat. Vesa (Fiji) is another short-duration food preservation method for fish (Table 10.2). Fish are cleaned by removing the scales and guts, staple food such as breadfruit are not peeled, after this, the food item is then added to a hot plate or directly above the fire.

10.5 TEK and Sustainable Food System

Integrating traditional ecological knowledge into modern agricultural practices presents a promising path for developing sustainable food systems. TEK, based on indigenous and local knowledge systems, offers valuable insights into sustainable resource management, biodiversity conservation and climate resilience. By

Table 10.2 Traditional food preservation techniques in the South Pacific

Method	Examples	Regions
Fermentation	Kokoda (raw fish in citrus/coconut milk) Ma (fermented breadfruit/taro paste)	Fiji, Micronesia, Tonga
Sun-drying/smoking	Maso (sun-dried fish) Smoked meat/fish with aromatic woods	Samoa, Papua New Guinea
Pit storage	Poi Pits (fermented taro/breadfruit) Underground yam/taro storage	Hawai'i, Polynesia, Vanuatu, Fiji
Coconut milk/oil	Palusami (taro leaves + coconut milk) Preserved fish in coconut cream	Samoa, Tonga, Fiji
Salt curing	Salted fish jerky Seawater brine for meats	Coastal regions across the South Pacific
Leaf wrapping	Laulau (meat/fish in taro leaves) Banana-leaf-wrapped starches	Hawai'i, Samoa, Fiji
Natural preservatives	Citrus-marinated fish (like ceviche) Seawater brining	Widely practiced
Hangi/Umu/Lovo	Slow-cooked meats/vegetables in earth ovens	Māori (NZ), Fiji, Samoa, Polynesia
Voyage preservation	Dried breadfruit (mā) Fermented poi (taro paste)	Micronesia, Hawai'i

harmonising traditional practices with contemporary agricultural technologies, it is possible to improve environmental sustainability and community resilience. This approach not only preserves biodiversity but also enhances community identity and empowers local populations (Adefila et al. 2024). To build more sustainable food systems, several approaches can be taken, including bolstering local practices, integrating traditional knowledge and addressing systemic issues while also recognising the importance of food safety and the need to reduce food loss and waste (Ferguson et al. 2022; Vilariño et al. 2017).

Food security is achieved when all individuals, at all times, possess physical, social and economic access to adequate, safe and nutritious food that meets their dietary needs and preferences for an active and healthy lifestyle (Ferguson et al. 2022; Rabbi et al. 2021). A food system encompasses production, processing, distribution and consumption and includes social, economic, political, institutional and environmental activities (Ferguson et al. 2022). Resilience is the ability of a system to overcome disturbance and reorganise while maintaining its function, structure and identity. Shocks can provide opportunities to enhance and transform food systems (Ferguson et al. 2022). Re-localising food systems requires understanding local dynamics, their actors and their connections in networks while also addressing power imbalances and systemic issues (Guell et al. 2021). Implementing a circular economy in the food system can contribute to environmental, economic and social sustainability. This includes reducing food waste, reusing byproducts and focusing on the food chain for potential solutions and current issues (Elena Segneanu 2018).

Supporting local food production for local consumption is essential for food safety and sustainability. It can help in reducing the reliance on imported foods, which can be a source of food insecurity (Kuhnlein 2015). Traditional agroforestry systems and polyculture methods should also be promoted for their ecological and nutritional benefits (Marrero and Mattei 2022). Organic farming practices can improve soil health, reduce dependence on external sources and contribute to biodiversity. Establishing community seed banks can preserve valuable landraces and provide access to various locally adapted crops (Nunn et al. 2024). Developing links between farms and the tourism sector can create economic opportunities for local food producers (Shah et al. 2018). Reducing food loss and waste is essential for enhancing food safety and sustainability. Strategies include improving storage and transportation, smart and innovative packaging and improved communication across the food supply chain. The use of scientific and technological systems in food production and processing can increase shelf life and enable the distribution to expand further. However, these should be implemented in ways that do not undermine or threaten local systems and must be evaluated for their environmental and social impacts.

By recognising the interactions among plants, animals, humans and the environment, the One Health approach can address food safety, security and sustainability issues by recognising the interactions among plants, animals and the environment (Garcia et al. 2020). It is also essential to address trade-offs among food security, safety and sustainability to achieve a balance based on evidence and implement solutions that do not compromise any of these three factors. Promoting nutrition education, mainly when it is based on local and traditional knowledge, enhances the capacity to improve food choices and nutrient density of food resources (Lyons et al. 2020). Policies that promote local farmers promote traditional knowledge, and create an accessible and inclusive environment for neo-traditional approaches are necessary. Land use models should be based on a combination of traditional and modern farming practices (Shah et al. 2018).

10.6 TEK for Climate-Impacted Food Security

Traditional ecological knowledge (TEK) is a vital resource for ensuring food security in the face of climate change, particularly for Indigenous communities who have developed these practices through generations of interaction with their environments (Lugo-Morin 2020; Weber 2012). TEK provides specific strategies for adapting to the impacts of climate change. This includes knowledge of weather patterns, optimal planting and harvesting times and strategies for dealing with extreme events such as cyclones and droughts (Lako 2021; Nunn et al. 2024). Indigenous communities often use conventional lunar calendars to guide their agricultural and marine activities, incorporating them with natural seasonal changes and promoting crop rotation. For example, the iTaukei traditional calendar in Fiji divides the year into seasons and sub-seasons linked with specific climatic conditions and crop cycles, helping them in crop harvest and sustainable fishing. This calendar also enables them to ensure enough food supply, water, shelter, medicines and other necessary resources during climate disasters. (Lako 2021; Shah et al. 2018). TEK contains knowledge of climate-resilient crops, like giant swamp taro, dalo ni tana, tivoli and yams, which can withstand waterlogging, saline conditions and drought. These wild crops can be vital for food security during extreme weather events (Shah et al. 2018). TEK promotes sustainable agricultural practices that are less dependent on external inputs and more aligned with climate, such as mixed cropping and agroforestry techniques, creating diverse and resilient ecosystems that act as natural germplasm conservatories, conserving valuable landraces. The traditional agroforestry systems also kept the people of the Pacific Islands more self-sufficient and well-nourished in the world (Lako 2021; Shah et al. 2018). TEK evolved as place-based knowledge; it is well suited to managing any change in that region. Combining TEK with modern scientific approaches can lead to more effective and sustainable climate adaptation strategies.

10.7 Conclusion

The traditional ecological knowledge of wild foods in the South Pacific is a testimony to Indigenous communities' ingenuity, resilience and deep environmental wisdom. These knowledge systems have sustained human populations for millennia, maintaining biodiversity, adapting to environmental pressure and integrating cultural identity within the natural world. Practices such as pit fermentation, lunar-guided fishing and harvesting and community-based marine stewardship reveal a sophisticated understanding of ecological interdependencies, where sustainability is not merely a concept but a living goal. These systems emphasise that food security is inseparable from cultural continuity, spiritual reverence and communal governance—principles increasingly relevant in a globalised world coping with climate crises and biodiversity loss.

However, with the first colonisation, followed by globalisation, the survival of TEK faces unprecedented challenges, from the erosion of intergenerational knowledge transmission to the invasion of Western dietary norms. Preserving this knowledge demands more than documentation; it requires centring Indigenous sovereignty in conservation policies, revitalising land and sea rights and developing equitable partnerships between TEK holders and scientific institutions. By integrating Indigenous practices with modern ecological science, we can co-create adaptive strategies that honour both cultural heritage and environmental sustainability. TEK of wild foods of the South Pacific offers not just solutions for food security but a profound reminder that humanity's future depends on rekindling our relationship with the Earth as stewards, not conquerors. In safeguarding these traditions, we protect a legacy of resilience that holds timeless lessons for nurturing prosperous, biodiverse and culturally vibrant societies.

References

Aalbersberg WGL, Lovelace CEA, Madhoji K, Parkinson SV (1988) Davuke, the traditional Fijian method of pit preservation of staple carbohydrate foods. Ecol Food Nutr 21:173–180. https://doi.org/10.1080/03670244.1988.9991030

Adefila AO, Ajayi OO, Toromade AS, Sam-Bulya NJ (2024) Integrating traditional knowledge with modern agricultural practices: a sociocultural framework for sustainable development. World J Biol Pharm Health Sci 20:125–135. https://doi.org/10.30574/WJBPHS.2024.20.2.0850

Allen MG (2015) Framing food security in the Pacific Islands: empirical evidence from an island in the Western Pacific. Regional Environmental Change, 15(7), 1341–1353. https://doi.org/10.1007/s10113-014-0734-5

Borelli T, Hunter D, Powell B et al (2020) Born to eat wild: an integrated conservation approach to secure wild Food plants for Food security and nutrition. Plants 9:1299. https://doi.org/10.3390/plants9101299

Bua B, Onang C (2017) Validating the role of African indigenous vegetables for food and nutrition security in Uganda. J Food Sci Eng 7(6). https://doi.org/10.17265/2159-5828/2017.06.005

Ezzati M, Lopez AD, Rodgers A et al (2002) Selected major risk factors and global and regional burden of disease. Lancet 360:1347–1360. https://doi.org/10.1016/S0140-6736(02)11403-6

Ferguson CE, Tuxson T, Mangubhai S et al (2022) Local practices and production confer resilience to rural Pacific food systems during the COVID-19 pandemic. Mar Policy 137:104954. https://doi.org/10.1016/j.marpol.2022.104954

Garcia SN, Osburn BI, Jay-Russell MT (2020) One health for food safety, food security, and sustainable food production. Front Sustain Food Syst 4:1. https://doi.org/10.3389/fsufs.2020.00001

Gough KV, Bayliss-Smith T, Connell J, Mertz O (2010) Small Island sustainability in the Pacific: introduction to the special issue. Singap J Trop Geogr 31:1–9. https://doi.org/10.1111/j.1467-9493.2010.00382.x

Guell C, Brown CR, Iese V et al (2021) "We used to get food from the garden." Understanding changing practices of local food production and consumption in small island states. Soc Sci Med 284:114214. https://doi.org/10.1016/j.socscimed.2021.114214

Gupta S, Srivastava A, Lal EP (2017) Indigenous leafy vegetables for food and nutritional security in two districts of Jharkhand, India. Int J Curr Microbiol Appl Sci 6:901–909

IFAD (2016) The Traditional Knowledge Advantage: Indigenous peoples' knowledge in climate change adaptation and mitigation strategies

Imathiu S (2021) Neglected and underutilized cultivated crops with respect to indigenous African leafy vegetables for Food and nutrition security. J Food Secur 9:115–125. https://doi.org/10.12691/jfs-9-3-4

Kuhnlein HV (2015) Food system sustainability for health and well-being of Indigenous Peoples. Public Health Nutr 18(13):2415–2424. https://doi.org/10.1017/S1368980014002961

Lako J (2021) Eco-cultural practices related to sustainable food systems in coastal communities in Fiji. In: Park W (ed) Maritime living heritage: coastal communities in the Asia-Pacific region and their traditional Food system. ICHCAP, pp 20–29

Lock K, Pomerleau J, Causer L et al (2005) The global burden of disease attributable to low consumption of fruit and vegetables: implications for the global strategy on diet. Bull World Health Organ 83:100–108. https://doi.org//S0042-96862005000200010

Lugo-Morin DR (2020) Indigenous communities and their food systems: a contribution to the current debate. J Ethn Foods 7:1–10. https://doi.org/10.1186/s42779-019-0043-1

Lugo-Morin DR (2022) Building an ethnic food ethic: the case of the Ngigua indigenous People of Southern Puebla, Mexico. Food Ethics 7:3. https://doi.org/10.1007/s41055-021-00095-8

Lyons G, Dean G, Tongaiaba R, Halavatau S, Nakabuta K, Lonalona M, Susumu G (2020) Macro- and micronutrients from traditional food plants could improve nutrition and reduce non-communicable diseases of islanders on atolls in the South Pacific. Plants 9(8):942. https://doi.org/10.3390/plants9080942

Mabhaudhi T, Chibarabada TP, Chimonyo VGP et al (2019) Mainstreaming underutilized indigenous and traditional crops into Food systems: a South African perspective. Sustainability 11: 172. https://doi.org/10.3390/su11010172

Magiri R, Gaundan S, Singh S, Pal S, Bakare A, Choongo K, Zindove T, Okello W, Mutwiri G, Iji PA (2022) The Role of Agricultural Institutions in Providing Support Towards Sustainable Rural Development in South Pacific Island Countries. J Agric Sci 14(2):104. https://doi.org/10.5539/jas.v14n2p104

Marrero A, Mattei J (2022) Reclaiming traditional, plant-based, climate-resilient food systems in small islands. Lancet Planet Health 6:e171–e179. https://doi.org/10.1016/S2542-5196(21)00322-3

Maseko I, Mabhaudhi T, Tesfay S et al (2018) African leafy vegetables: a review of status, production and utilization in South Africa. Sustainability 10:16. https://doi.org/10.3390/su10010016

Meldrum G, Padulosi S, Lochetti G et al (2018) Issues and prospects for the sustainable use and conservation of cultivated vegetable diversity for more nutrition-sensitive agriculture. Agriculture-Basel 8:112. https://doi.org/10.3390/agriculture8070112

Mibei EK, Ojijo NKO, Karanja SM, Kinyua JK (2011) Compositional attributes of the leaves of some indigenous African leafy vegetables commonly consumed in Kenya. Ann Food Sci Technol 2:146–154

Mungofa N, Sibanyoni JJ, Mashau ME, Besw D (2020) Prospective role of indigenous leafy vegetables as functional food ingredients. Molecules 25:377–391. https://doi.org/10.1016/b978-0-12-811808-5.00020-9

Naika A (2020) Preserve your own food. Promoting healthy eating through home food processing and preservation. FAO. https://www.fao.org/documents/card/en/c/CB2113EN

Njume C, Goduka NI, George G (2014) Indigenous leafy vegetables (imifino, morogo, muhuro) in South Africa: a rich and unexplored source of nutrients and antioxidants. Afr J Biotechnol 13: 1933–1942. https://doi.org/10.5897/ajb2013.13320

Nnamani CV, Oselebe HO, Agbatutu A (2009) Assessment of nutritional values of three underutilized indigenous leafy vegetables of Ebonyi State, Nigeria. Afr J Biotechnol 8:2321–2324

Nunn PD, Kumar R, Barrowman HM et al (2024) Traditional knowledge for climate resilience in the Pacific Islands. WIREs Clim Change 15:e882. https://doi.org/10.1002/wcc.882

Okeno JA, Chebet DK, Mathenge PW (2003) Status of indigenous vegetable utilization in Kenya. Acta Hortic. https://doi.org/10.17660/ActaHortic.2003.621.10

Otang-Mbeng W, Mashabela MN (2020) A review of beneficial phytochemicals and postharvest studies on some indigenous leafy vegetables from the Mpumalanga province of South Africa. Med Plants 12:533–544. https://doi.org/10.5958/0975-6892.2020.00066.0

Rabbi MF, Hasan M, Kovács S (2021) Food security and transition towards sustainability. Sustainability 13:12433. https://doi.org/10.3390/su132212433

Ranum P, Peña-Rosas JP, Garcia-Casal MN (2014) Global maize production, utilization, and consumption. Ann N Y Acad Sci 1312:105–112. https://doi.org/10.1111/nyas.12396

Segneanu AE (2018) Food security into a circular economy. Food Sci Nutr 4:1–4. https://doi.org/10.24966/FSN-1076/100038

Shah S, Moroca A, Bhat JA (2018) Neo-traditional approaches for ensuring food security in Fiji Islands. Environ Dev 28:83–100. https://doi.org/10.1016/j.envdev.2018.11.001

Sievert K, Lawrence M, Naika A, Baker P (2019) Processed Foods and Nutrition Transition in the Pacific: Regional Trends, Patterns and Food System Drivers. Nutrients

Umerah NN, Nnam NM (2019) Nutritional composition of neglected underutilized green leafy vegetables and fruits in South East eo-political zone of Nigeria. Asian Food Sci J 11:1–17. https://doi.org/10.9734/afsj/2019/v11i230058

Vilariño MV, Franco C, Quarrington C (2017) Food loss and waste reduction as an integral part of a circular economy. Front Environ Sci 5. https://doi.org/10.3389/fenvs.2017.00021

Vogliano C, Raneri JE, Coad J et al (2021) Dietary agrobiodiversity for improved nutrition and health outcomes within a transitioning indigenous Solomon Island food system. Food Secur 13: 819–847. https://doi.org/10.1007/s12571-021-01167-7

Weber E (2012) Climate and environmental change and food security—some conceptual considerations. J Pac Stud 32:91–102

WHO (2003) Diet, nutrition, and the prevention of chronic diseases. Report of a joint WHO/FAO expert consultation, WHO technical report series 916. WHO

Xikombiso G, Mbhenyane (2017) Indigenous foods and their contribution to nutrient requirements. S Afr J Clin Nutr 30(4):5–7

Chapter 11
From Field to Future: Green Revolution 2.0—Biochar Innovations for Sustainable Agri-Health

Vikash Kumar, Nikhil Raghuvanshi, Nikhil Chandel, Alok Rai, Kshitij Parmar, Eid Sadun Alotaibi, and Ajay Tomar

Abstract The increasing accumulation of agricultural waste worldwide has exacerbated greenhouse gas emissions, prompting the exploration of sustainable solutions such as biochar. Biochar, a carbon-rich material produced through pyrolysis of organic matter under limited oxygen, offers versatile applications in agriculture, energy production, and environmental management. Its unique properties, including a porous structure, abundant functional groups, and stable carbon components, make it a promising tool for enhancing soil fertility, mitigating greenhouse gases, and improving crop yields. Recent research highlights biochar's potential in moisture retention, nutrient absorption, and carbon sequestration, while also reducing the need for chemical fertilizers and alleviating plant stress. However, challenges such as economic viability, technological limitations, and long-term soil impacts hinder its widespread adoption. This chapter delves into the properties of biochar, its multifaceted roles in agriculture—ranging from improving soil physical and biological properties to addressing heavy metal pollution and its contribution to sustainable farming practices. It also examines the limitations and future prospects of biochar, emphasizing the need for expanded research and innovative applications to fully harness its potential in the Green Revolution 2.0. By integrating scientific insights and practical considerations, this chapter provides a comprehensive overview of

V. Kumar · N. Raghuvanshi
Faculty of Agricultural Sciences, GLA University, Mathura, Uttar Pradesh, India

N. Chandel · K. Parmar · A. Tomar (✉)
Amity Institute of Organic Agriculture, Amity University Uttar Pradesh, Noida, Uttar Pradesh, India

A. Rai
Department of Agriculture, Deen Dayal Upadhyaya Gorakhpur University, Gorakhpur, Uttar Pradesh, India

E. S. Alotaibi
Department of Mathematics and Statistics, AlKhurmah University College, Taif University, Taif, Saudi Arabia

M. Thakur, T. Belwal (eds.), *Food Biodiversity and Consumer Engagements*,
Sustainable Development and Biodiversity 46,
https://doi.org/10.1007/978-981-95-7777-4_11

biochar's transformative role in modern agriculture and its implications for sustainable development.

Keywords Sustainable agriculture · Soil health · Carbon sequestration · Plant growth

11.1 Introduction

Increasing greenhouse gas emissions is the primary environmental threat posed by the worldwide accumulation of waste from agriculture. Researchers have thus sought out numerous ways in which these materials can be repurposed, including as soil mulch, organic manure, and biochar (El–Metwally et al. 2022). Biochar is a renewable energy resource that stands out due to its versatility; it can be used to produce heat, electricity, and liquid biofuels (Wang et al. 2021). This is because biochar is produced through thermal processes of transformation with restricted oxygen availability at temperatures below 700 °C (Ulusal et al. 2021). Biochar, which is created through the pyrolysis of organic matter, is characterized by a distinct porous structure, an abundance of functional groups, a wide range of mineral nutrients, and stable carbon components (Wang et al. 2021).

Charcoal, also known as pyrolytic biomass, has been used by ancient communities for agricultural purposes for thousands of years. Biochar application has been hailed as a promising strategy for increasing yields while decreasing environmental impact and the need for chemical fertilizers in recent years (Wainaina et al. 2020). Soil fertility, greenhouse gas mitigation, moisture retention, carbon sequestration, and nutrient absorption are all areas that recent research has shown biochar to have the potential to improve (Liu et al. 2021). Adding biochar to the soil has been shown empirically to increase crop yields and reduce plant stress caused by biotic or abiotic factors. There are many obstacles to incorporating biochar into sustainable agriculture, despite the growing awareness of its benefits (Awasthi et al. 2020).

Certain properties of biochar, such as its ability to hold water, its surface area, and its carbon-to-nitrogen ratio, have been shown to have a major impact on the health of soil microorganisms (Hale et al. 2015). As stated in the aforementioned reference (Parmar et al. 2014), biochar can be made from a wide variety of processed and unprocessed biomass materials. The economic value of agricultural and agro-industrial residues in biochar synthesis is especially high (Hardy et al. 2016). To better utilize biochar for agricultural improvement, studies are currently underway to better understand its long-term effects on soil properties (Kamali et al. 2022). Biochar is a technology with pros and cons that must be weighed against issues of economics, technology, and sustainable development (Homagain et al. 2016). New studies emphasize the need for expanding biochar's spectrum of applications in agriculture (Mylavarapu et al. 2013). This chapter comprehensively explores biochar's properties, agricultural roles, limitations, and prospects for the future.

11.1.1 Biochar Properties

The main raw material conditions and the pyrolysis process, which vary in physical and chemical properties, have the greatest impact on biochar (e-Gedien 2020). Biochar produced from wood with a high lignin content is more carbon dense than biochar produced from herbaceous raw materials (Pituello et al. 2014), but it is deficient in nitrogen (N). Carbonization breaks down biomass into its component parts while preserving a high concentration of carbon. Biochar's properties can be modified to make the final product more carbonaceous and amenable to technical use (Weber and Quicker 2018). Charcoal is the byproduct of the incomplete combustion of plant material, and its chemical composition depends on the type of fuel used and the burn conditions, which include the availability of oxygen, the temperature, and the length of time that the material is exposed to fire (Fig. 11.1). Complete combustion in an oxygenated environment results in the loss of all carbon and leaves behind only white ash. Carbon (C) remains as charcoal when plant tissue is burned in low-oxygen conditions, while oxygen, nitrogen, and hydrogen escape as gas (Briggs et al. 2012). Due to its heterogeneous nature, biochar's composition includes both volatile and stable elements. Characterizing biochar involves looking at its surface area, C content, and mineral matter. Because of its high carbon content and stability (which includes resistance to both microbial and abiotic decay in the soil), biochar can be used for long-term carbon storage (Singh et al. 2020).

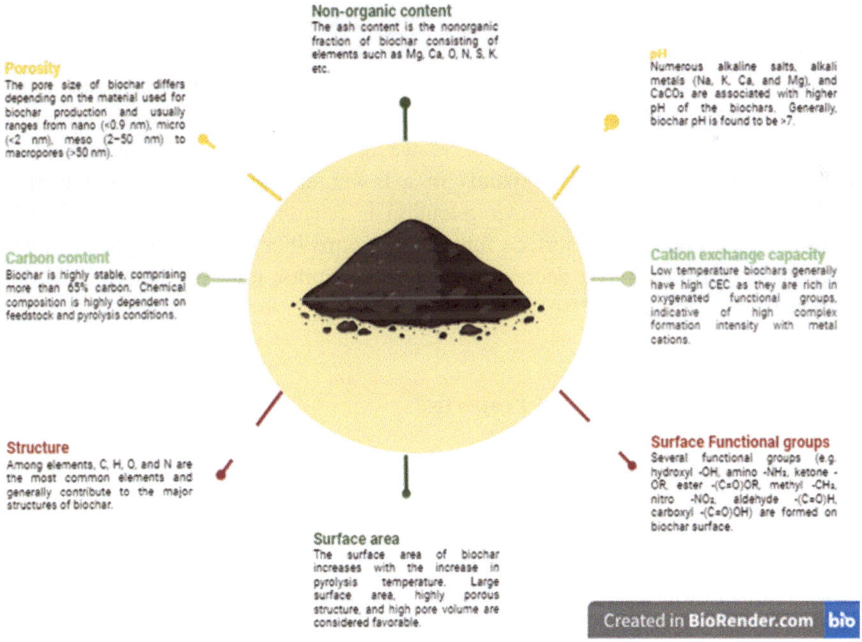

Fig. 11.1 Physio-chemical and functionality properties of biochar (Yadav et al. 2023)

11.1.1.1 Physical Properties of Biochar

Density, porosity, surface area, hydrophobicity, pore size, pore volume, and water-holding capacity are all examples of physical properties. Biochar density has been variously referred to as particle density, bulk density (BD), true density, skeletal density, and envelope density (Weber and Quicker 2018). The BD of biochar is typically less than 0.6 g cm^{-3} (Blanco-Canqui 2017). Biochar's porosity improves as pyrolysis temperatures rise. (Somerville and Jahanshahi 2015) found that the porosity of biochar made from wood materials at atmospheric pressure varied from about 50% at 300 °C to 70% at 850 °C. Hemicellulose is abundant in grass materials and can be broken down at room temperature. At temperatures between 350 °C and 700 °C, the porosity of grass biochar increases (Brewer et al. 2014). The amount of gas given off during carbonization is proportional to the surface area exposed. Biochar surface area grows in proportion to the temperature of pyrolysis. Furthermore, surface area grows even more as residence time grows. Biochar made from sewage sludge has a surface area of less than 100 m^2 g^{-1}. While the surface area of biochar derived from some biomasses can exceed 800 m^2 g^{-1} (Weber and Quicker 2018), the majority of biochar falls into the 100 m^2 g^{-1} range. Biochar can have macropores (pore diameter > 1000 nm), mesopores (pore diameter—0.05 nm to 0.002 nm), or micropores (pore diameter—0.05 nm to 0.0001 nm) (Brewer et al. 2014). Up to 80% of the total pore volume may be made up of micropores. There is a positive correlation between pore volume and surface area, and an increase in total pore volume as the pyrolysis temperature is raised. Pore volume has been measured in a variety of ways, and the results may vary greatly depending on the method used. The surface functional groups are responsible for hydrophobicity, while the bulk porosity of biochar's determines its water-holding capacity. An increase in the pyrolysis temperature has an effect not only on the surface functional group of biochar but also on its water retention. However, biochar's ability to absorb water will improve as its porosity expands in a hotter environment. The connection between the two in terms of water retention is crucial, but it has received little attention. The reduction of surface functional groups that occurs at higher temperatures has been linked to a decrease in water adsorption onto the inner surface of biochar (Pimchuai et al. 2010).

11.1.1.2 Biochar's Chemical Properties

General chemical features include the hydrogen ion concentration, oxygen to carbon and hydrogen to carbon ratios, and the elemental make-up of a substance. The average pH of biochar is 8.9, but it can range from 5.7 to 12.3 (Ahmad et al. 2014). Biochar pH and cation exchange capacity both increase with higher pyrolysis temperatures (Table 11.1), while O:C and H:C ratios decrease (Wang et al. 2015). Hydrothermal carbonization yields acidic hydrochars rather than the neutral chars produced by dry pyrolysis (Fang et al. 2015). Biochar's fundamental components

Table 11.1 Application of biochar for crop growth and yield

Types of biochar	Crop	Amount	Result	References
Rice straw biochar	Lettuce	18 t ha^{-1}	Increase 15% in yield	Zheng et al. (2017)
Eucalyptus wood, bamboo, and rice husk	Rice	20 t ha^{-1}	Si-rich biochars in enhancing plant growth	Shetty and Prakash (2020)
Sewage sludge, cow dung, and poultry manure	Water spinach	40 t ha^{-1}	Biochar did not improve the growth of the plant	Khan et al. (2018)
Sugarcane bagasse biochar	Maize and groundnut cropping system	10 t ha^{-1} biochar with 100% RDF	Increase yield of both crops	Rahman et al. (2021)
Pine chips, poultry litter, swine solids, switchgrass	Carrot, lettuce, soybean, and sweet corn	20 t ha^{-1}	Swine solids and mixture of pine chips and poultry litter were the most effective in increasing shoot and root biomass	Olszyk et al. (2020)
Prosopis juliflora biochar with RDF	Maize	10 t ha^{-1} with 50 and 100% RDF	Initially growth increase but decline after 3 years	Kamau et al. (2019)
Agricultural waste biochar	Cowpea	Apply 5 t ha^{-1} in spot and ring methods	Enhanced growth, nodule number and grain yield	Yeboah et al. (2020)
Lantana camara	Field pea	12 t ha^{-1} and 18 ha^{-1}	Maximum germination, recorded higher growth attributes, seeds/pod, and grain yield	Berihun et al. (2017)
Composting biochar	Oat	0–50 t ha^{-1}	Plant biomass, yield, total organic carbon increased with amount of biochar	Fedeli et al. (2023)
Peanut shell biochar	Rice	33.75 t ha^{-1}, 67.5 t ha^{-1}, and 101.25 t ha^{-1}	Increasing rice yield with increase in amount of biochar in saline-sodic soil	Ran et al. (2020)

are the elements. Different types of biochars are produced due to the variety of physiochemical reactions that take place during pyrolysis (Xiao et al. 2014). Biochar contains varying amounts of trace elements such as Cu, Fe, Zn, and Mn in addition to the four major elements of carbon, hydrogen, oxygen, and nitrogen (Wu et al. 2016). The carbon (C) in biochar can be found in a wide range of forms, including functionalized carbon, aliphatic carbon, aromatic carbon, and organic compounds (Sun et al. 2014). Hydrogen is essential to biochar's ability to sorb and attract ionizable molecules, and it also serves as a building block. Both organic and inorganic forms of O exist. Proteins, amino acids, pyrrolic acid, and pyridine are all forms of nitrogen that are particularly abundant in plants. This is especially true of leaves and herbaceous species. Some amide-N within biochar was reduced as pyrolysis temperature increased, and peptide-N bonds were converted to N-heteroaromatic carbon compounds. By altering the electronic band structures, N doping has the potential to enhance the sorption and catalytic capacities of N-doped biochar (Omondi et al. 2016). Plants have a wide range of P content, from 0.1% to 1.0%. Under 700 °C, P will not be volatilized during pyrolysis. Biochar is enriched in phosphorus (P) because pyrolysis converts organic P to inorganic P (Cao et al. 2011). Plants may be able to get some of the phosphorus they need from the biochar. Also, P in biochar has the potential to precipitate and bind certain heavy metals (Inyang et al. 2012). Total sulphur and phosphorus increased as well during the process of pyrolysis and gasification of oak and corn biomass. Gasification yields 73–100% organosulfur, while pyrolysis yields sulphate, organosulfur, and sulphide (Cheah et al. 2014). Biochar that has been enriched with sulphur could be used as a fertilizer or soil amendment (Zhang et al. 2012). Silicon is essential for the development of many different types of crops (Epstein 1994), including rice, barley, maize, and wheat. The Si content of pyrolyzed biochar derived from these plants is typically quite high. Biochar made from rice straw, for instance, saw an increase in silicon content from 6.16% when pyrolyzed at 150 °C to 18.29% when subjected to 700 °C (Xiao et al. 2014). It has also been demonstrated that the presence of silicon in biochar contributes to the solubilization and absorption of heavy metals like Al and Cd (Xu and Chen 2015). As the temperature of pyrolysis rises, biochar contains a greater quantity of elements such as P, K, Si, Ca, S, Mg, Zn, Fe, and Mn; however, these elements become less bioavailable. This is because at higher pyrolysis temperatures, these elements become incorporated into the biochar's highly aromatic structure. The bioavailability of these nutrients may be increased by pyrolysis at a lower temperature. Biochar made from maize straw that was pyrolyzed at 300 and 400 °C had higher concentrations of exchangeable K, making them more bioavailable to plant roots than biochar made at 500 °C (Naeem et al. 2016).

11.1.1.3 Biochar's Functional Properties

Biological and chemical processes interact with biochar at its surface. Structure-related functionality, free radicals, surface charge, and surface functional groups are all thought to play a role in these activities (Zhang et al. 2017). Biochar's surface features a wide variety of functional groups beyond just aliphatic and aromatic ones, such as epoxy, hydroxyl, carboxyl, carbonyl, acyl, ether, ester, azyl, amido, and sulfonic acid. X-ray photoelectron spectroscopy (XPS), X-ray diffraction (XRD), Fourier transform infrared spectroscopy (FTIR), Raman spectroscopy (RS), and solid-state 13C nuclear magnetic resonance (NMR) have all been used to measure these classes (Wang et al. 2017). The phenolic carboxyl group, the OH group, the carbonyl group, and the ester group all occur frequently. Hydrogen bond-induced sorption, biochar alkalis, capacitance, hydrophilicity/hydrophobicity, pH buffering, surface charge alterations, and cation exchange capacity (CEC) improvement are all possible because these groups can associate/dissociate with protons in the environment (Xiao et al. 2018). For additional chemical and surface chemistry modifications of biochar, such as oxidation, amination, and sulfonation, the functional groups are ideal sites (Wang et al. 2016). Biochar's surface charge is strongly correlated with the pH of the solution because of the aliphatic/aromatic nature of its surface and the dissociation of functional groups like carboxyl groups, which serve as major ion-exchange sites for the adsorption and desorption of nutrients and heavy metals (Chen et al. 2014). During the carbonization process, free radicals are produced. Due to the electron-rich aromatic surface, pyrolysis increases the temperature at which radical species transition from oxygen-centered radicals to combined carbon-centered and oxygen-centered radicals, which stabilize the surface free radicals. Biochar has been shown to reduce plant growth due to the presence of free radicals (Amin and Eissa 2017). Other research has shown that free radicals' reactions with hydrogen peroxide and persulfate contribute significantly to the degradation of organic contaminants (Xiao et al. 2018).

11.1.2 Harnessing Biochar in Agriculture

11.1.2.1 Enhancing Plant Growth with Biochar

Depending on the soil type, the variation in the raw materials, and the pyrolysis situations during biochar production, the application of biochar either increased or decreased crop yield. Increases in dry biomass yield and quality of medicinal plant material (Senna; *Cassia angustifolia*) were observed after biochar was applied to deeply weathered acidic soils (Basak 2021). The authors of a recent study on the Senna plant hypothesized that adding biochar to the soil increased the plant's efficiency in using nutrients and absorbing them. The rice yield was greater on MgO-rich biochar-amended saline soil than on unamended saline soil (Wu et al. 2019). A black dermosol soil saw increased yields of ginger (*Zingiber officinale*) after biochar was applied to it (Farrar et al. 2019). By decreasing the SAR, ESP, and sodium adsorption by plants, biochar enriched with MgO and MnO significantly increased the root and shoot growth of safflower (*Carthamus tinctorius*) in saline conditions (Ran et al. 2020). Problems with root development are a major challenge for plants growing in soil. The function of a plant's roots cannot be overstated. Biochar amendment increased root volume 29%, root biomass 32%, root length 52%, surface area 39%, number of root tips 17%, and root diameter 9.9% (Xiang et al. 2017). These findings imply that root establishment and growth are both significantly enhanced in soils amended with biochar. A better ability to grow and capture more nutrients from the soil could be facilitated by an increase in these root parameters. Improved CEC, C, N, and P content, and decreased hydraulic conductivity were all factors in a study showing a significant increase in maize grain yield when biochar was applied at a rate of 15 t ha^{-1} to sandy soil (Uzoma et al. 2011). High crop yield under salt stress conditions was also attributed to biochar's ability to improve nutrient contents in plant tissues by facilitating greater uptake of the minerals and assisting in the maintenance of optimal tissue water content (Ran et al. 2020). Wu et al. (2019) reported similar results after applying biochar composite to a salty soil, where increased P availability in the soil led to a notable increase in rice yield. The aromatic hydrocarbons, phenols, and organic acids in biochar could be replaced by a much larger portion of medium and long-chain in alkanoic acids and alkanes through torrefaction with minerals and other organic fertilizers, which could play a crucial role in enhancing the germination of seeds, shifting the microbial population, and creating systemic plant defence against stress, leading to high crop yields (Amin and Eissa 2017). Blackwell et al. (2015) added biochar to compost, they saw a boost in microbial activity, a reduction in nitrogen loss during decomposition, and a synergistic increase in pak choi (*Brassica rapa*) plant height, fresh biomass weight, and dry biomass weight compared to when biochar was used alone. High application rates (7.5 t ha^{-1}) of biochar-based fertilizer resulted in higher maximum photosynthesis, plant relative growth rate, and rates of photosynthesis at saturation, as evidenced by a higher stem height, weight, and yield of aboveground biomass than with other organic fertilizer treatments. Increased crop yields have been linked to increased biochar application, which is thought to have enhanced soil properties such as nutrient availability and nutrient use efficiency (Gunes et al. 2015).

11.1.2.2 Enhancing Soil Health with Biochar

Enhancing the Physical Properties of Soil

Soil Bulk Density

Bulk density, which is directly related to soil compactness, is a useful indicator of physical soil characteristics. Soil porosity was improved and bulk density was greatly decreased after biochar application. When added to soil, biochar increases porosity while decreasing bulk density because of the inverse relationship between bulk density and porosity (Kakaire et al. 2015). The magnitude of biochar's effects on these two soil properties, however, is sensitive to its physical properties, application rate, and soil texture (which depend on feedstock types and pyrolysis temperatures). Soil bulk density decreases as the rate of biochar application increases. The reduction in soil bulk density brought on by biochar application is also influenced by the texture of the soil. According to (Blanco-Canqui 2017), bulk density decreased more in coarse-textured soils after biochar application than in fine-textured soils. By mixing in biochar, the bulk density of sandy soil dropped to 1.42 g cm^{-3}, while that of silty soil dropped to 1.38 and that of clay soil dropped to 1.34 g cm^{-3} (Qin et al. 2016). The difference in particle density and size between biochar particles and sand versus clay may account for this reaction.

Soil Porosity

The soil pore structure, which affects how water is used, stored, and transformed as well as how much oxygen and space the soil's fauna and flora have, is greatly influenced by the biochar's particle size, hydraulic connectivity, and pore distribution. Biochar's high surface area promotes microbial activity in soil pores, which in turn encourages plant root development and a greater diversity of soil microorganisms. Biochar can increase soil saturated hydraulic, permeability, and porosity, according to research by Oguntunde et al. (2010). Biochar increases crop yields by 13% in medium-textured soils and by 10% in coarse-textured soils (Jeffery et al. 2011). Incorporating biochar into the soil may be to blame for this phenomenon, as doing so has been shown to improve crop yields by shrinking soil pores (Dokoohaki et al. 2017). Biochar increases soil porosity from 5 to 10–25 m, as determined by Rasa et al. (2018). In general, the addition of biochar to soil can have an effect on its fertility, ability to retain water, degree of compaction, ability to let air and water pass through it, pore distribution, pore connectivity and porosity.

Soil Hydraulic Properties

Biochar's high porosity and significant specific surface area allow it to alter water retention and available soil water capacity when applied to the soil (Abrol et al. 2016). Adding biochar to the soil has been shown in studies to increase the soil's ability to hold water, which in turn increases crop yield (Verheijen et al. 2019). In addition, biochar can increase plant water use efficiency, so it's a good idea to use it in places where water is scarce (Ventura et al. 2012). Water retention and soil's mechanical strength can be altered by modifying its porosity, pore size, distribution, connectivity, and shape after biochar has been applied. Water retention may be enhanced in biochar with a larger pore volume and larger average pore diameter. This shows that the soil's hydraulic conductivity can be improved by adding biochar. By incorporating biochar into soil, saturated hydraulic conductivity can be improved (Trifunovic et al. 2018). When the soil has a high-water conductivity, it can speed up drainage and infiltration (Abel et al. 2013). Rapid drainage of the soil can reduce runoff risks while decreasing the number of agrochemicals and nutrients that can be filtered and dissolved in water (Li et al. 2013). The amount of biochar added and the surface and pore structures of the various types of biochar account for the variation in biochar's effects on saturated soil hydraulic conductivity (Uzoma et al. 2011).

Soil Temperature and Colour

Soil heat can be affected by biochar addition because of changes in both soil temperature and surface reflectance (Briggs et al. 2012). Soil temperature can be raised by incorporating biochar into the ground (Zhang et al. 2013). In addition, because biochar is a black particulate matter, it can be applied to soil to make it appear darker.

Table 11.2 Potential effect of biochar on properties of the soil

Biochar type	Amount	Soil impact	Reference
Straw biochar	40 t ha^{-1}	Decrease bulk density between 3.9% and 5.1%	Burrell et al. (2016)
Rice straw	2 g 10 g^{-1} soil	Sucrase and urease activities and increased in the microbial population	Li et al. (2023)
Corn	30 t ha^{-1} and 60 t ha^{-1}	Reduce runoff and nutrients loss but increase surface erosion	Yin et al. (2022)
Manure, wood, and plant residue	40–80 t ha^{-1}	Raised soil pH by 46%; CEC by 20%; SOC by 27%	Singh et al. (2022)
Hardwood	10, 20, and 30 t ha^{-1}	Reduced soil erosion by increase amount of biochar 7.4, 20, and 73.5%, respectively	Adekiya et al. (2020)
Pinewood-based biochar	5, 10, 20%	20% biochar enhanced porosity by 50%	Chang et al. (2021)
Wheat straw	0, 20, and 40 t ha^{-1}	N_2O emissions decreased and soil organic carbon increased.	Li et al. (2015)
Household garbage	2% by weight	Increase in total available N, P, K and decrease in Al saturation	Schulz and Gla-ser (2012)
Swine manure biochar	25 t·ha^{-1}	Improved carbon source utilization	(Ayaz et al. 2022)
Red gram, mesquite wood, maize stalk and cotton	5 t ha^{-1} with 10 t FYM and RDF	The CEC increased by 1.4 cmol^{+} kg^{-1} and the C buildup by 4.4 t ha^{-1} when the pH was raised from 5.7 to 6.3	Pandian et al. (2016)

Enhancing Nutrient Availability

Important micronutrients for rhizobacterial nodulation include molybdenum, phosphorus, potassium, calcium, magnesium, iron, bromine, and biochar, which may be a source of these elements. When biochar is used in conjunction with crop residues, nutrient supply is greatly enhanced, leading to increased crop yields. Tan (2023) showed a 25% increase in crop yield compared to conventional method, with corresponding increases in soil organic matter content (from 0.93% to 1.25%), soil moisture content (from 18% to 23%), etc. Soil quickly loses its chemical fertilizer after application due to leaching and debasing (Dhaliwal 2022). Composting, which can lead to less soil, can therefore increase costs for farmers. Polluting the environment can be accomplished through the draining of vital plant nutrients like P, K, and N. In such instances, it may be best to use biochar in conjunction with fertilizers or organic manure. Another team came to a similar conclusion, finding that groundnut yield went up alongside improvements in soil pH, available N, P, and CEC (Luo 2017). Numerous studies have demonstrated that biochar improves soil fertility, productivity and crop productivity for a wide variety of crops (Table 11.2).

Fostering Biological Properties and Enzymatic Activity

Multiple studies have looked into how biochar affects soil microbes and found that it has stimulating positive effects on microbial activity, diversity, and soil enzymes (Tan 2022). Researchers have found that adding even a small percentage of biochar to soil improves microbial (bacterial and fungal) growth and activity, but that adding more than about 5% of biochar gradually reduces the microbial community. Other experiments supported the idea that adding biochar to the microbiota dominated by bacteria would increase the number of 16 S rRNA replicas (Derycke 2016). However, both studies found that compared to a control, adding 5% biochar decreased microbial gene replicas by 74% and 25%, respectively. One theory suggests that the pH of the environment played a crucial role. In contrast to an acidic setting, bacteria and fungi thrived in slightly alkaline and neutral pH soil. Soil pH was between 7 and 7.5 when 1–2% biochar was added; however, when 5% biochar was added, the pH increased to 8.5, which was found to be unfavourable to microbial activity and, consequently, impeded bacterial and fungal transcription mechanisms (Sutar 2021). Biochar has been shown to positively affect both Arbuscular and Ectomycorrhizal fungi in soil, according to a few studies. Another team of researchers speculated that the presence of biochar may have provided an ecosystem in which microorganisms could protect themselves from predators while also meeting their individual requirements for mineral nutrients, carbon, and energy. The addition of biochar is not universally well received by bacteria. The bacterial denitrification process could be slowed down by proposals involving biochar. Soil N_2 availability may decrease and N_2O emissions may increase due to denitrifying microbes. Thus, biochar application increases soil N_2 richness by decreasing denitrified abundance. The utilization of biochar leads to the proliferation of iron (Fe)-reducing microorganisms, contributing notably to the heightened emissions of methane (CH_4). This phenomenon is particularly prominent in rice cultivation soils where the incorporation of biochar serves as a stable substrate for the establishment of biofilms. This, in turn, facilitates the effective transfer of charged particles between various microbial communities and insoluble acceptors such as iron oxides (Fe). Consequently, the presence of Fe-reducing microbes spurred by biochar amendment plays a substantial role in the cycling of nitrogen (N) and carbon (C) within these soils. Many groups studied biochar's effect on Fe-reducing bacteria and came to the conclusion that it could act as a kind of ecosystem for microbes, enhancing soil properties like pH buffering, nutrient availability, and water-holding capacity (Roopnarain 2021). Large organic compounds in soil need to be broken down into simpler monomer units before they can be used by plants, and this is the job of the soil microorganisms. To aid in the reprocessing of N, P, and C, microbes produce exoenzymes such as urease, phosphatase, glucosidase, invertase, cellulase, laccase, and galactosidase (Piotrowska-Długosz 2020). Increased nutrient supply and availability lead to a greater microbial population and, in turn, enhanced exoenzyme synthesis and activity, which can be achieved through biochar incorporation. Hydrolytic enzymes, alkaline phosphatase, and urease are all examples of exoenzymes that have been shown to benefit from the addition of about 2% biochar. As biochar concentrations were raised, enzyme levels were lowered (Palansooriya 2019). Only a few researchers have noticed an uptick in alkaline phosphomonoesters and alkaline phosphatase activities with such low concentrations of biochar (roughly 2.5%), revealing enzyme activity related to phosphate cycling. Exoenzymes like *-1,4-N acetylglucosaminidase*, *-1,4-glucosidase*, and *-D-cellobiohydrolase* were found to benefit from the addition of biochar, while phosphatase and -1,4-glucosidase activities were negatively affected (Pokharel 2020). Activated carbon, according to several sources, also inhibited invertase and alkaline phosphatase activity. According to these studies, the introduction of biochar could have a negative effect on microbial growth and enzymatic activities, and high pH could have impacted the metabolic functions due to biochar addition (Palansooriya 2019; Pokharel 2020).

Heavy Metal Polluted Soils

Heavy metal (HM) contamination has been known to negatively affect microbiological diversity, groundwater, soil quality, and agricultural productivity. Biochar is a bio-residue material that can be used to treat HM-contaminated soil sustainably. Although biochar amendment can be used in conjunction with other methods to address heavy metal-contaminated soils (Mosa et al. 2016), it does not completely eliminate the presence of heavy metals in soil but rather stabilises them by converting the heavy metals into less bio accessible and less soluble forms. Similar to chemical stabilization, in which chemicals like phosphate and carbonate are applied to polluted soils to cause a reaction with the heavy metal contaminants therein and turn them into precipitates (Lian and Xing 2017), this procedure also results in the formation of a precipitate. Heavy metals in treated soils have their bioavailability and ecotoxicity reduced to below the risk level, allowing for the restoration of the soil's desirable function in order to grow safe crops. Biochar-facilitated soil remediation involves the application of specific biochar products at optimal rates to heavily metal-polluted sites, followed by a thorough mixing of the biochar with the contaminated soil. For various heavy metal pollutants, biochar employs a variety of remediation mechanisms. Furthermore, the adsorption mechanism for the same heavy metal ion varies depending on the biochar (Qian and Chen 2014). Heavy metals can be removed through physical adsorption when their ions diffuse into the pores of the sorbent. Biochar applied to soil can absorb metal ions on pore surfaces, where they can be transformed into hydroxide, carbonate, and phosphate precipitates. Heavy metal uptake and bioaccumulation by soil organisms (including plant roots) are reduced as the bioactive fraction and water-soluble forms of heavy metals in soil (Ahmad et al. 2014). Heavy metal toxicity in biochar-amended soils should be regularly checked for metals removal accuracy and their immobilization effect may wear off with time. Biochar removes heavy metals from soil by physical sorption, complexation, ion exchange, precipitation, electrostatic attraction, and other mechanisms. That may help lessen their leachability as well (Ding et al. 2016). Biochar possessing abundant oxygen and featuring various acidic functional groups, hydroxyl moieties, carbonyl, including phenolic, lactonic, and carboxylic, has the capacity to intricately bind (complex) heavy metals and metalloids both on its surface and within its internal pore structure. Electrostatic repulsion between metal ions and the surface charge of biochar traps the heavy metals. The electrostatic force is at work here. Biochar can also fix metals by facilitating their ion exchange from the soil matrix to its surface. In the ion exchange process, the residual efficiency of heavy metals is most affected by the size and metal group on the surface of biochar (Fang et al. 2015). Heavy metals are immobilized by biochar, and the mechanism most commonly attributed to it is precipitation (Xiao et al. 2018). Precipitation refers to the process of forming a solid in solution or on a surface during the sorption process.

11.1.2.3 Biochar's Role in Carbon Sequestration

Soil plays an important role in the global carbon cycle and acts as a major carbon sink. According to Bond-Lamberty and Thomson (2010), this has an immediate effect on global warming. Carbon sequestration (SC) has been proposed as a means of reducing carbon dioxide emissions in soil, as reported by Wang and Wang (2019). Biochar's CS cycle in soil and crop plants is depicted in Fig. 11.2. Biochar's high resistance to biodegradation stems from its highly condensed aromatic structure (Kuhlbusch 1998). The soil benefits from biochar because it helps to sequester carbon. (Yin et al. 2022) claim that biochar's labile-C can offset and even reverse

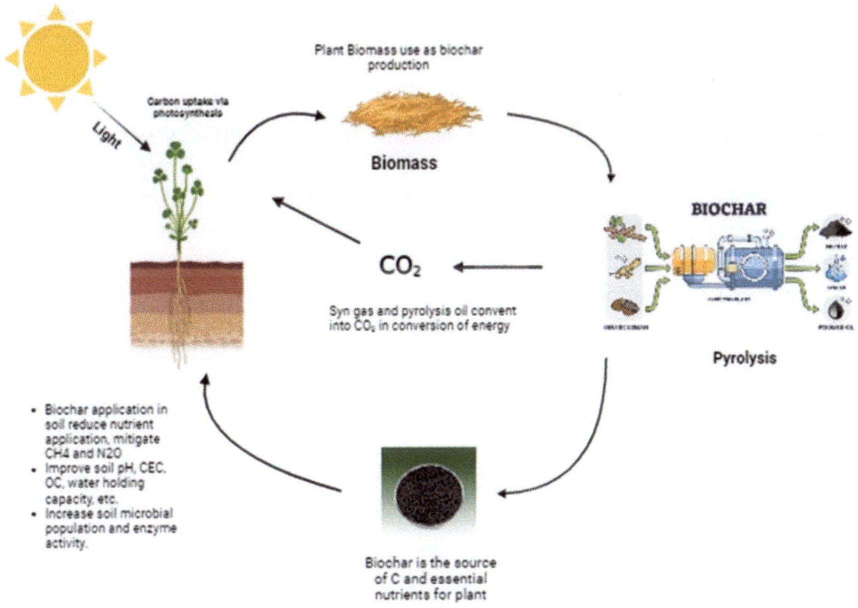

Fig. 11.2 Carbon sequestration cycle with biochar application

the long-term effects of microbial decomposition on soil climate. Photosynthesis stores CO_2 in biomass, and its decay releases the gas into the atmosphere. By slowing down or stopping the decomposition of biomass, biochar keeps carbon locked up in a stable form and out of the air. This is the "carbon-negative effect" of biochar. Soil with biochar significantly reduced the amount of carbon dioxide released, as shown by research by Kim Oanh et al. (2018). Biochar is a win–win strategy for lowering greenhouse gas emissions and improving food security around the world. Soil amendment with biochar has two potential advantages: long-term carbon sequestration and beneficial soil change.

11.1.3 Limitations and Challenges of Biochar Implementation

In comparison to controlled environments like greenhouses, outdoor applications of biochar materials generally lead to lower levels of pollutant sorption. The direct implementation of biochar in natural field conditions seems to have minimal impact on soil optimization. The potential for biochar to manage nutrients and have other positive effects has raised some concerns among researchers. Soil microbial abundance and diversity measurements may be impacted by factors like extractable carbon, pyrolysis temperature, and incubation time (Xu et al. 2013). Biochar applied to soil can bind and inactivate agrochemicals like pesticides and fertilizers (Zhou et al. 2014). The net oxidation of carbon hydrogen in the soil was reduced and the production of soil and N_2O was lowered in many of the biochars evaluated here (Mohammadi 2021). There are concerns that biochar's presence of phytotoxic substances could inhibit seed germination (Kuppusamy et al. 2016). Soil biological processes and seed germination are affected by the excess biochar applied to it because of changes in EC and pH (Spokas and Reicosky 2009). When applied to agricultural soil, dusty nature, and biochar's poor mechanical strength can be problematic because of the frequent loss of significant volumes to wind and rain due to the soil's application conditions (Mylavarapu et al. 2013). Biochar compounds may be pushed deeper into the soil or larger particles may be carried away during heavy precipitation and runoff events, decreasing their proximity to impurities and their ability to adsorb and interact (Xu and Wang 2015). Short-term performance of carbon-rich solids can be impacted by bacterial degradation of biochar particles in the soil. Bacterial growth causes decomposition, which can release pollutants at higher depths. Biochar matrices interact strongly, making it difficult for microbes to access the compounds they contain (Goswami et al. 2016). Biochar, once applied to the soil, becomes permanently adsorbent in the soil (Spokas and Reicosky 2009). The presence of heavy metals in biochar creates a risk of release (Mylavarapu et al. 2013). Unfortunately, not all biodegradable waste can be recycled into a form of biochar that is useful for agricultural purposes. According to Kuppusamy et al. (2016), biochar that is produced using particular processes and components is vulnerable to microbial degradation and possesses a low capacity for the storage of nutrients.

11.1.4 The Future of Biochar: Prospects and Conclusion

Recent years have witnessed an increase in studies focusing on the potential of biochar in agricultural use. Despite this, the application of biochar for soil remediation remains relevant. While most biochar's exhibit significant sorption outcomes of over 45–50% when integrated into agricultural fields without prior investigation, certain biochar's might not effectively remove soil pollutants, particularly in the context of soil remediation (Logan et al. 2007). The interaction between biochar and soil composition is influenced by several factors, including cation exchange capacity (CEC), biochar types, the formation of complexes, and microbial activity. Factors such as the materials used for biochar production, pyrolysis temperature, and the properties of contaminants, such as polarity, must all be considered. Numerous laboratory-based studies (Das et al. 2023) have explored the amalgamation of biochar within soil. However, there has been a relative scarcity of such field-based studies. Further research into the reclamation of contaminated soil under optimal conditions is necessary to comprehensively grasp the dynamics of biochar in such environments (Jin et al. 2013). Attaining precise details about the biochar in use, including feedstock type, proposed application rate, and recovery process, requires considerable effort for each unique case. Anand et al. (2022) underscore the need for more investigation to ascertain the quantity and type of biochar that a given soil type can endure without disrupting its inherent function. While understanding the mechanisms of biochar with soil physicochemical properties is crucial, comprehending the interactions between these mechanisms holds equal importance (Choi and Park 2015). Despite their notable resistance to biological degradation, there is a lack of knowledge about the long-term fate of individual biochar particles (Das et al. 2023). The effectiveness of trials hinges on their alignment with each specific use case (Ran et al. 2020). In conclusion, it can be inferred that the use of biochar enhances soil physicochemical and biological properties, along with improving crop growth and yield.

References

Abel S, Peters A, Trinks S, Schonsky H, Facklam M, Wessolek G (2013) Impact of biochar and hydrochar addition on water retention and water repellency of sandy soil. Geoderma 202(203): 183–191

Abrol V, Ben-Hur M, Verheijen F, Keizer JJ, Martins M, Tenaw H, Tchehansky L, Graber ER (2016) Biochar effects on soil water infiltration and erosion under seal formation conditions: rainfall simulation experiment. J Soils Sediments 16(12):2709–2719

Adekiya AO, Agbede TM, Olayanju A, Ejue WS, Adekanye TA, Adenusi TT, Ayeni JF (2020) Effect of biochar on soil properties, soil loss, and cocoyam yield on a tropical Sandy loam Alfisol. Sci World J 2020:e9391630. https://doi.org/10.1155/2020/9391630

Ahmad M, Rajapaksha AU, Lim JE, Zhang M, Bolan N, Mohan D, Vithanage M, Lee SS, Ok YS (2014) Biochar as a sorbent for contaminant management in soil and water: a review. Chemosphere 99:19–33

Amin AE-EAZ, Eissa MA (2017) Biochar effects on nitrogen and phosphorus use efficiencies of zucchini plants grown in a calcareous sandy soil. J Soil Sci Plant Nutr 17:912–921

Anand A, Kumar V, Kaushal P (2022) Biochar and its twin benefits: crop residue management and climate change mitigation in India. Renew Sust Energ Rev 156:111959. https://doi.org/10.1016/j.rser.2021.111959

Awasthi MK, Sarsaiya S, Patel A, Juneja A, Singh RP, Yan B, Awasthi SK, Jain A, Liu T, Duan Y, Pandey A, Zhang Z, Taherzadeh MJ (2020) Refining biomass residues for sustainable energy and bio-products: an assessment of technology, its importance, and strategic applications in circular bioeconomy. Renew Sustain Energy Rev 127:109876

Ayaz M, Feizienė D, Feiza V, Tilvikienė V, Baltrėnaitė-Gedienė E, Khan A (2022) The impact of swine manure biochar on the physical properties and microbial activity of loamy soils. Plants 11(13):1729. https://doi.org/10.3390/plants11131729

Basak BB (2021) Repurposing distillation waste biomass and low-value mineral resources through biochar-mineral-complex for sustainable production of high-value medicinal plants and soil quality improvement. Sci Total Environ 760:143319. https://doi.org/10.1016/j.scitotenv.2020.143319

Berihun T, Tolosa S, Tadele M, Kebede F (2017) Effect of biochar application on growth of garden pea (*Pisum sativum* L.) in acidic soils of Bule Woreda Gedeo zone Southern Ethiopia. Int J Agron 2017:e6827323. https://doi.org/10.1155/2017/6827323

Blackwell P, Joseph S, Munroe P, Anawar HM, Storer P, Gilkes RJ (2015) Influences of biochar and biochar-mineral complex on mycorrhizal colonisation and nutrition of wheat and sorghum. Pedosphere 25(686):695. https://doi.org/10.1016/S1002-0160(15)

Blanco-Canqui H (2017) Biochar and soil physical properties. Soil Sci Soc Am J 81(4):687–711. https://doi.org/10.2136/sssaj2017.01.0017

Bond-Lamberty B, Thomson A (2010) Temperature-associated increases in the global soil respiration record. Nature 464:579–582

Brewer CE, Chuang VJ, Masiello CA, Gonnermann H, Gao X, Dugan B, Driver LE, Panzacchi P, Zygourakis K, Davies CA (2014) New approaches to measuring biochar density and porosity. Biomass Bioenergy 66:176–185. https://doi.org/10.1016/j.biombioe.2014.03.059

Briggs C, Breiner JM, Graham RC (2012) Physical and chemical properties of Pinus ponderosa charcoal : implications for soil modification. Soil Sci 177:263–268. https://doi.org/10.1097/SS.0b013e3182482784

Burrell LD, Zehetner F, Rampazzo N, Wimmer B, Soja G (2016) Long-term effects of biochar on soil physical properties. Geoderma 282:96–102. https://doi.org/10.1016/j.geoderma.2016.07.019

Cao X, Ma LN, Liang Y, Gao B, Harris W (2011) Simultaneous immobilization of lead and atrazine in contaminated soils using dairy-manure biochar. Environ Sci Technol 45:4884–4889

Chang Y, Rossi L, Zotarelli L, Gao B, Shahid MA, Sarkhosh A (2021) Biochar improves soil physical characteristics and strengthens root architecture in Muscadine grape (Vitis rotundifolia L.). Chem Biol Technol Agric 8(1):7. https://doi.org/10.1186/s40538-020-00204-5

Cheah S, Malone SC, Feik CJ (2014) Speciation of sulfur in biochar produced from pyrolysis and gasification of oak and corn stover. Environ Sci Technol 48:8474–8480

Chen Z, Xiao X, Chen B, Zhu L (2014) Quantification of chemical states, dissociation constants and contents of oxygen-containing groups on the surface of biochars produced at different temperatures. Environ Sci Technol 49:309–317

Choi YK, Park SJ (2015) Hydrogen storage capacity of highly porous carbons synthesized from biomass-derived aerogels. Carbon Lett 16(2):127–131. https://doi.org/10.5714/CL.2015.16.2.127

Das SK, Ghosh GK, Avasthe R (2023) Application of biochar in agriculture and environment, and its safety issues. Biomass Convers Biorefinery 13:1359–1369. https://doi.org/10.1007/s13399-020-01013-4

Derycke S (2016) Coexisting cryptic species of the Litoditis marina complex (Nematoda) show differential resource use and have distinct microbiomes with high intraspecific variability. Mol Ecol 25:2093–2110

Dhaliwal SS (2022) Biofortification—a frontier novel approach to enrich micronutrients in field crops to encounter the nutritional security. Molecules 27:1340

Ding ZH, Wan YS, Hu X, Wang SS, Zimmerman AR, Gao B (2016) Sorption of lead and methylene blue onto hickory biochars from different pyrolysis temperatures: importance of physicochemical properties. J Ind Eng Chem 37:261–267

Dokoohaki H, Miguez FE, Laird D, Horton R, Basso AS (2017) Assessing the biochar effects on selected physical properties of a sandy soil: an analytical approach. Commun Soil Sci Plant Anal 48(12):1387–1398

e-Gedien, B (2020) Dependence of pyrolysis temperature and lignocellulosic physical-chemical properties of biochar on its wettability. Biomass Convers Biorefinery 2020:1–19. https://doi.org/10.1007/s13399-020-00711-

El–Metwally I, Geries L, Saudy H (2022) Interactive effect of soil mulching and irrigation regime on yield, irrigation water use efficiency and weeds of trickle–irrigated onion. Arch Agron Soil Sci 68:1103–1116

Epstein E (1994) The anomaly of silicon in plant biology. Proc Natl Acad Sci 91:11–17

Fang J, Gao B, Chen J, Zimmerman AR (2015) Hydrochars derived from plant biomass under various conditions: characterization and potential applications and impacts. Chem Eng J 267: 253–259

Farrar MB, Wallace HM, Xu CY, Nguyen TTN, Tavakkoli E, Joseph S (2019) Short-term effects of organo-mineral enriched biochar fertiliser on ginger yield and nutrient cycling. J Soils Sediments 19:668–682. https://doi.org/10.1007/s11368-018-2061-9

Fedeli R, Alexandrov D, Celletti S (2023) Biochar improves the performance of Avena sativa L. grown in gasoline-polluted soils. Environ Sci Pollut Res 30:28791–28802. https://doi.org/10.1007/s11356-022-24127-w

Goswami D, Thakker JN, Dhandhukia PC (2016) Portraying mechanics of plant growth promoting rhizobacteria (PGPR): a review. Cogent Food Agric 2(1):1127500. https://doi.org/10.1080/23311932.2015.1127500

Gunes A, Inal A, Sahin O, Taskin M, Atakol O, Yilmaz N (2015) Variations in mineral element concentrations of poultry manure biochar obtained at different pyrolysis temperatures, and their effects on crop growth and mineral nutrition. Soil Use Manag 31:429–437

Hale L, Luth M, Crowley D (2015) Biochar characteristics relate to its utility as an alternative soil inoculum carrier to peat and vermiculite. Soil Biol Biochem 81:228–235. https://doi.org/10.1016/j.soilbio.2014.11.023

Hardy JEDB, Cornelis JT, Houben D, Leifeld J, Lambert R (2016) Evaluation of the long-term effect of biochar on properties of temperate agricultural soil at pre-industrial charcoal kiln sites in Wallonia, Belgium. Eur J Soil Sci:1–10. https://doi.org/10.1111/ejss.12395

Homagain K, Shahi C, Luckai N, Sharma M (2016) Life cycle cost and economic assessment of biochar-based bioenergy production and biochar land application in Northwestern. For Ecosyst 3:1–10. https://doi.org/10.1186/s40663-016-

Inyang M, Gao B, Yao Y, Xue Y, Zimmerman AR, Pullammanappallil P, Cao X (2012) Removal of heavy metals from aqueous solution by biochars derived from anaerobically digested biomass. Bioresour Technol 110:50–56

Jeffery S, Verheijen FGA, Velde M, Bastos AC (2011) A quantitative review of the effects of biochar application to soils on crop productivity using meta-analysis. Agric Ecosyst Environ 144(1):175–187

Jin H, Wang X, Gu Z, Polin J (2013) Carbon materials from high ash biochar for supercapacitor and improvement of capacitance with HNO3 surface oxidation. J Power Sources 236(285):292. https://doi.org/10.1016/j.jpowsour.2013.02.088

Kakaire J, Makokha GL, Mwanjalolo M, Mensah AK, Menya E (2015) Effects of mulching on soil hydro-physical properties in Kibaale Sub-catchment, South Central Uganda. Appl Ecol Environ Res 3(5):127–135. https://doi.org/10.12691/aees-3-5-1

Kamali M, Sweygers N, Al-salem S, Appels L, Aminabhavi TM, Dewil R (2022) Biochar for soil applications-sustainability aspects, challenges and future prospects. Chem Eng J 428:131189. https://doi.org/10.1016/j.cej.2021.131189

Kamau S, Karanja NK, Ayuke FO, Lehmann J (2019) Short-term influence of biochar and fertilizer-biochar blends on soil nutrients, fauna and maize growth. Biol Fertil Soils 55:661–673

Khan T, Salma M, Hossain S (2018) Impacts of different sources of biochar on plant growth characteristics. Am J Plant Sci 9:1922–1934. https://doi.org/10.4236/ajps.2018.99139

Kim Oanh NT, Permadi DA, Dong NP, Nguyet DA (2018) Emission of toxic air pollutants and greenhouse gases from crop residue open burning in Southeast Asia. In: Vadrevu KP, Ohara T, Justice C (eds) Land-atmospheric research applications in South and Southeast Asia, Springer Remote sensing/photogrammetry. Springer, Cham, pp 47–66

Kuhlbusch TAJ (1998) Black carbon and the carbon cycle. Science 280:1903–1904. https://doi.org/10.1126/science.280.5371.1903

Kuppusamy S, Thavamani P, Megharaj M, Venkateswarlu K, Naidu R (2016) Agronomic and remedial benefits and risks of applying biochar to soil: current knowledge and future research directions. Environ Int 87:1–12. https://doi.org/10.1016/j.envint.2015.10.018

Li Y, Ru G, Rui Y, Wei H, Yong L, Xiao H, Wu J (2013) Using a simple soil column method to evaluate soil phosphorus leaching risk. Acta Hydrochim Hydrobiol 41(11):1100–1107

Li B, Fan CH, Xiong ZQ, Li QL, Zhang M (2015) The combined effects of nitrification inhibitor and biochar incorporation on yield-scaled N_2O emissions from an intensively managed vegetable field in southeastern China. Biogeosciences 12(6):2003–2017. https://doi.org/10.5194/bg-12-2003-2015

Li X, Shi Y, Li Z, Yang X, Xiao J (2023) Influence of biochar-triggered bottom-up effects on soil enzymes, soybean defenses, and the fitness of Spodoptera litura and its parasitoid. Biol Control 185:105309. https://doi.org/10.1016/j.biocontrol.2023.105309

Lian F, Xing B (2017) Black carbon (biochar) in water/soil environments: molecular structure, sorption, stability, and potential risk. Environ Sci Technol 51:13517–13532

Liu J, Jiang B, Shen J, Zhu X, Wu J (2021) Contrasting effects of straw and straw-derived biochar applications on soil carbon accumulation and nitrogen use efficiency in doublerice cropping systems. Agric Ecosyst Environ 311:107286

Logan B, Cheng S, Watson V, Estadt G (2007) Graphite fiber brush anodes for increased power production in air-cathode microbial fuel cells. Environ Sci Technol 41(9):3341–3346. https://doi.org/10.1021/es062644y

Luo X (2017) Use of biochar-compost to improve properties and productivity of the degraded coastal soil in the Yellow River Delta, China. J Soils Sediments 17:780–789

Mohammadi A (2021) Overview of the benefits and challenges associated with pelletizing biochar. Processes 9:1591

Mosa A, El-Banna MF, Gao B (2016) Biochar filters reduced the toxic effects of nickel on tomato (Lycopersicon esculentum L.) grown in nutrient film technique hydroponic system. Chemosphere 149:254–262

Mylavarapu R, Nair V, Morgan K (2013) An introduction to biochars and their uses in agriculture. EDIS 2013:1–4

Naeem MA, Khalid M, Ahmad Z, Naveed M (2016) Low pyrolysis temperature biochar improves growth and nutrient availability of Maize on Typic Calciargid. Commun Soil Sci Plant Anal 47: 41–51

Oguntunde PG, Abiodun BJ, Ajayi AE, Nick V (2010) Effects of charcoal production on soil physical properties in Ghana. J Plant Nutr Soil Sci 171(4):591–596

Olszyk D, Shiroyama T, Novak J, Cantrell K, Sigua G, Watts D, Johnson MG (2020) Biochar affects growth and shoot nitrogen in four crops for two soils. Agrosyst Geosci Environ 3(1): e20067. https://doi.org/10.1002/agg2.20067

Omondi MO, Xia X, Nahayo A, Liu X, Korai PK, Pan G (2016) Quantification of biochar effects on soil hydrological properties using meta-analysis of literature data. Geoderma 274:28–34

Palansooriya KN (2019) Response of microbial communities to biochar-amended soils: a critical review. Biochar 1:3–22

Pandian K, Subramaniayan P, Gnasekaran P, Chitraputhirapillai S (2016) Effect of biochar amendment on soil physical, chemical and biological properties and groundnut yield in rainfed Alfisol of semi-arid tropics. Arch Agron Soil Sci 62(9):1293–1310. https://doi.org/10.1080/03650340.2016.1139086

Parmar A, Nema PK, Agarwal T (2014) Biochar production from agro-food industry residues: a sustainable approach for soil and environmental management. Curr Sci 107:1673–1682

Pimchuai A, Dutta A, Basu P (2010) Torrefaction of agriculture residue to enhance combustible properties. Energy Fuel 24:4638–4645

Piotrowska-Długosz A (2020) Carbon and nitrogen cycling in soil. In: Significance of the enzymes associated with soil C and N transformation. Springer, pp 399–437

Pituello C, Francioso O, Simonetti G, Pisi A, Torreggiani A, Berti A (2014) Characterization of chemical – physical, structural and morphological properties of biochars from biowastes produced at different temperatures. J Soils Sediments:1–13. https://doi.org/10.1007/s11368-014-0964-7

Pokharel P (2020) Biochar increases soil microbial biomass with changes in extra-and intracellular enzyme activities: a global meta-analysis. Biochar 2:65–79

Qian L, Chen B (2014) Interactions of aluminum with biochars and oxidized biochars: implications for the biochar aging process. J Agric Food Chem 62:373–380

Qin X, Li Y, Hong W, Chong L, Li J, Wan Y, Gao Q, Fan F, Liao Y (2016) Long-term effect of biochar application on yield-scaled greenhouse gas emissions in a rice paddy cropping system: a four-year case study in South China. Sci Total Environ 569–570:1390–1401

Rahman MM, Mostofa MG, Keya SS, Siddiqui MN, Ansary MMU, Das AK, Rahman MA, Tran LS-P (2021) Adaptive mechanisms of halophytes and their potential in improving salinity tolerance in plants. Int J Mol Sci 22(19):10733. https://doi.org/10.3390/ijms221910733

Ran C, Gulaqa A, Zhu J, Wang X, Zhang S, Geng Y (2020) Benefits of biochar for improving ion contents, cell membrane permeability, leaf water status and yield of rice under saline–sodic paddy field condition. J Plant Growth Regul 39:370–377

Rasa K, Heikkinen J, Hannula M, Kai A, Hyvluoma J (2018) Why does willowbiochar increase a clay soil water retention capacity? Biomass Bioenergy 119:346–353

Roopnarain A (2021) Unravelling the anaerobic digestion 'black box': biotechnological approaches for process optimization. Renew Sustain Energy Rev 152:111717

Schulz H, Glaser B (2012) Effects of biochar compared to organic and inorganic fertilizers on soil quality and plant growth in a greenhouse experiment. J Plant Nutr Soil Sci 175(3):410–422. https://doi.org/10.1002/jpln.201100143

Shetty R, Prakash NB (2020) Effect of different biochars on acid soil and growth parameters of rice plants under aluminium toxicity. Sci Rep 10:12249. https://doi.org/10.1038/s41598-020-69262-x

Singh BK, Liu H, Trivedi P (2020) Eco-holobiont: a new concept to identify drivers of host-associated microorganisms. Environ Microbiol 22(2):564–567. https://doi.org/10.1111/1462-2920.14900

Singh H, Northup BK, Rice CW, Prasad PVV (2022) Biochar applications influence soil physical and chemical properties, microbial diversity, and crop productivity: a meta-analysis. Biochar 4(1):8. https://doi.org/10.1007/s42773-022-00138-1

Somerville M, Jahanshahi S (2015) The effect of temperature and compression during pyrolysis on the density of charcoal made from Australian eucalypt wood. Renew Energy 80:471–478

Spokas KA, Reicosky DC (2009) Impacts of sixteen different biochars on soil greenhouse gas production, impacts Sixt. Differ Biochars Soil Greenh Gas Prod 3:179–193

Sun Y, Gao B, Yao Y, Fang J, Zhang M, Zhou Y, Chen H, Yang L (2014) Effects of feedstock type, production method, and pyrolysis temperature on biochar and hydrochar properties. Chem Eng J 240:574–578

Sutar SS (2021) Recent advances in biochar-based dye remediation. In: Biochar and its application in bioremediation, pp 135–167

Tan S (2022) A perspective on the interaction between biochar and soil microbes: a way to regain soil Eminence. Environ Res 214:113832

Tan M (2023) Conversion of agricultural biomass into valuable biochar and their competence on soil fertility enrichment. Environ Res 234:116596

Trifunovic B, Gonzales HB, Ravi S, Sharratt BS, Mohanty SK (2018) Dynamic effects of biochar concentration and particle size on hydraulic properties of sand. Land Degrad Dev 29(4): 884–893

Ulusal A, Varol EA, Bruckman VJ, Uzun BB (2021) Opportunity for sustainable biomass valorization to produce biochar for improving soil characteristics. Biomass Convers Biorefinery 11: 1041–1051

Uzoma K, Inoue M, Andry H, Fujimaki H, Zahoor A, Nishihara E (2011) Effect of cow manure biochar on maize productivity under sandy soil condition. Soil Use Manag 27:205–212

Ventura F, Salvatorelli F, Piana S, Pieri L, Pisa PR (2012) The effects of biochar on the physical properties of bare soil. Earth Sci 103(1):5–11

Verheijen F, Zhuravel A, Silva FC, Amaro A, Ben-Hur M, Keizer JJ (2019) The influence of biochar particle size and concentration on bulk density and maximum water holding capacity of sandy vs sandy loam soil in a column experiment. Geoderma 1(347):194

Wainaina S, Awasthi MK, Sarsaiya S, Chen H, Singh E, Kumar A, Ravindran B, Awasthi SK, Liu T, Duan Y, Kumar S, Zhang Z, Taherzadeh MJ (2020) Resource recovery and circular economy from organic solid waste using aerobic and anaerobic digestion technologies. Bioresour Technol 301:122778

Wang J, Wang S (2019) Preparation, modification and environmental application of biochar: a review. J Clean Prod 227:1002–1022. https://doi.org/10.1016/j.jclepro.2019.04.282

Wang L, Cui XF, Cheng HG, Chen F, Wang JT, Zhao XY, Lin CY, Pu X (2015) A review of soil cadmium contamination in China including a health risk assessment. Environ Sci Pollut Res Int 22:16441–16452

Wang P, Wu C, Guo Y, Wang C (2016) Experimental and theoretical studies on methylene blue and methyl orange sorption by wheat straw-derived biochar with a large surface area. Phys Chem Chem Phys 18:30196–30203

Wang B, Gao B, Fang J (2017) Recent advances in engineered biochar productions and applications. Crit Rev Environ Sci Technol 47:2158–2207

Wang L, Rinklebe J, Sik Y, Daniel O, Daniel CWT (2021) Biochar composites: emerging trends, field successes and sustainability implications. Soil Use Manag 14:1–25. https://doi.org/10.1111/sum.12731

Weber K, Quicker P (2018) Properties of biochar. Fuel 217:240–261

Wu HL, Che XD, Ding ZH, Hu X, Creamer AE, Chen H, Gao B (2016) Release of soluble elements from biochars derived from various biomass feedstocks. Environ Sci Pollut Control Ser 23: 1905–1915

Wu L, Wei C, Zhang S, Wang Y, Kuzyakov Y, Ding X (2019) MgO-modified biochar increases phosphate retention and rice yields in saline alkaline soil. J Clean Prod 235(901):909. https://doi.org/10.1016/j.jclepro.2019.07.043

Xiang Y, Deng Q, Duan H, Guo Y (2017) Effects of biochar application on root traits: a meta-analysis. GCB Bioenergy 9:1563–1572

Xiao X, Chen B, Zhu L (2014) Transformation, morphology, and dissolution of silicon and carbon in rice straw-derived biochars under different pyrolytic temperatures. Environ Sci Technol 48: 3411–3419

Xiao X, Chen B, Chen Z, Zhu L, Schnoor JL (2018) Insight into multiple and multilevel structures of biochars and their potential environmental applications: a critical review. Environ Sci Technol 52(9):5027–5047. https://doi.org/10.1021/acs.est.7b06487

Xu Y, Chen B (2015) Organic carbon and inorganic silicon speciation in rice-bran-derived biochars affect its capacity to adsorb cadmium in solution. J Soils Sediments 15:60–70

Xu F, Wang D (2015) Analysis of lignocellulosic biomass using infrared methodology. In: Pretreatment of biomass. Elsevier, pp 7–25

Xu X, Cao X, Zhao L (2013) Comparison of rice husk-and dairy manure-derived biochars for simultaneously removing heavy metals from aqueous solutions: role of mineral components in biochars. Chemosphere 92(8):955–961. https://doi.org/10.1016/jchemosphere.2013.03.009

Yadav SPS, Bhandari S, Bhatta D, Poudel A, Bhattarai S, Yadav P, Ghimire N, Paudel P, Paudel P, Shrestha J, Oli B (2023) Biochar application: a sustainable approach to improve soil health. J Agric Food Res 11:100498. https://doi.org/10.1016/j.jafr.2023.100498

Yeboah E, Asamoah G, Ofori P, Amoah B, Agyeman KOA (2020) Method of biochar application affects growth, yield and nutrient uptake of cowpea. Open Agric 5(1):352–360. https://doi.org/10.1515/opag-2020-0040

Yin J, Zhao L, Xu X, Li D, Qiu H, Cao X (2022) Evaluation of long-term carbon sequestration of biochar in soil with biogeochemical field model. Sci Total Environ 822:153576. https://doi.org/10.1016/J.SCITOTENV.2022.153576

Zhang A, Bian R, Pan G, Cui L, Hussain Q, Li L, Zheng J, Zheng J, Zhang X, Han X (2012) Effects of biochar amendment on soil quality, crop yield and greenhouse gas emission in a Chinese rice paddy: a field study of 2 consecutive rice growing cycles. Field Crop Res 127:153–160

Zhang Q, Wang Y, Wu Y, Wang X, Du Z, Liu X, Song J (2013) Effects of biochar amendment on soil thermal conductivity, reflectance, and temperature. Soil Sci Soc Am J 77(5):1478

Zhang Y, Chen XP, Ma WQ, Cui ZL (2017) Elucidating variations in nitrogen requirement according to yield, variety and cropping system for Chinese rice production. Pedosphere 27: 358–363

Zheng R, Sun G, Li C, Reid BJ, Xie Z, Zhang B (2017) Mitigating cadmium accumulation in greenhouse lettuce production using biochar. Environ Sci Pollut Control Ser 24:6532–6542

Zhou Y, Gao B, Zimmerman AR, Chen H, Zhang M, Cao X (2014) Biochar-supported zerovalent iron for removal of various contaminants from aqueous solutions. Bioresour Technol 152:538–542. https://doi.org/10.1016/j.biortech.2013.11.021

Part II
Food Biodiversity: Innovations in Safe, Sustainable, Nutrient-Rich Foods

Chapter 12
Food Biodiversity: BIA in Community Nutrition—Assessing Impact of Nutrient-Rich, Sustainable Foods

Manya Gupta, Karuna Singh, Mayuri Rastogi, and Aditi Rikhari

Abstract Bioelectrical impedance analysis (BIA) offers a non-invasive approach to assess nutritional status by measuring tissues' opposition to electrical current flow. This technique evaluates body composition as: fat mass, fat-free mass, and total body water, making it a key tool for identifying malnutrition risks, which affect 20–62% of hospitalized patients. Established screening methods like NRS-2002 from ESPEN and SGA provide initial risk detection, but BIA delivers precise data on fluid distribution and bioelectrical properties such as resistance and reactance. This investigation highlights BIA's effectiveness in generating detailed insights into nutritional health. By analyzing these parameters, the study aims to create a robust predictive model for early identification of deficiencies at the community level. Such a framework supports timely interventions and tailored dietary strategies, enhancing patient outcomes through proactive care.

Keywords Bioelectrical impedance analysis · Body composition · Phase angle · Fat-free mass · Singular frequencies BIA

12.1 Introduction

A non-invasive technique for determining nutritional status and the composition of the body is called impedance bioelectrical analysis, or BIA (Dias et al. 2024). It offers significant data on the body's water, fat, or lean mass balance by measuring the electrical current's resistance as it travels through various bodily tissues. Because it is straightforward, inexpensive, and simple to use, BIA has become a more often used prediction method in nutrition state analysis (Campa et al. 2022).

Even though "impedance bioelectrical analysis" has a sophisticated tone, BIA instruments are made of straightforward technologies. These assess how quickly a

M. Gupta · K. Singh · M. Rastogi (✉) · A. Rikhari
Department of Nutrition and Dietetics, Sharda University, Greater Noida, Uttar Pradesh, India
e-mail: mayuri.rastogi@sharda.ac.in

M. Thakur, T. Belwal (eds.), *Food Biodiversity and Consumer Engagements*, Sustainable Development and Biodiversity 46,
https://doi.org/10.1007/978-981-95-7777-4_12

small amount harmless current of electricity circulates throughout your entire body (Abasi et al. 2022).

Globally, malnutrition is a serious issue for public health. According to surveys, 40–90% of older persons in community and health care settings are malnourished, and up to 50% of seniors admitted to hospitals arrive malnourished (Isautier et al. 2019). It starts by outlining the fundamentals of bioelectrical impedance analysis (BIA) and how it measures the composition of the body, particularly fat content, muscle mass, and level of hydration, by using the electrical characteristics of tissues (Ward 2019). The authors next discuss the limits of conventional evaluation methods and the significance of evaluating nutritional status in both clinical and community settings.

Impedance bioelectric measurement (BIA) is a precise, non-invasive, and cost-effective method for measuring body water compartments in both healthy and pathologically diseased individuals (Moonen and Van Zanten 2021). These estimates will be used to obtain important information about body composition and nutritional status. Since a 15% reduction in TBW could be a serious health risk, TBW measurement is critical to health status assessment. Furthermore, TBW is useful for accurately monitoring the effects of medications (Zheng et al. 2022).

The incidence as degree of nutritional risk is influenced by the patient population and the evaluation procedures used to estimate nutritional risk. In rich nations, this hazard of nutrition varies between 20% and 62% of hospitalized patients (Kyle et al. 2013). These are nutritional screening methods and nutritional assessments, respectively. Albumin is a marker of nutritional risk. Several screening tools have been developed to assess nutritional risk, such as the Healthy Risk Testing Tool 2002 (NRS-2002) developed by the European Society for Enteral and Parenteral Nutrition (ESPEN) and the Subjective Global Evaluation (SGA) questionnaire.

Impedance bioelectrical analysis (BIA) is a valuable method for body composition study that allows three important body compartments to be evaluated: fat mass, fat-free mass, and water. A critical review has been carried out to examine the theories, practises, kinds, limitations, and main applications of this approach.

12.2 History Context

The eighteenth century is the first time to invent the machine to calculate the overall body weight and fat, they are in the greater scope of frequency on a high range of tissue, and it was completely developed into the twentieth century, was to the calculation of all body fat, water, muscle, and tissues.

Tissue electrical properties have been characterized since 1872. These features are then extended to a wider range of frequency upon a more diverse set of cells, particularly injured or changing tissues post-mortem (Dylke and Ward 2021). Used the impedance of electricity to measure a total body water (TBW) indication to conduct the initial research. In the end of the 1970s, the principles for impedance bioelectrical analysis (BIA) had been well established, including the relationship between impedance and the body's water content (Krishnan 2021). Numerous multi-frequency BIA analyzers were also available on the market following the 1990s release of single-frequency BIA analyzers. The results are generated fast and consistently, the procedure is simple and non-invasive, and the equipment is safe and lightweight (Chan 2023).

It is reasonable to assess the state of bioimpedance electrical research in 2021. Philipson's "Les lois de la résistance électrique des tissus vivants" (which is the laws for the electrical resistance that govern living tissues) published in 1921 marked 100th anniversary in 2021, indicating, it is impossible to pinpoint a precise beginning time for research in this field (Ward 2021).

12.3 Bioelectrical Impedance Analysis's Operating Principles

The process in nutrition evaluation uses the body composition to determine the total body fat. There are several methods for determining the composition of the body. Many established techniques have been used to quantify the mass of fat (FM) or free mass (FFM), the two primary body sections, with accuracy (Bonilla et al. 2022). Within the fatty tissue area, there are tiny quantities of necessary lipids or glucose, within or external fluid, physical and non-physical minerals, and all body amino acids, assuming a stable composition (Lyons-Reid et al. 2020). The measurement of mass free of fat, mass of fat, cells in the body volume, total water in the body, internal fluid, or external fluid may all be done with great consistency when using BIA frequently (Joshi and Bagal 2014). Several compositions of the body models are separated into many parts in the models (Fig. 12.1). The sections have distinct methodologies. According to the conventional understanding of both components of body composition, the body is split into two divisions (2-C). The remainder of cells are collected or referred to as mass free of fat (FFM), with body fat making up a single portion of both sections (Fedewa et al. 2021). Apart from ascertaining the mass devoid of fat, the three-component model of the body's composition (3-C) isolates the residual solids, which mostly consist of minerals and proteins, from the liquid component (water) (Bipin et al. 2023). The four components of body composition (4-C), like the mitochondrial method, which is more comprehensive than the previous models, divide the mass that is free of fat into the three fundamental or physiological compartments: the mass of cells in the body (BCM), fluid, or extra cell fluid (ECF), and outside the cell solids (ECS). Thus, mass free of fat (FFM) = BCM + ECF + ECS, and total fat mass = FFM—weight of the body (mass) (Marroni et al. 2017).

Then, the mass of cells in the body (BCM) and the mass of fat (FM) may be measured since the mass free of fat (FFM) is a constant component of a person's body mass (Subhedar et al. 2014). When they make contact with skin, they enable the measurement of reactance (Xc) and resistance (R) (Francisco et al. 2020). The

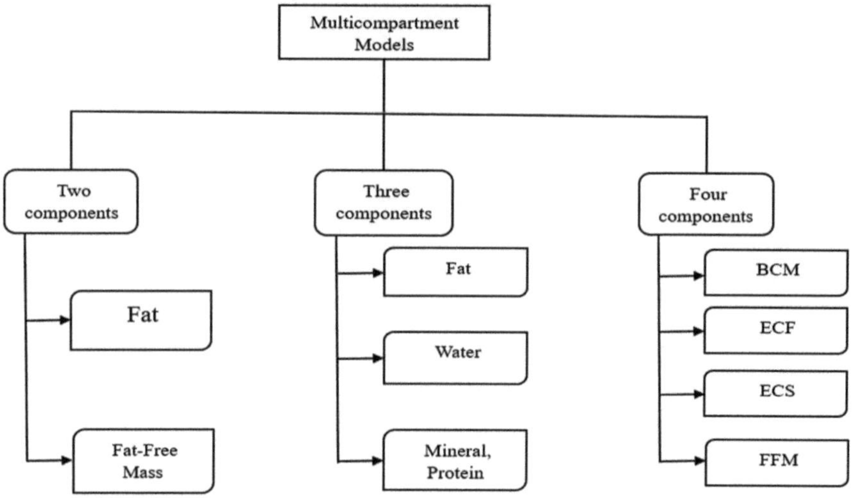

Fig. 12.1 A three-model multicompartment model consists of two basic components that evaluate the mass body components that are composed of fat and fat-free material. Another one has three parts that the body uses in turn, depending on what it needs for protein, fat, fluid, and minerals. The information found in the mass of body cells (BCM), extra cell fluid (ECF), external solids (ECS), as well as mass free of fat (FFM), is evaluated in the finished model

quantity of water in the external cell water (ECW) and internal cell water (ICW) phases is then calculated, and the R and Xc values are used to approximate the total amount of water in the body (TBW). The results of these computations also include the impedance (Z) and phase angle (φ) (Davydov et al. 2021).

The process of passing a low-amplitude, lower and high-frequency, painful current of electricity. Resistivity (R) corresponds to the width (L) or inversely corresponding on the area of the cross section (A) in a length of homogeneous, evenly conductive material (Fig. 12.2). Even though the human being was not a perfect cylinder or the conductivity of it varies, the impedance ratio (Length2/R) or the amount of fluid, that comprises chemicals which carry electricity throughout the body's tissues, may be connected experimentally (de Miranda et al. 2024). In practical terms, determining the conductivity length, which typically ranges as hand to ankle, is more challenging than determining height. Therefore, there is an empirical association between height2/R and lean body mass, which typically includes 73% water. Because of the body's intrinsic field inhomogeneity, the term height2/R denotes an analogous cylinder that needs to be matched to the actual geometries by a suitable value. One of the several factors influencing this number is the architecture of the segments under study (Sørensen et al. 2020).

When BIA is evaluated, the phase angle between the direct measurements it provides and the death rates in various conditions manifests itself as a predictive indication of its applicability. In a later part of this chapter, liver problems are covered. The phase angle (PA) is the proportion of intra- to outside fluid and can be influenced by nutritional and hydration levels (Iqbal 2013). It is normally between

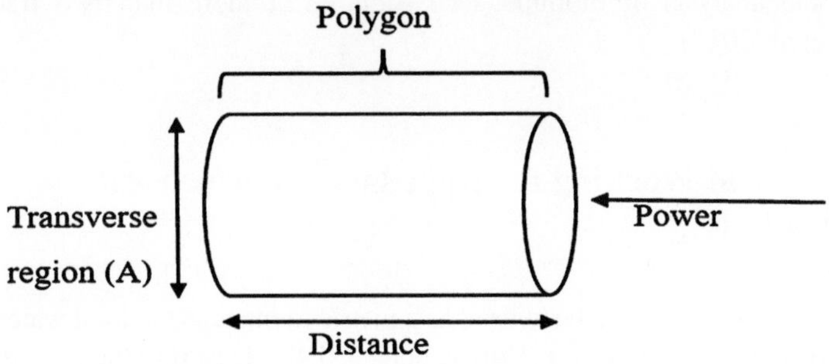

Fig. 12.2 An explanation of the relationship among geometries and efficiency using a cylinder model. Resistance in a homogeneous conductor length with an equal area of cross-section is proportional to the long (L) and inversely relates to its lengthwise area (A)

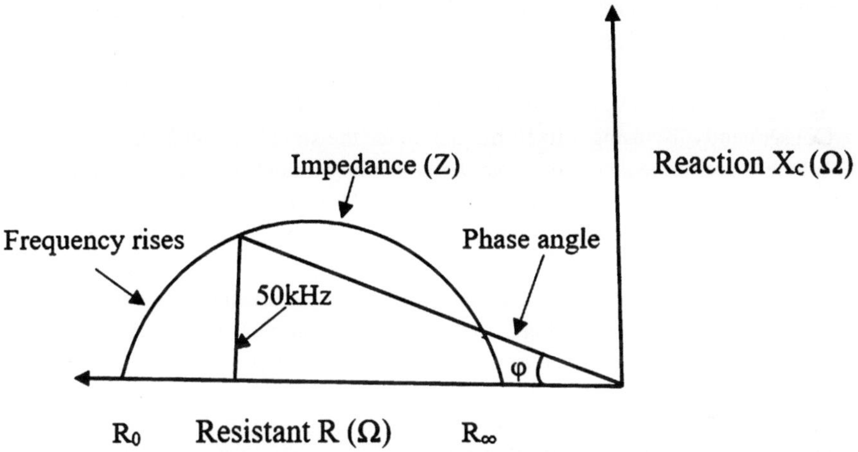

Fig. 12.3 Illustrating the visual derivation of the phase angle and its relationship to the voltage of the applied current, impedance (Z), resistivity (R), and response (Xc)

6° and 7° in healthy subjects and up to 8.5° in competitors. A small PA (<5°) means that the integrity of cells has been lost (Fig. 12.3). In comparison to impedance, which PA appears to offer an additional accurate indication of dietary habits and is strongly correlated with cell honesty (Marra et al. 2019). That was the total body electrical current circulating on only resistance (great broadcast amplitude), which has a φ of 0°, and pure response (cell membranes devoid of fluid), which has a φ of 90° (Vujicic et al. 2019). Single-frequency bioimpedance measurements (SF-BIA) is one of the most common bioimpedance analysis techniques. It analyses bioimpedance data determined that 50 KHz electric current, and the electrodes are placed in the hands as well as feet (Ward and Brantlov 2023). The results of SF-BIA can be used to estimate mass free of fat and body water content, but not intracellular water variations. The term "multiple-frequency bioimpedance analysis" (MF-BIA) refers to the analysis of bioimpedance acquired at more than two frequencies (Marroni et al. 2017).

12.4 Physical Working for Impedance Bioelectrical Analysis

The body's composition has been used as a metric to manage the total water content of the body and mass free of fat (Vujicic et al. 2019). Measuring the different blood types that travel through the organ is crucial. Clinicians are used to employing impedance techniques to measure variations in the heart ventricular volume (Joshi and Bagal 2014). Since its introduction over 30 years ago, BIA for whole body impedance assessment of the total water in the body (TBW) has become a widely used method (Fig. 12.4) (Marroni et al. 2017). The connection mentioned previously could be repeated by L/L since any fraction stays constant when multiplied by an amount that equals one. It is possible to solve for V by rearranging the equation since A * L equals volume (V, cm^3) (de Miranda et al. 2024).

Consequently, by using a fixed frequency for the signal and analyzing a conductor's length and impedance, one may solve for an unidentified factor, or area. The theory behind calculating the "electricity amount" for a body space, such as the FFM or TBW, was that the electrical volumes of various biological tissues vary (Joshi and Bagal 2014).

The BIA has been shown to be a useful tool when evaluating body compartments in a range of therapeutic settings, such as malnourishment, trauma, pre- or post-operative treatment, compensatory diseases of the liver, dialysis for the kidneys, and cancers. The phase angle (PA) is the impedance bioelectrical characteristic that was clinically established in light of these fluctuations (Lee et al. 2024). The phase angle is calculated using the formula that takes resistance (R) and reactance (Xc) into consideration (de Miranda et al. 2024). Although this assumption is somewhat incorrect at first, the geometrical form of a human being more closely resembles an array with five tubes: a trunk, two arms, and two legs (Fig. 12.5), except for the head. Since R is inversely correlated with the area of the cross-section, the highest or bottom areas, which include the lowest cross-sectional area, are going to have the biggest influence on whole-body R values. About 50% of body mass is made up by the trunk, which contributes 5–12% to the overall body weight (Jin et al. 2019).

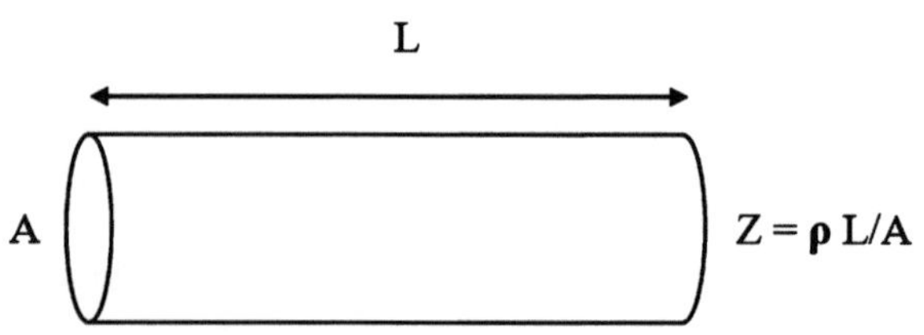

Fig. 12.4 The impedance (Z) of a geometrical conductor is related to its length (L), area of cross-section (A), and specific resistance (ρ)

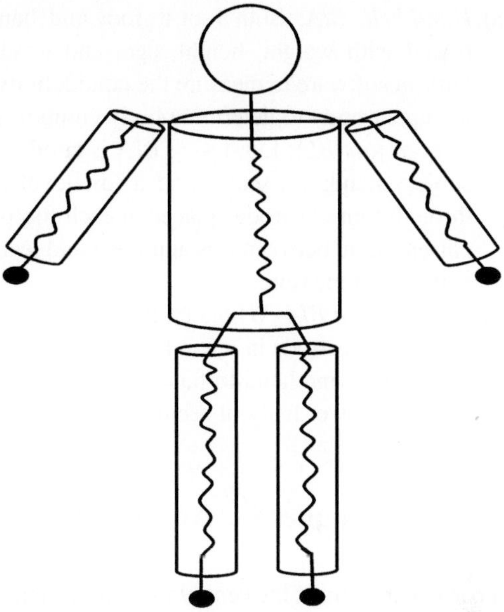

Fig. 12.5 A more accurate depiction of the body would consist of five electrically linked tubes: a trunk, two arms, and two legs

Using cables that are fastened to electrodes or to conductive surfaces that come into contact with skin, low-amplitude, lower or high-energy, painful electrical power is applied throughout the life cycle in the bimpedance bioelectric analysis (BIA). The person being treated maintains the dorsal decubitus position with their hands and legs remaining parallel to their bodies. One electrode is located at the joint of the wrist on the right side, while the other is situated at the center of the fingers level to the back of the hand. A second set of wires is placed on the right side in the middle-toe level of the medial feet and the ankle joint (Marroni et al. 2017).

12.5 Types of Bioelectrical Impedance Analysis (BIA)

BIA devices come in a variety of variety, but they always require two contact sites. Your two hands are the two points of a portable device (called hand-hand BIA). The two locations for contact on the typical BIA scale are represented by your two feet (called foot-foot bioimpedance measurement). BIA has two distinct points of touch (also called a hand-to- foot BIA).

BIA, beginning with the previously stated single-frequency BIA (SF-BIA). This process uses a 50 kHz flow that is transferred between the hand and foot surface electrodes (Marroni et al. 2017). Certain BIA devices employ other locations, such foot–foot, hand-to-hand, and hand-to- foot electrodes. These are the three mains of electrodes that help to collect the data of human body.

(a) *Hand-held BIA*: Both foot-to-foot and hand-to-hand technologies, once calibrated with weight, height, age, and gender, hand-to-hand BIA devices use built-in software to measure the conductivity of a little alternating current across the upper body to determine body composition.
(b) *Foot-to-foot BIA*: Leg-to-leg BIA is similar to hand-hand procedures in that the subject stands on scales and a low-level current is sent to their lower body through four electrodes placed at each footplate. The difference in the electrical current route between this approach and technique may have an impact on body component measure.
(c) *Hand-to-foot BIA*: Whole body measure is obtained using hand-to-foot BIA by use of an electrode in a fixed foot-plate and electrodes in hand grips. Given that foot-to-foot and hand-to-hand approaches might not take into consideration the upper, or lower body, it seems sense to believe that hand-to-foot measures.

12.6 Techniques for Analyzing Bioelectrical Impedance

In particular, cells of the central nervous system favor a greater frequency of electric current, while muscle and fat respond to a lower voltage (Fig. 12.6). This suggests that BIA uses different frequencies to function. The methods used in BIA are described below (Aldobali and Pal 2021).

(a) *Singular frequencies BIA (SF-BIA)*: This type of BIA measures impedance using a singular frequency (usually 50 kHz) of electrical current. It is frequently utilized in portable devices and home scales for fundamental body composition analysis (Orsso et al. 2022). SF-BIA sensors were utilized to measure FFM as well as TBW for secure, hydrated persons by applying derivative parameters independently; generally speaking, SF-BIA is imperfect for physical wellness without safely maintained health and safely maintained hydrated level (Aldobali and Pal 2021).

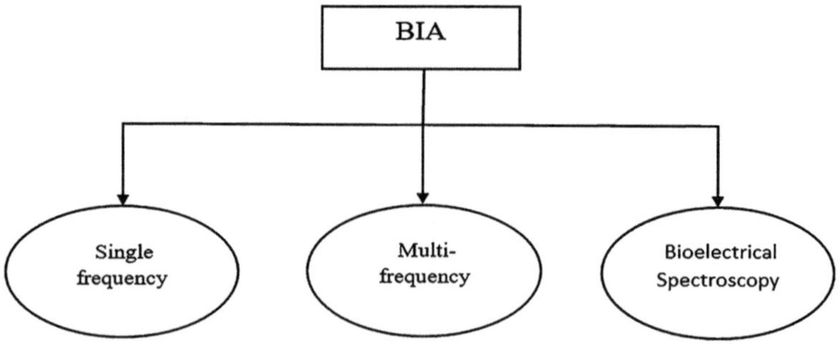

Fig. 12.6 Impedance bioelectrical analysis (BIA) may be performed in a variety of way to split frequency based on how the tissue responds to certain frequency

(b) *Multiple frequencies BIA (MF-BIA)*: MF-BIA uses experimental regression models based on linearity, just as SF-BIA, but it additionally accounts for impedances at different frequencies. At different frequencies, MF-BIA assesses FFM, TBW, ICW, and ECW (0, 1, 5, 50, 100, 200, and 500 kHz) (de la Cruz Marcos et al. 2021). For frequencies exceeding 200 kHz and below 5 kHz, limited reproduction capability was observed, particularly for response times at small frequencies (Aldobali and Pal 2021). Resistance measurement is done throughout a variety of frequencies ranges, being that ECW is detected. Total body water is higher frequency current flows through cell membranes and both ICW and ECW at low frequency current are blocked by them (Mylott et al. 2014).

(c) *Bioelectrical spectroscopy (BIA)*: Bioelectrical spectroscopy is an advanced form of BIA that utilizes a wide range of frequencies and measures impedance at multiple points across the body. In contrast to MF-BIA, BIS first creates prediction equations based on empirical data, then uses mathematical modeling and mixed equations to anticipate R0 and R∞ or establish connections among R as well as body fluid chambers (Silva et al. 2023). The Cole model is used in the process referred to as bioelectrical impedance spectroscopy (BIS). This method also measures body impedance throughout a frequency range (Mylott et al. 2014).

12.7 Sections of the Body

(a) *Composition of the body (BC)*: The human body is composed of several fluid and tissue compartments, as shown in the body composition graph, like mass of fat, FFM, TBW, ICW, ECW, BCM, abdominal protein molecules, or mineral content of bones (Fig. 12.7) (Subhedar et al. 2014). Body composition (BC) is a study of the human body's fluid, mass, and clinical tissue measurements

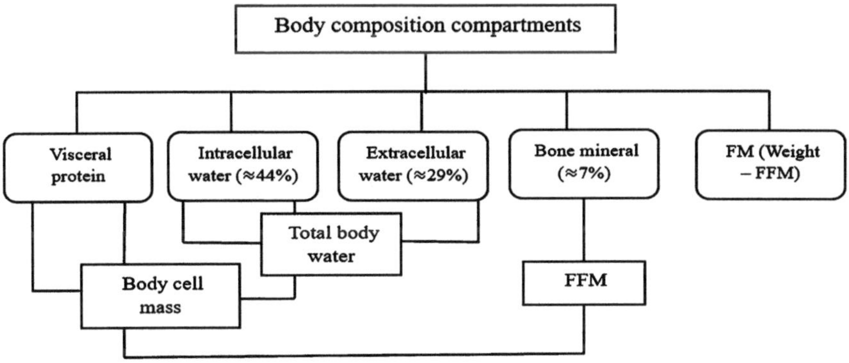

Fig. 12.7 Figure displaying the total body water content (TBW), mass free of fat (FFM), extracellular water content (ECW), water within cells (ICW), and mass of body cells (BCM)

(Aldobali and Pal 2021). Found that MF-BIA and SF, the two BIA methods currently in use, are not reliable enough to assess TBW when hydrated variations are present.

It has not been demonstrated that equations created of hydrated populations are applicable to those altered hydration (Jethe et al. 2019).

Everything that is not body fat is included in the area known as mass free of fat, which also contains visceral protein, extracellular fluid ($\approx$29%), intracellular fluid ($\approx$44%), and bone mineral content ($\approx$7%). The compartment known as the total water in the body (TBW) is composed of extracellular fluid (ECW) and intracellular fluid (ICW) (Marroni et al. 2017).

(b) *Mass free of fat (FFM)*: Everything that is not body fat is called FFM. In the literature, many BIA formulae forecast FFM. The numerous regression solutions' variables and the people to whom they apply differ across these equations (Jethe et al. 2019). The whole quantity for non-fat (lean) muscle mass is known as fat-free mass (FFM), also often called lean body mass (LBM). It is made up of around 73% fluid, 20% protein, 6% nutrients, or 1% centigrade (Aldobali and Pal 2021).

(c) *Body cell mass (BCM)*: Measurements to "cells substance," "BCM," and "physiologically energetic muscle," and "ICW," are controversial in terms of their physiologic value; in contrast, FFM includes anything that does not constitute fat in the body (Jethe et al. 2019). Therefore, BCM is made up of components of "living" fat cells rather than processed fat lipids. Living cells in BCM also contain water. Intracellular water is the term for this water (ICW) (Aldobali and Pal 2021).

12.8 Bioelectrical Impedance Analysis (BIA) Application

As this chapter's introduction noted, there are a number of BIA instrument types available today that differ in terms of cost, electrode appearance, and measuring style. A variety of uses for devices can be satisfied, such as research, use in specialized hospital departments such as the kidney unit, transplantation of the liver, and cancer, to name a few, as well as utilization in medical centers, healthcare facilities, medical wellness centers, fitness centers, and clients' homes (Naranjo-Hernández et al. 2019). As such, it is essential for understanding this item and its intended application. This is the body composition measure in BIA, based on weight measurements and changes in body composition. Detecting the illness and excess body fat percentage was beneficial. The observed tissue content was associated with certain situations and frequencies (Peart et al. 2019).

(a) *SF-BIA Nutritional assessment to measurement of BC in weight-stable subjects*: The nutritional evaluation of body composition for weight-stable human subjects using single-frequency bioimpedance. This highly commercialized approach is reasonably priced, accessible, and offered by medical facilities, clinics, physicians' places of work, health clubs, and training centers (Naranjo-Hernández et al. 2019). Due to the singular 50 kHz frequencies of this instrument, regression models may be used to calculate TBW and FFM. The empirical connection produced in the equation, as they tested TBW for healthy people, was verified by the original research. With an average frequency of 100 kHz and a 100 μA current used with an overall tetrapolar resistance approach, H^2/Z was the most accurate predictor of TBW, accounting for 84% of the overall variability (R^2) (Peart et al. 2019). This instrument's single frequency might make it difficult to understand the results of tests. Since height (cm) was easier to get than route length, it was patient as the employed changed their body composition.

(b) *MF-BIA monitoring of ill patients and measure of changes in BC*: Monitoring sick patients' multi-frequency bioimpedance and measuring variations in their body composition. As this approach uses currents with frequencies spanning from 5 to 100 kHz, it provides additional resources for assessment, such as the determination of ICW (Naranjo-Hernández et al. 2019). Because lower frequencies are membrane-permeable, the measurement is possible. Because different frequencies resulted in a more exact for determining total body weight (TBW). All of the studies that have been analyzed thus far have demonstrated the effectiveness of using BIA in people who have weight stability. The application of this technique to evaluate changes in the makeup of the body has been investigated by a number of researchers. BIA studies examining changes in the composition of the body that are acute (within a day), middle (1–2 days), and ongoing (monthly) have been classified (Peart et al. 2019).

(c) *Application of BIA in community nutrition assessment framework*: Accuracy, accessibility, and cost are important things to think about when selecting an instrument to measure body composition in a community framework. To get the most accurate and reliable results, it's important to choose the right method to detect tissue, diseases, or overall health among participants. BIA has proven to be an excellent method for identifying the future health challenges by assessing the participants' physiological parameters. This is especially have a significance with people who are malnourished, women at menopausal age, or populations at risk in diverse community settings. BIA strengthens evidence-based public health initiatives and focused nutritional strategies by supporting extensive screening, tracking nutritional transitions, and evaluating community-based nutrition interventions (Ng et al. 2023).

12.9 Conclusion

Bioelectrical impedance analysis (BIA) is used in body composition analysis because it offers a rapid and non-invasive evaluation of several body components, including water content, fat mass, and muscle mass. Because various tissues (such as muscle and fat) have differing electrical conductivities, BIA measures the resistance to electrical passage across the body. It helps measure body fat, it changes to a diet to the action. This method is often used in fitness and healthcare to evaluate changes in body composition, monitor progress, and tailor fitness or nutrition plans.

References

Abasi S, Aggas JR, Garayar-Leyva GG, Walther BK, Guiseppi-Elie A (2022) Bioelectrical impedance spectroscopy for monitoring mammalian cells and tissues under different frequency domains: a review. ACS Meas Sci Au 2:495–516

Aldobali M, Pal K (2021) Bioelectrical impedance analysis for evaluation of body composition: a review. Int Congr Adv Technol Eng 1(10):1–10

Bipin C, Rao CV, Sridevi PV (2023) Speckle aware spatial search based segmentation algorithm for crop classification in SAR images using a three component K-NN model. J Appl Remote Sens 17(4):048503–048503

Bonilla DA, De León LG, Alexander-Cortez P, Odriozola-Martínez A, Herrera- Amante CA, Vargas-Molina S, Petro JL (2022) Simple anthropometry-based calculations to monitor body composition in athletes: scoping review and reference values. Nutr Health 28(1):95–109

Campa F, Gobbo LA, Stagi S, Cyrino LT, Toselli S, Marini E, Coratella G (2022) Bioelectrical impedance analysis versus reference methods in the assessment of body composition in athletes. Eur J Appl Physiol 122:561–589

Chan AY (2023) Biomedical device technology: principles and design. Charles C Thomas Publisher

Davydov DM, Boev A, Gorbunov S (2021) Making the choice between bioelectrical impedance measures for body hydration status assessment. Sci Rep 11(1):7685

De la Cruz Marcos S, Redondo del Río MP, de Mateo SB (2021) Applications of bioelectrical impedance vector analysis (BIVA) in the study of body composition in athletes. Appl Sci 11(21):97801

de Miranda AC, de Oliveira Coelho GM, de Oliveira Cattem MV, Koury JC (2024) Fat-free mass predictive equation using bioelectrical impedance and maturity offset in adolescent athletes: development and cross-validation. Nutrition 123:112415

Dias CV, Dias JC, Laranjo C, Cardoso P, De Sousa-Coelho AL (2024) On the run—comparing bioimpedance analysis (BIA) using portable devices. Proceedings 99:11

Dylke ES, Ward LC (2021) Three decades of bioelectrical impedance spectroscopy in lymphedema assessment: an historical perspective. Lymphat Res Biol 19(3):206–214

Fedewa MV, Nickerson BS, Tinsley GM, Esco MR, Dunbar EG, Boucher AG, DeLeon RM (2021) Examining race-related error in two-compartment models of body composition assessment: a systematic review and meta- analysis. J Clin Densitom 24(1):156–168

Francisco R, Matias CN, Santos DA, Campa F, Minderico CS, Rocha P, Heymsfield SB, Lukaski H, Sardinha LB, Silva AM (2020) The predictive role of raw bioelectrical impedance parameters in water compartments and fluid distribution assessed by dilution techniques in athletes. Int J Environ Res Public Health 17(3):759

Iqbal SR (2013) Physics of bio-electrical impedance analysis: phase angle and its application. Adv Life Sci Technol 9:4–12

Isautier JM, Bosnić M, Yeung SS, Trappenburg MC, Meskers CG, Whittaker AC, Maier AB (2019) Validity of nutritional screening tools for community- dwelling older adults: a systematic review and meta-analysis. J Am Med Dir Assoc 20:1351–1e13

Jethe JV, Deshpande AK, Ananthakrishnan T, Jindal G (2019) Bioelectrical impedance analysis and its clinical application. MGM J Med Sci 6:42–47

Jin M, Du H, Zhang Y, Zhu H, Xu K, Yuan X, Pan H, Shan G (2019) Characteristics and reference values of fat mass index and fat free mass index by bioelectrical impedance analysis in an adult population. Clin Nutr 38(5):2325–2332

Joshi MSR, Bagal PUR (2014) Development of bioelectrical impedance analyzer for body composition analysis. IOSR-JEEE 9(5):53–62

Krishnan S (2021) Biomedical signal analysis for connected healthcare. Academic

Kyle UG, Genton L, Pichard C (2013) Low phase angle determined by bioelectrical impedance analysis is associated with malnutrition and nutritional risk at hospital admission. Clinical nutrition, 32(2), 294–299

Lee SW, Daud ZAM, Lim JH, Ibrahim I, Chan YM, Zakaria NF (2024) Bioelectrical impedance analysis-derived phase angle as a pragmatic screening tool for assessing sarcopenia in multiethnic peritoneal dialysis patients. Clin Nutr Open Sci 57:26–38

Lyons-Reid J, Ward LC, Kenealy T, Cutfield W (2020) Bioelectrical impedance analysis—an easy tool for quantifying body composition in infancy. Nutrients 124:920

Marra M, Sammarco R, De Lorenzo A, Iellamo F, Siervo M, Pietrobelli A, Donini LM, Santarpia L, Cataldi M, Pasanisi F, Contaldo F (2019) Assessment of body composition in health and disease using bioelectrical impedance analysis (BIA) and dual energy X-ray absorptiometry (DXA): a critical overview. Contrast Media Mol Imaging. https://doi.org/10.1155/2019/3548284

Marroni CA, Miranda D, Boemeke L, Fernandes SA (2017) Phase angle bioelectrical impedance analysis (BIA) as a biomarker tool for liver disease. In: Biomarkers in liver disease (biomarkers in disease: methods, discoveries and applications). Springer Science, Berlin, pp 735–751

Moonen HPFX, Van Zanten ARH (2021) Bioelectric impedance analysis for body composition measurement and other potential clinical applications in critical illness. Curr Opin Crit Care 27(4):344–353

Mylott E, Kutschera E, Widenhorn R (2014) Bioelectrical impedance analysis as a laboratory activity: at the interface of physics and the body. Am J Phys 82(5):521–528

Naranjo-Hernández D, Reina-Tosina J, Min M (2019) Fundamentals, recent advances, and future challenges in bioimpedance devices for healthcare applications. J Sens. https://doi.org/10.1155/2019/9210258

Ng JKC, Lau SLF, Chan GCK, Tian N, Li PKT (2023) Nutritional assessments by bioimpedance technique in dialysis patients. Nutrients 16(1):15

Orsso CE, Gonzalez MC, Maisch MJ, Haqq AM, Prado CM (2022) Using bioelectrical impedance analysis in children and adolescents: pressing issues. Eur J Clin Nutr 76(5):659–665

Peart DJ, Balsalobre-Fernández C, Shaw MP (2019) Use of mobile applications to collect data in sport, health, and exercise science: a narrative review. J Strength Cond Res 33(4):1167–1177

Silva AM, Campa F, Stagi S, Gobbo LA, Buffa R, Toselli S, Silva DAS, Gonçalves EM, Langer RD, Guerra-Júnior G, Machado DR (2023) The bioelectrical impedance analysis (BIA) international database: aims, scope, and call for data. Eur J Clin Nutr 77(12):1143–1150

Sørensen TI, Frederiksen P, Heitmann BL (2020) Levels and changes in body mass index decomposed into fat and fat-free mass index: relation to long-term all-cause mortality in the general population. Int J Obes 44(10):2092–2100

Subhedar R, Subhedar V, Dave P, Mishra P, Kaur A (2014) Bioelectrical impedance technology for evaluating human body composition parameters: "an advanced diagnostic technology for body composition and physical fitness analysis". Int J Physiother Res 2(6):824–830

Vujicic B, Dorcic G, Babic V, Rundic A, Devcic B, Simac DV, Zaputovic L, Racki S (2019) Comparison of clinical assessment and multifrequency bioimpedance analysis as methods of estimating volume status in peritoneal dialysis patients–a single-center experience. Clin Nephrol 91(6):334

Ward LC (2019) Bioelectrical impedance analysis for body composition assessment: reflections on accuracy, clinical utility, and standardization. Eur J Clin Nutr 73:194–199

Ward LC (2021) Electrical bioimpedance: from the past to the future. J Electr Bioimpedance 12(1): 1–2. https://doi.org/10.2478/joeb-2021-0001. PMID: 34413916; PMCID: PMC8336311

Ward LC, Brantlov S (2023) Bioimpedance basics and phase angle fundamentals. Rev Endocr Metab Disord 24(3):381–391

Zheng K, Lu J, Liu X, Ji W, Liu P, Cui J, Li W (2022) The clinical application value of the extracellular-water-to-total-body-water ratio obtained by bioelectrical impedance analysis in people with advanced cancer. Nutrition 96:111567

Chapter 13
Health-Promoting, Gluten-Free Breakfast Cereals from Common Buckwheat: A Nutrient-Dense Innovation and Consumer Acceptability Study

Aneesha Nalin Kumar and Pratibha Singh

Abstract The growing hunger demands, water stress among various countries, challenging environmental disasters, and reliance only on few crops for consumption and use in food industries are leading to food insecurity. This study evaluates the acceptability of products made from Buckwheat as a Breakfast cereal, such as rolled flakes and honey coated muesli with nuts. Changing climatic conditions impose the need for more diverse crop utilization for sustainability. Buckwheat is one of the sustainable options to tackle global hunger and food insecurity.

Keywords Breakfast cereals · Buckwheat · Sustainable foods

13.1 Introduction

Breakfast cereals are usually made from grains like Wheat, Corn, and Oats, are popular, quick to prepare, and available for consumption easily. Buckwheat, having a varied nutrient profile and health benefits, is underutilized in the food industry, particularly in ready to eat (RTE) cereal production.

13.1.1 Buckwheat as a Sustainable Crop

The most challenging aspect of global food insecurity is the fastest growing world's population, resulting from dependency on only on few grains. Buckwheat is an ancient pseudocereal that can be cultivated in conditions like extreme cold

A. N. Kumar · P. Singh (✉)
Department of Nutrition and Dietetics, Manav Rachna International Institute of Research and Studies (Deemed to be University), Faridabad, Haryana, India
e-mail: aneesha0316@gmail.com; pratibha.fas@mriu.edu.in

M. Thakur, T. Belwal (eds.), *Food Biodiversity and Consumer Engagements*, Sustainable Development and Biodiversity 46,
https://doi.org/10.1007/978-981-95-7777-4_13

conditions, scarcity of water, dry soil, and in different seasons, which makes it a sustainable crop (Luitel et al. 2017; Thakur and Tiwari 2019; Nalinkumar and Singh 2020; Pirzadah and Ul Rehman 2021; Rana et al. 2012; Zhang et al. 2023; Zhao et al. 2023).

In times of a shortage of food supply, it can be used as an emergency food source due to its ability to grow rapidly and mature early. It has a strong potential to fight poverty, hunger, and malnutrition. Because of its numerous health benefits, it has the potential to be a superfood. Owing to its rich nutritional profile, it has been identified by the Food and Agriculture Organization (FAO) that it can have a significant contribution to food security, improving nutrition and health (Pirzadah and Malik 2020; Thakur and Belwal 2023; Thakur 2024).

13.1.2 Buckwheat's Role in Food Industry

Buckwheat is rich in essential nutrients that promote Health but, yet it is underutilized even in food industries, especially in the production of RTE Breakfast cereals (Farooq et al. 2016; Babu et al. 2018; Bekkering and Tian 2019). There is an increasing demand for RTE foods as well as gluten-free Cereal products. This study aims to evaluate the overall acceptability of Common Buckwheat based RTE Breakfast cereals.

13.1.3 Buckwheat-Based RTE Breakfast Cereals

Rolled flakes and Honey-coated Muesli with nuts were developed from Buckwheat. The main objective of the study was to study its acceptability as a Breakfast cereal.

13.1.4 Buckwheat Rolled Flakes

Local variety (Kuttu) of Common Buckwheat (*Fagopyrum Esculentum* Moench) was procured from the local market of Haryana, India. It was cleaned and Buckwheat rolled flakes were prepared using a roller flaking machine. Figure 13.1 shows the process of preparation of Buckwheat rolled flakes. It was served with hot milk. It was compared with commercially available Cornflakes as a control for overall acceptability.

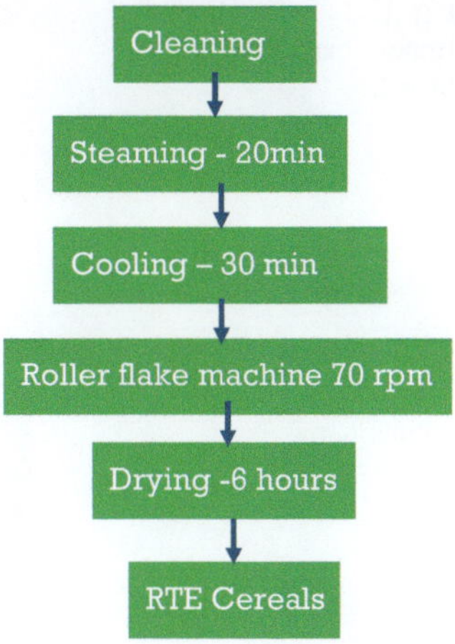

Fig. 13.1 Process of preparing rolled Buckwheat flakes

Table 13.1 Proportion of ingredients used for the preparation of Honey-coated Buckwheat Muesli

Ingredients	Quantity for 100 grams
Buckwheat flakes	70 grams
Almonds	5 grams
Raisins	7 grams
Flax seed	3 grams
Honey	20 grams
Sugar	5 grams

13.1.5 Honey-Coated Buckwheat Muesli with Nuts

Honey-coated Buckwheat Muesli was prepared by using Buckwheat flakes, roasted Flax seeds, Honey, Almonds, and Raisins. Table 13.1 shows the proportion of ingredients used for the preparation of Honey-coated Buckwheat Muesli. Figure 13.2 shows the preparation of Buckwheat rolled flakes. It can be served with hot milk. It was compared with commercially available Honey-coated Muesli with nuts as a control for overall acceptability.

13.1.6 Sensory Evaluation of Developed Products

Overall acceptability of products was evaluated by 30 semi-trained panellists using a 9-point Hedonic scale with a score of 5 being a limit of acceptability. Figure 13.3 shows the scores of overall acceptability of the developed products along with their control. Table 13.2 shows the mean scores of overall acceptability of the developed products.*significant at 5% level of significance

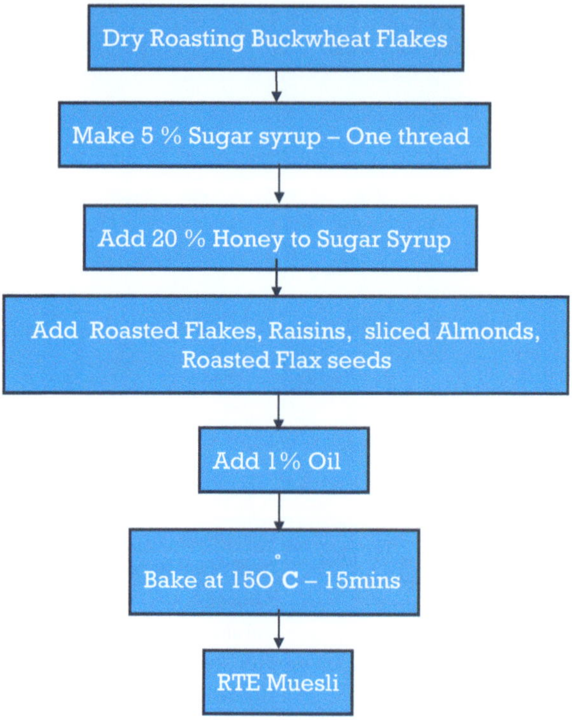

Fig. 13.2 Preparation of Honey coated Buckwheat Muesli

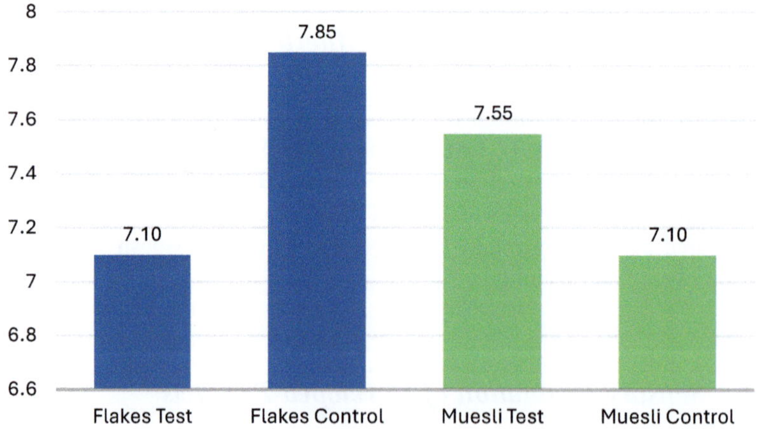

Fig. 13.3 The scores of overall acceptability of the developed products along with their control

The mean acceptability score for the plain Buckwheat flakes (7.1) was lower than the control (7.85). This difference was statistically significant ($p = 0.0232$). This suggests that the plain Buckwheat flakes were less acceptable than the control product. The mean acceptability score for the Buckwheat Muesli (7.55) was slightly

Table 13.2 Means scores of overall acceptability of the developed products

	Test Mean ± S.D	Control Mean ± S.D	U value	$p < 0.05$
Plain flakes with milk	7.1 ± 1.02	7.85 ± 0.81	115.5	0.0232*
RTE muesli with milk	7.55 ± 1.27	7.1 ± 1.11	141	0.1141

*significant at 5% level of significance

higher than the control (7.1). However, this difference was not statistically significant ($p = 0.1141$). This suggests that the Buckwheat Muesli was perceived as similar in acceptability to the control product.

13.2 Benefits of Incorporating Buckwheat as a RTE Breakfast Cereal

13.2.1 Increases the Utilization and Encourages Cultivation

The current approach in processing cereal-based products in the market highly considers staple crops like Rice, Wheat, and Maize. Buckwheat has been grown since ancient years but its cultivation is less common when compared to other cereals, thereby its utilization in the food industry(Joshi et al. 2019). Thus, consumers have fewer choice options to incorporate Buckwheat into their daily diet. Even though Buckwheat can be utilized by preparing homemade traditional recipes, the need for instant RTE foods is high in the current fast-moving life and both males and females have dual roles of homemakers and breadwinners.

13.2.2 Nutritive Value

Buckwheat has all the essential macro-nutrients and is rich in dietary fiber, protein, vitamins, and minerals. Most cereals lack the essential amino acid lysine, but Buckwheat contains lysine.

13.2.3 Health Benefits

Many studies show that Buckwheat's nutritional properties, its bioactive flavonoids like Rutin and Quercetin has many health benefits such as preventing Cancer, Type II Diabetes Mellitus, Coronary Artery Diseases, and Cardiovascular Diseases. It also has antiviral, antibacterial, and antifungal properties, thereby enhancing immune function. Buckwheat contains no gluten and is ideal for individuals with Coeliac Sprue and gluten food allergies (Rauf et al. 2020).

13.2.4 Food Market Expansion

By incorporating Buckwheat in the manufacture of various RTE foods like Breakfast cereals, baked products and snacks, there is a scope for new market expansion for target consumers who seek healthier and sustainable food options.

13.2.5 Crop Diversity and Reduced Reliance on Traditional Grains

Incorporating Buckwheat in daily diet can increase the demand for its cultivation and can help diversify crop rotation and reduce the reliance on few staple crops like wheat, maize, and rice. Due to its quick-growing capacity, it also supports farmers and promotes crop diversity (Babu et al. 2018; Rangappa et al. 2023).

13.3 Buckwheat's Role in Sustainability and Food Security

The most challenging aspect for food insecurity is the fastest growing world population resulting from dependency on only on few grains. One in three people across the world is malnourished and many countries are also burdened with hunger and undernutrition (IFPRI 2016). Buckwheat is a versatile crop that can thrive in various conditions drought and cold temperatures. In times of a shortage of food supply, it can meet the needs of food demand as it grows rapidly. Because of its numerous health benefits, it has the potential to be a superfood. Owing to its rich nutritional profile, it has been identified by Food and Agricultural Organization (FAO) that it can have a significant contribution to food security, improving nutrition, and health (Babu et al. 2018; Pirzadah and Malik 2020; Rana et al. 2012). It shows diversity in its genetic profile that aids in improving the quality of other crops (Zamaratskaia et al. 2023).

13.3.1 Climate Changes

Climate change, such as heatwaves, cold climate, droughts, and floods, will triple by 2040, causing food price spikes and potential civil unrest (FAO 2023; IFPRI 2016).

Buckwheat can be grown in high temperatures, droughts, and cold temperatures. It is also resistant to UV wave exposure, salty waters, and contamination with heavy metals (Babu et al. 2018; Jha et al. 2024; Luitel et al. 2017).

13.3.2 Greenhouse Gas Emissions

There are many reasons for the increase in greenhouse gas emission but one of them is through the agrifood systems, which have increased by 10% from 2000 up to 2021. Buckwheat requires less fertilizer and its farming methods include organic methods. Buckwheat, with its unique metabolites that manage weeds and decompose biomass, contributes to biomass production (Björkman and Shail 2013; Pinski et al. 2023; Vieites-Álvarez et al. 2024).

13.3.3 Water Scarcity for Agriculture

The biggest challenge in the world is that there are water stress levels around or above 100% in many Middle East countries and North Africa, followed by 30–60% stress levels in the rest of the world. This indicates there will be food scarcity. Buckwheat has a deep tap root structure system, which takes water for its growth even from the deepest layers of water meagreness. Thus, Buckwheat can be used as a sustainable crop even in conditions like drought or water scarcity(FAO 2023).

13.3.4 Land Availability for Agriculture

A recent FAO report shows that there is a decline in agricultural land by 86 million hectares between 2000 and 2021 (FAO 2023). By 2050, estimates indicate that the increasing need for food will necessitate an extra 320–850 million hectares of agricultural land (Fróna et al. 2019). Further increase in population and food needed per head might create an increase in demand for food of 40% by 2030 and 70% by 2050 (Senker 2011).

Buckwheat can be grown to meet the urgent demands of food as it grows rapidly and matures early in times of food scarcity. In a study, it is observed that the yield of Buckwheat when cultivated in October was 9.7 q/ha, followed by November as 11.0 q/ha (Rangappa et al. 2023).

13.3.5 Challenges in Incorporating Buckwheat in Breakfast Cereals

Processing Challenges: Buckwheat processing can be more complex and expensive compared to other grains, which may impact production costs.

Availability and Cost: Since Buckwheat cultivation has decreased overtime and based on its availability, depending on the region, its supply might be limited, and its cost could fluctuate, impacting product pricing.

13.3.6 Opportunities for Buckwheat's Incorporation in Breakfast Cereals

- Health and Wellness Trend: The growing consumer interest in healthy and sustainable food aligns perfectly with Buckwheat's nutritional profile and environmental benefits. Placing Buckwheat-based cereals as premium health-beneficial products can increase their utilization and contribute to health.
- Product Differentiation: Incorporating Buckwheat can help breakfast cereal manufacturers differentiate their products and attract a niche market.
- Sustainability: By using Buckwheat, food industries can improve their sustainability credentials and appeal to health conscious consumers.
- Existing Products Variations: Blending Buckwheat with other grains or flavorings to create a more familiar taste profile.
- Research and Development: More research is needed in food technological applications to optimize Buckwheat utilization in different food products.
- Consumer Education: Buckwheat has not gained popularity like quinoa, oats, millets, and thus its utilization is less (Bekkering and Tian 2019).
- Adopting New Technology: There is a need to adopt advanced processing technologies to improve efficiency and reduce costs in manufacturing Buckwheat based new food products.
- Industry Collaboration with Research Institutes: To design and develop innovative Buckwheat-based food products, there is a need for combined efforts from the industrial food technologists and research institutes for providing specialized processing solutions.
- Livelihood for Farmers: Farmers can earn better if they are given contract farming to cultivate buckwheat as it has versatile growing conditions and yield strength. Thus, considering approaches like contract farming would enable the secure a consistent supply of buckwheat.
- Cost Optimization: Explore ways to optimize production processes and reduce costs.

13.3.7 Other Culinary Uses of Buckwheat

Buckwheat can be incorporated into traditional recipes like Buckwheat roti, Buckwheat dosa, Buckwheat based khichidi. Bakery products likes biscuits and cookies can also be prepared. In the food industry, Buckwheat can be used to prepare RTE foods like muesli, flakes, chocos, etc. Buckwheat can be incorporated in making noodles, vermicelli, and pasta.

13.4 Conclusion

Food insecurity is a significant issue, with the fastest-growing world's population relying on few grains. Global food insecurity has increased, with many people struggling to access adequate food regularly. Buckwheat, a versatile crop, can thrive in various conditions and can be used as an emergency crop during food shortages. It has the potential to make sufficient food available through various seasons, reduce hunger, poverty, and malnutrition, thus making it a sustainable crop.

References

Babu S, Yadav GS, Singh R, Avasthe RK, Das A, Mohapatra KP, Tahashildar M et al (2018) Production technology and multifarious uses of buckwheat (Fagopyrum spp.): a review. Indian J Agron 63(4):415–427

Bekkering CS, Tian L (2019) Thinking outside of the cereal box: breeding underutilized (pseudo)-cereals for improved human nutrition. Front Genet 10.1289. https://doi.org/10.3389/fgene.2019.01289

Björkman T, Shail JW (2013) Using a buckwheat cover crop for maximum weed suppression after early vegetables. HortTechnology 23(5):575–580. https://doi.org/10.21273/HORTTECH.23.5.575

FAO (2023) World food and agriculture—statistical yearbook 2023. FAO, Rome

Farooq S, Ul Rehman R, Pirzadah TB, Malik B, Ahmad Dar F, Tahir I (2016) Cultivation, agronomic practices, and growth performance of buckwheat. In: Molecular breeding and nutritional aspects of buckwheat, pp 299–313. https://doi.org/10.1016/B978-0-12-803692-1.00023-7

Fróna D, Szenderák J, Harangi-Rákos M (2019) The challenge of feeding the world. Sustainability 11(20):5816. https://doi.org/10.3390/SU11205816

International Food Policy Research Institute (IFPRI) (2016) Global nutrition report 2016: from promise to impact: Ending malnutrition by 2030. https://doi.org/10.2499/9780896295841

Jha R, Zhang K, He Y, Mendler-Drienyovszki N, Magyar-Tábori K, Quinet M, Germ M et al (2024) Global nutritional challenges and opportunities: buckwheat, a potential bridge between nutrient deficiency and food security. Trends Food Sci Technol 145:104365. https://doi.org/10.1016/J.TIFS.2024.104365

Joshi DC, Chaudhari GV, Sood S, Kant L, Pattanayak A, Zhang K, Fan Y, Janovská D, Meglič V, Zhou M (2019) Revisiting the versatile buckwheat: reinvigorating genetic gains through integrated breeding and genomics approach. Planta 250(3):783–801. https://doi.org/10.1007/S00425-018-03080-4

Luitel DR, Siwakoti M, Jha PK, Jha AK, Krakauer N (2017) An overview: distribution, production, and diversity of local landraces of buckwheat in Nepal. Adv Agric 2017:1–6. https://doi.org/10.1155/2017/2738045

Nalinkumar A, Singh P (2020) An overview of buckwheat (Fagopyrum spp)—an underutilized crop in India—nutritional value and health benefits. Int J Med Res Health Sci 9(7):39–44. https://www.ijmrhs.com/medical-research/an-overview-of-buckwheat-emfagopyrum-sppeman-underutilized-crop-in-indianutritional-value-and-health-benefits-47686.html

Pinski A, Zhou M, Betekhtin A (2023) Editorial: advances in buckwheat research. Front Plant Sci 14:1190090. https://doi.org/10.3389/FPLS.2023.1190090

Pirzadah TB, Malik B (2020) Pseudocereals as super foods of 21st century: recent technological interventions. J Agric Food Res 2:100052. https://doi.org/10.1016/j.jafr.2020.100052

Pirzadah TB, Ul Rehman R (2021) Buckwheat: forgotten crop for the future: issues and challenges. doi:https://doi.org/10.1201/9781003089100

Rana J, Chauhan R, Sharma T, Gupta N (2012) Analyzing problems and prospects of buckwheat cultivation in India. Eur J Plant Sci Biotechnol 6(2):50–56

Rangappa K, Kumar A, Choudhury BU, Moirangthem P, Layek J, Rajkhowa D, Sarma AK, et al. (2023) Buckwheat: Potential stress-tolerant crop for mid-hills of Eastern Himalaya under changing climate. https://doi.org/10.5772/INTECHOPEN.112096

Rauf M, Yoon H, Lee S, Hyun DY, Lee MC, Oh S, Choi YM (2020) Evaluation of *Fagopyrum esculentum* Moench germplasm based on agro-morphological traits and the rutin and quercetin content of seeds under spring cultivation. Genet Resour Crop Evol. https://doi.org/10.1007/s10722-020-00899-3

Senker P (2011) Foresight: the future of food and farming, final project report. Prometheus 29(3). https://doi.org/10.1080/08109028.2011.628564

Thakur M (2024) Millets: the multi-cereal paradigm for food sustainability, World Sustainability Series. Springer, Cham. https://doi.org/10.1007/978-3-031-64237-1_1

Thakur M, Belwal, T. (2023) Bioactive components. Springer, Singapore doi:https://doi.org/10.1007/978-981-19-2366-1_34

Thakur M, Tiwari P (2019) Millets: the underutilized nutritious functional foods. Plant Arch 19(1): 875–883

Vieites-Álvarez Y, Reigosa MJ, Sánchez-Moreiras AM (2024) A decade of advances in the study of buckwheat for organic farming and agroecology (2013–2023). Front Plant Sci 15:1354672. https://doi.org/10.3389/FPLS.2024.1354672

Zamaratskaia G, Gerhardt K, Knicky M and Wendin K (2023) Buckwheat: an underutilized crop with attractive sensory qualities and health benefits. Critical Reviews in Food Science and Nutrition. https://doi.org/10.1080/10408398.2023.2249112

Zhang K, He Y, Lu X, Shi Y, Zhao H, Li X, Li J et al (2023) Comparative and population genomics of buckwheat species reveal key determinants of flavor and fertility. Mol Plant 16(9): 1427–1444. https://doi.org/10.1016/J.MOLP.2023.08.013

Zhao H, He Y, Zhang K, Li S, Chen Y, He M, He F et al (2023) Rewiring of the seed metabolome during Tartary buckwheat domestication. Plant Biotechnol J 21(1):150–164. https://doi.org/10.1111/PBI.13932

Chapter 14
Vegan Millet Milk: A Nutrient-Rich, Allergy-Friendly and Eco-Sustainable Alternative to Dairy

Ananya Mohan, Vasudha Sharma, and Rahul Kumar

Abstract Millets are gaining global attention as resilient and nutritious staple crops capable of addressing food insecurity in regions vulnerable to climate change, such as India and sub-Saharan Africa. Their ability to thrive in arid, low-input environments makes them suitable alternatives to traditional cereals, contributing to sustainable agricultural systems that can withstand climatic stresses and reduce dependence on irrigation and high-fertility soils. Recognizing this potential, the United Nations General Assembly declared *2023 the International Year of Millets* to elevate awareness of their role in enhancing food and nutritional security, promote sustainable production, and stimulate policy attention and investment in research and development.

In parallel, dietary challenges related to dairy consumption are increasing in many developed regions. *Lactose intolerance*, caused by a deficiency in the enzyme lactase, impairs lactose digestion and can exacerbate gastrointestinal conditions, necessitating the reduction or elimination of dairy from affected individuals' diets. Additionally, *milk allergy*, an immune response to milk proteins, requires strict avoidance of dairy products to prevent adverse reactions. These conditions have driven demand for plant-based dairy alternatives that are free from lactose and dairy proteins. Millets offer an effective option for such alternatives. They can be used to formulate *dairy-free products*, such as millet-based milk, which serve as nutritious substitutes for conventional dairy. Beyond meeting the needs of individuals with lactose intolerance and milk allergies, millet-derived foods provide essential macro- and micronutrients, supporting diversified diets and contributing to global efforts to improve food security and nutrition under changing climatic conditions.

Keywords Allergy · Intolerance · Lactose · Milk · Millet · Vegan

A. Mohan (✉) · V. Sharma
Department of Food Technology, Jamia Hamdard (Deemed to be University), New Delhi, India

R. Kumar
Department of Food and Nutritional Sciences, University of Reading, Reading, UK

M. Thakur, T. Belwal (eds.), *Food Biodiversity and Consumer Engagements*, Sustainable Development and Biodiversity 46,
https://doi.org/10.1007/978-981-95-7777-4_14

14.1 Introduction

The concept of a vegan diet is expanding and is being acknowledged as a positive and healing dietary option. As per the Mediterranean Diet Score and the Healthy Eating Index 2010, the diet quality score of a vegan diet is the highest, whereas an unsupervised omnivorous diet has the lowest score. Using food frequency questionnaires, a survey was conducted recently where out of 1475 participants, 104 were vegans. To control a plethora of medical diseases such as cancer and cardiovascular diseases, veganism is becoming more popular. Research suggests that plant-based or vegan dietary patterns may help reduce the risk of chronic inflammatory conditions such as rheumatoid arthritis, and they could serve as effective medical nutrition therapy for managing components of metabolic syndrome, obesity, type 2 diabetes, and associated cardiovascular risk factors—by improving inflammatory profiles and metabolic health outcomes (Glick-Bauer and Yeh 2014).

In the past few years, the number of people who wanted to adopt a vegan diet has increased exponentially (Radnitz et al. 2015). Veganism can be a part of an individual's identity and/or lifestyle, or it can be related to health as a part of a medical treatment plan (Cramer et al. 2017).

14.1.1 Significance of Vegan Dietary Patterns

In recent research, it has been observed that, as compared to the non-vegetarians, the vegans and lacto-ovo-vegetarians had a BMI 3–5 points lesser. In comparison to the non-vegetarians, the vegans and lacto-ovo-vegetarians had 75% and 55% lesser risk of hypertension, respectively. Both of these are also associated to lower risk of type-2 diabetes. For lacto-ovo-vegetarians, the risk of diabetes decreased by a range from 38% to 61%, and for vegans, the range was between 47% and 78%. Compared to non-vegetarians, vegans showed a lower risk for all cardio-metabolic-related outcomes than lacto-ovo-vegetarians (Le and Sabate 2014).

The randomized clinical trials have shown that vegan diets protect against diabetes and reduce weight, blood pressure, glycosylated haemoglobin, and low-density lipoprotein cholesterol (Wang et al. 2023).

In recent years, an increasing trend for the consumption of vegan food products is seen in people since they demand functional meals that enhance the health and well-being of a person, thereby the demand for ready-to-drink beverages is increasing. Due to the sedentary and hectic lifestyle of people, they wish to consume food that can be prepared in lesser time and/or is ready to eat. In most of the developed countries, the demand for vegan milk is increasing. Plant-based milk appears similar to dairy-based milk, however it is an extracted micro particle suspension of a selected raw material (Sandhya and Anandakumar 2021).

14.2 Importance of Millets in Enhancing National Security

Millets have a characterization of "yesterday's coarse grains and today's *nutri-cereals*" since they are resistant to pests as well as diseases and can withstand the hard conditions prevalent in Asia and Africa's arid and semi-arid regions. Millets can be referred to as "future crops". (Gowda et al. 2022). They have an enhanced nutritional profile as compared to wheat and rice. This is because of to enhanced protein content and a more balanced amino acid composition. Millets also include a number of phytochemicals that, due to their anti-inflammatory and antioxidant properties, have therapeutic properties. They also consist of several important amino acids, notably sulfur-containing amino acids such as methionine and cysteine (Prasad 2023).

According to the FAO, the global production of millets is 89.17 million metric tons from 74 million hectares. Our country is the world's leading millet producer, contributing roughly 19% of total global output. Tamil Nadu, Rajasthan, Gujarat, Uttarakhand, Karnataka, Maharashtra, Haryana, Uttar Pradesh, Madhya Pradesh, and Andhra Pradesh are India's leading millet producers. These 11 states currently produce about 98% of India's millets (Dharmaraj et al. 2023).

Millets are composed of plenty of minerals like ß-carotene and vitamins. Nowadays, millets solve the problem of malnutrition and obesity, which is highly prominent in Indian people. Hundred grams of dietary fiber is present in millets. The bran consists of essential antioxidants, minerals, and vitamins like zinc, iron, magnesium, copper, B vitamins, fibre, and phytonutrients (Thakur and Tiwari 2019; Thakur 2024). The germ is known as the embryo that, post-pollination, forms into a new plant. The germ provides E vitamins, vitamin B, antioxidants, unsaturated fats, and phytonutrients. If the endosperm is allowed to develop, it gives energy to the plant and is known as the food source of the germ. The endosperm contains carbohydrates full of starch, proteins, and trace amounts of vitamins as well as minerals (Ambati and Sucharitha 2019). There are three distinct millets—major millets, minor millets, and pseudo millets. Their nutritional profile is mentioned in Table 14.1.

14.3 Vegan Millet Milk (Sustainable Approach)

This type of milk is formulated by the extraction of millet grains in the water that consists of iron, calcium, and magnesium. Choosing plant-based milk substitutes has its own ecological advantages as milk derived from cows is associated with high emissions of greenhouse gases, water, and land usage. In contrast, the manufacturing methods of plant-based milk replacements are characterized by fewer resource requirements and a decreased environmental imprint (Pandey and Singh 2024). There are a number of processing techniques involved in the formation of vegan millet milk, such as the ultrasound technique, which is a non-thermal food

Table 14.1 Type of millets

English Name	Local name	Energy (per 100 g)	Protein (per 100 g)	Carbohydrates (per 100 g)	Fats (per 100 g)	Total dietary fiber (per 100 g)
Sorghum (M)	Jowar	333 kcal	10.5 g	67.9 g	1.7 g	10.3 g
Pearl (M)	Bajra	362 kcal	11.7 g	67 g	5 g	11.5 g
Finger (M)	Ragi	338 kcal	7.4 g	72 g	1.4 g	11.3 g
Buckwheat (P)	Kuttu	116.5 kcal	4.2 g	21.5 g	1.2 g	2.1 g
Foxtail (m)	Kangani	332 kcal	12.4 g	60.9 g	4.5 g	8.4 g
Kodo (m)	Koden	355 kcal	8.5 g	65.9 g	1.9 g	6.3 g
Barnyard (m)	Sanwa	304 kcal	11.9 g	65.5 g	2.3 g	12.6 g
Proso (m)	Chena	340 kcal	12.2 g	70.4 g	3.2 g	8.5 g
Little millet (m)	Kutki	330 kcal	8.6 g	67.0 g	5.4 g	6.2 g
Amaranth (P)	Rajgira	375 kcal	14.45 g	66.1 g	6.8 g	9.5 g

M major millet, *m* minor millet, *P* pseudo-millet

processing methodology that is used independently or in combination with other methods like microwave, dehydration, supercritical carbon dioxide, etc. In solids, liquids, and gases (elastic media), sound travels as longitudinal waves. Longitudinal sound waves are pressure waves that oscillate between mean positions, leading to compressions and rarefactions. High-intensity ultrasound is used in several food and dairy processing operations for extraction, homogenization, emulsification, and degassing. (Sarangapany et al. 2022).

Another example includes sonication, to sterilize liquid meals, much greater acoustic energy is used; the procedures are classified as sonication alone, mono-sonication (pressure and ultrasound), thermos sonication (heat and ultrasound), or mono-thermo-sonication (heat, pressure, and ultrasound) (Tiwari and Mason 2012).

A common processing that is used in pasteurization and commercial sterilization is the thermal processing. Even though it is effective, it has a harmful effect on the organoleptic and nutritional properties. Heat energy is delivered to the target material either directly or indirectly to obtain the necessary temperature for lowering the microbial burden (Vanga 2018).

Homogenization enhances color, flavor consistency, creaminess, creaming stability, and mouthfeel. High-pressure homogenization works by pressurizing an emulsion premix to several hundred bars before pushing it through a disruption system, similar to full-stream homogenization (Köhlera and Schuchmann 2011).

New homogenization valves, together with ceramic seats and needles, as well as diamond covering, can sustain pressures ranging from 300 MPa to 400 MPa, enabling homogenization processing in the pharmaceutical, dairy, and cosmetic industries to reduce particle sizes below the micron (Dumay et al. 2013).

While comparing dairy milk with non-dairy milks, UHT milk had greater values when compared to all other plant-based milk alternatives in terms of color brightness (L), titratable acidity, yellowness (b), density, and pH. All plant-based beverages had higher redness (a) than UHT milk, but soy drinks had higher b and chroma (C), and almond and soy drinks had higher pH values. The almond drink possessed the highest a value, whereas the soy drink had the highest b and C values. The highest pH level was seen in the almond drink, whereas most titratable acidity was observed in soy drinks and coconut drinks. Rice beverage was denser than a beverage made up of oat, almond, and soy. The beverage made from rice possessed higher nutritional values and UFA/SFA, MUFA/SFA, and DFA/ OFA ratios. In the coconut drink and dairy milk, it was observed that the fat percentage was lower when compared to other beverages. Soy drink had the highest PUFA/SFA ratio and overall EFA content, while almond drink had the highest n-6/n-3 PUFA ratio (Daszkiewicz et al. 2024).

14.3.1 Comparative Analysis of the Nutritional Composition of Dairy-Based Milk to the Vegan Millet Milk

Millet milk has higher calories (383 kcal) and a similar carbohydrate content (78%), compared to dairy milk. The millet milk contains more crude fiber (0.9%). A high percentage of ash (0.35%) is found in the millet milk due to the presence of mineral content. There was minimal reduction in the content of protein (8.5–7.8%) as well as fat content (0.74–0.6%) as there as low temperature processing hen compared to the unprocessed milk made of millets (9.1%), however even after its processing, it was higher than the normal dairy milk (1.8–2.4% protein and 1.1–1.2% fat). At lower temperatures, TSS increased, while viscosity decreased. (Nair et al. 2020). The milk had 4.8 mg of dietary fiber per 100 g, 16.3 mg of calcium, and 7.1 mg of iron. Vegan millet milk production involves several basic steps. Some of these include straining, grinding, homogenization, UHT sterilization, degassing, and so on. The millet milk had a shelf life of 5 days, which was longer than the milk derived from cows. Overall, the milk made from millet is a suitable alternative to milk and other soft drinks.

According to a study by Akshaya et al. (2020) it was observed that there was great acceptance among the people for their Millet milk prepared by using Kodo, Proso, Barnyard as well as Foxtail millets (Akshaya et al. 2020). Using a 9-point hedonic scale, the fresh milk samples were examined by a panel containing ten semi-trained individuals. Before the evaluation of the sample using sensory analysis, the panelists were recommended to thoroughly rinse their mouth. The rating takes into account color, flavor, taste, scent, and general acceptability (Sunny et al. 2019).

14.4 Future Prospects

Ultra-high-pressure homogenization (UHPH) and high-pressure homogenization (HPH) are innovative food processing techniques that stabilize emulsions and food components at pressures from 60 to 400 MPa. In addition, they offer an increased nutritional profile, preservation of food, and functional enhancement. (Sherman et al. 2024).

The food articles that can substitute more than one to two adequate meals and may be utilized as a single nutritional source are called a meal replacement. It is an efficient and easy way to contribute to an adequate diet full of enriched proteins, carbohydrates, etc. It has been observed that the majority of vitamins, as well as minerals, are found in meal replacements. For overweight and obese citizens, people working in the health industry also suggest beverages as meal replacements. Shakes, bars, and beverages are used as meal substitutes (Kaur G, et al. 2024).

It is anticipated that the market will change and move in the direction of alternative plant proteins as per the demand of the consumer with respect to enhanced health, flavor as well as creative and unique products. Therefore, understanding the ideology and concept behind allergy-free and gluten-free food is becoming necessary. Research projects are addressing the nutritional quality of plant-based dairy alternatives, such as enhancing protein quality and glycemic characteristics (Jeske et al. 2018).

Millets are drought-resistant grains that possess a unique nutritional composition and a remarkable agronomic performance has an extraordinary nutritional profile and unusual agronomic performance, making it a potential food source for future diets. However, its undesirable properties, bitterness, and astringency make it difficult to develop millet-based culinary products (Kaur R et al. 2024). Hence, it is extremely important to overcome its properties that are not so desirable in order to make it more palatable. This is a research challenge that has to be addressed.

14.5 Conclusion

Nowadays, an agrarian crisis is being observed, enforcing the development of crops with respect to the detrimental effects of global warming. It has been observed that rigorous farming of several crops for food contributes to poor nutrition, genetic deterioration, and the abandonment of traditionally accessible crops rich in nutrition. These are known as poor man's crops, because they cater to roughly one-third of the global population. The use of naturally stress-resistant plants (NSRPs) can significantly relieve climate change problems by maintaining yield stability, global food security, and health. Agronomically, millets are beneficial since they can withstand drought, heat, salt, and biotic stresses. They also persist on marginal soils under rain-fed conditions. The milk derived from millets is a good substitute for people suffering from lactose intolerance and are unable to digest the milk from dairy and/or its products. Barnyard millet milk, foxtail millet milk, and finger millet milk had the highest overall approval.

References

Akshaya AC, Maivli CT, Vinothaa L, Ramya K (2020) Extraction of Milk from the millets. Int Res J Eng Technol 7(3)

Ambati K, Sucharitha KV (2019) Millets-review on nutritional profiles and health benefits. Int J Rec Sci Res 10(7):33943–33948

Cramer H, Kessler CS, Sundberg T, Leach MJ, Schumann D, Adams J, Lauche R (2017) Characteristics of Americans choosing vegetarian and vegan diets for health reasons. J Nutr Educ Behav 49(7):561–567

Daszkiewicz T, Florek M, Murawska D, Jabłońska A (2024) A comparison of the quality of UHT milk and its plant-based analogs. J Dairy Sci 107:10299

Dharmaraj U, YN S, BV SR, Singh SA (2023) Millets for the future 60(4):NIScPR-CSIR, India

Dumay E, Chevalier-Lucia D, Picart-Palmade L, Benzaria A, Gràcia-Julià A, Blayo C (2013) Technological aspects and potential applications of (ultra) high-pressure homogenisation. Trends Food Sci Technol 31(1):13–26

Glick-Bauer M, Yeh MC (2014) The health advantage of a vegan diet: exploring the gut microbiota connection. Nutrients 6(11):4822–4838

Gowda NN, Siliveru K, Prasad PV, Bhatt Y, Netravati BP, Gurikar C (2022) Modern processing of Indian millets: a perspective on changes in nutritional properties. Foods 11(4):999

Jeske S, Zannini E, Arendt EK (2018) Past, present and future: the strength of plant-based dairy substitutes based on gluten-free raw materials. Food Res Int 110:42–51

Kaur, G., Ahmadzadeh-Hashemi, S., Amir, S., Khan, Z. S., Gulsunoglu-Konuskan, Z., Karimidastjerd, A., & Aijaz, T. (2024). From millet to marvels: An improved future food through innovative processing. Future Foods, 9, 100367

Kaur, R., Shekhar, S., & Prasad, K. (2024). Functional beverages: recent trends and prospects as potential meal replacers. Food Materials Research, 4(1)

Köhlera K, Schuchmann HP (2011) Homogenisation in the dairy process-conventional processes and novel techniques. Procedia Food Sci 1:1367–1373

Le LT, Sabate J (2014) Beyond meatless, the health effects of vegan diets: findings from the Adventist cohorts. Nutrients 6(6):2131–2147

Nair UKA, Hema V, Sinija VR, Hariharan S (2020) Millet milk: a comparative study on the changes in nutritional quality of dairy and non-dairy milks during processing and malting. J Food Process Eng 43(3):e13324

Pandey S, Singh A (2024) Millet milk: a potential cattle milk alternative—extraction and value addition. J Food Qual 2024(1):3964937

Prasad S (2023) Application of millets in milk-millet based composite food: a review. Int J Sci Res Eng Dev 6(2):512–519

Radnitz C, Beezhold B, DiMatteo J (2015) Investigation of lifestyle choices of individuals following a vegan diet for health and ethical reasons. Appetite 90:31–36

Sandhya K, Anandakumar S (2021) Development and quality evaluation of retort processed ready-to-drink blended vegan milk. Pharma Innov J 10:818–823

Sarangapany AK, Murugesan A, Annamalai AS, Balasubramanian A, Shanmugam A (2022) An overview on ultrasonically treated plant-based milk and its properties–A review. Appl Food Res 2(2):100130

Sherman IM, Mounika A, Srikanth D, Shanmugam A, Ashokkumar M (2024) Leveraging new opportunities and advances in high-pressure homogenization to design non-dairy foods. Compr Rev Food Sci Food Saf 23(1):e13282

Sunny JD, Kundu P, Mehla R (2019) Preparation and physicochemical evaluation of dairy free alternative based on coconut and millet milk. Int J Pharm Biol Sci 9:565–572

Thakur M (2024) Millets: the multi-cereal paradigm for food sustainability, World sustainability series. Springer, Cham. https://doi.org/10.1007/978-3-031-64237-1_1

Thakur M, Tiwari P (2019) Millets: the underutilized nutritious functional foods. Plant Arch 19(1): 875–883

Tiwari BK, Mason TJ (2012) Ultrasound processing of fluid foods. In: Novel thermal and non-thermal technologies for fluid foods. Academic, pp 135–165

Vanga SKK (2018) Effects of novel processing techniques on proteins in plant-based alternative milk beverages. McGill University, Montreal

Wang T, Masedunskas A, Willett WC, Fontana L (2023) Vegetarian and vegan diets: benefits and drawbacks. Eur Heart J 44(36):3423–3439

Chapter 15
Cold Plasma Technology: A Sustainable, Non-thermal Innovation for Safe Food Processing

Sellam Perinban, Bindvi Arora, Anamika Thakur, Alka Joshi, and Valerie Orsat

Abstract Cold plasma technology offers a sustainable, non-thermal strategy for food preservation that inactivates microbes, enzymes, allergens, pesticide residues, and toxins with minimal impact on nutritional and sensory quality. Utilizing reactive species, UV radiation, and radical chemistry, cold plasma ensures safe, minimally processed foods while preserving taste, texture, and nutrients. Its potential extends to active packaging for shelf-life extension and waste reduction. Optimal performance depends on carefully controlled parameters, gas composition, pressure, electric field, frequency, and temperature—and understanding substrate and microbial diversity. Though promising, cold plasma is not a universal fix; integrated, hybrid approaches are essential to effectively safeguard food safety and quality in complex systems.

Keywords Cold Plasma technology · Plasma · Microbial inactivation · Sustainable · Non-thermal

15.1 Introduction

With the growing emphasis on clean label technologies in food processing, cold plasma applications have gained significant attention. The conventional thermal disinfection methods of having significant impact on the food products' nutritional quality, while the chemical-based disinfection methods are less preferred due to residual effects in foods. The novel non-thermal method, cold plasma, on the other hand, has proven to be a clean-label technology which not only preserves the nutritional quality of the food product but also does not leave any residues on the

S. Perinban · B. Arora · A. Thakur · A. Joshi (✉)
Division of Food Science and Postharvest Technology, ICAR-IARI, New Delhi, India
e-mail: alka@iari.res.in

V. Orsat
Department of Bioresource Engineering, McGill University, Montreal, QC, Canada

M. Thakur, T. Belwal (eds.), *Food Biodiversity and Consumer Engagements*, Sustainable Development and Biodiversity 46,
https://doi.org/10.1007/978-981-95-7777-4_15

food material. Hence, this technology has been extensively researched for its microbial decontamination properties, degradation of pesticide residues, allergens, and antibiotic residues, and to modify the functional properties of the food (Thakur and Modi 2020).

Apart from food processing, cold plasma and plasma-activated liquids are widely researched in medicine and dentistry for wound healing, disinfection, and disintegration of cancer cells. Further, the recent reports also suggest the application of cold plasma-activated water (PAW) in agriculture for seed germination and disease control. The mechanism behind these applications of cold plasma is mainly due to the reactive species and its oxidative properties. Even though this leads to disinfection and other beneficial changes in the food material, due to the high susceptibility of food materials to oxidative stress, there are undesirable changes in the quality of food materials implicated by this oxidation mechanism. Hence, it is important to optimize the reactive species generation and its interaction to achieve the desirable function of cold plasma without affecting the quality of the food materials. Furthermore, the underlying reaction mechanisms and the specific effects of these reactive species on various food components are not yet completely explored, necessitating further in-depth research on these aspects (Thakur and Belwal 2023).

The characteristics and concentration of these reactive species are influenced by various operational parameters, including the applied voltage, system configuration, gas composition, treatment time, and method of application. As an emerging technology, cold plasma currently faces a significant hurdle: unlike established methods such as ozone or UV processing, it has not yet received Generally Recognized As Safe (GRAS) status from the FDA for food applications. Overcoming this regulatory roadblock will be essential for the widespread commercial adoption and integration of cold plasma technology within the food industry.

15.2 Classification of Plasma

Plasma is a fourth state of matter, which is produced by the ionization of gases. Plasma is abundantly present in the universe, and it is broadly classified into thermal and non-thermal plasma based on the thermal equilibrium of free electrons (Te) with neutral gas molecules (Tg), and negative ions (Ti). Further, based on the operating pressure conditions, NTP's further classified as atmospheric pressure plasmas and low-pressure plasmas (Fig. 15.1).

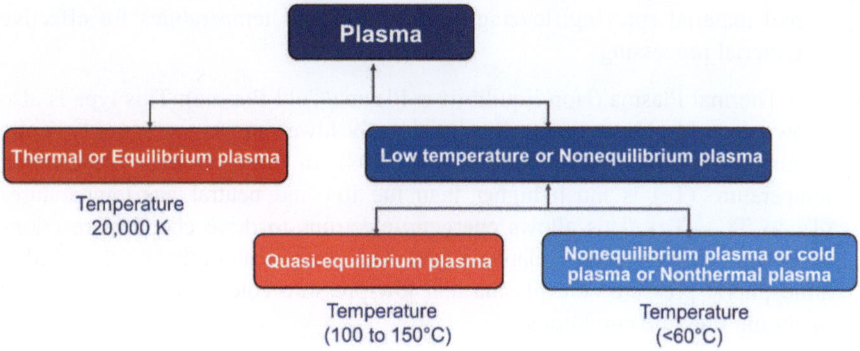

Fig. 15.1 Classification of plasma. (Subrahmanyam et al. 2024)

15.2.1 High-Temperature Plasma (Thermally Equilibrium Plasma)

This category, also known as thermally equilibrium plasma, is characterized by all particle species existing in thermodynamic temperature equilibrium at the same very high temperature (Te ≈ Ti ≈ Tg) (Afshari and Hosseini 2012). The gas temperature (Tp) of these plasmas typically ranges from 10^6 to 10^8 K (Huang and Tang 2007).

- Examples: The most prominent examples are fusion plasmas found in the core of stars and in experimental fusion reactors (like tokamaks). These plasmas are fully ionized and require extreme temperatures to sustain nuclear fusion reactions, necessitating powerful magnetic confinement.

15.2.2 Low-Temperature Plasma (Thermally Non-Equilibrium Plasma)

This broader category is known as thermally non-equilibrium plasma because not all particle species are at the same temperature. It is further subdivided based on the degree of thermal equilibrium:

- Thermal Plasma (Quasi-Equilibrium Plasma): In this subcategory, particle species are in a local thermal equilibrium (LTE) state. While the heavy particles (ions and neutrals) reach very high temperatures, there can be slight deviations from full equilibrium, where Te might be slightly higher than Tg. The gas temperature (Tg) for quasi-equilibrium plasma is about 2×10^4 K (Afshari and Hosseini 2012).
  - Examples: Common applications include plasma torches and arc welding plasmas. These are widely used in industrial processes for cutting, welding,

and material spraying, leveraging their high gas temperatures for effective material processing.

- Non-Thermal Plasma (Non-Equilibrium Plasma/Cold Plasma): This type is also known as cold plasma due to its significantly lower gas temperature, typically ranging from 300 to 1000 K (Fridman 2008). In these plasmas, the electron temperature (Te) is much higher than the ion and neutral gas temperatures (Te $\gg$ Ti $\approx$ Tg). This allows energetic electrons to drive chemical reactions while the bulk gas remains relatively cool. This plasma can further be classified as atmospheric pressure cold plasma and low-pressure cold plasma based on the operating pressure conditions.
  - Examples: Fluorescent lamps, dielectric barrier discharges (DBDs), and atmospheric pressure plasma jets are prime examples. Their low gas temperature makes them suitable for sensitive applications like medical sterilization, wound healing, food disinfection, surface modification of heat-sensitive materials, and environmental remediation.

This chapter focuses exclusively on the applications of cold plasma. Other plasma types fall outside the scope of this discussion.

15.3 Generation of Cold Plasma

Plasma is produced by the ionization of gas molecules by applying energy in the form of nuclear energy, electrical energy, mechanical energy, or thermal energy (Conrads and Schmidt 2000). Generating low-temperature plasma for technical uses commonly involves applying an electric field to a neutral gas. A few free electrons and ions, naturally present due to background radiation, are accelerated by this field. As they speed up, they collide with gas atoms, molecules, or electrode surfaces. These collisions are energetic enough to liberate more charged particles, creating an avalanche. This process continues until the rate at which new charged particles are formed equals the rate at which they are lost, establishing a stable, steady-state plasma (Conrads and Schmidt 2000).

In cold plasma, partially ionized gas is produced by relatively low power input as compared to thermal plasmas. This weakly ionized cold plasma has less than 0.1% ionization and the temperature will be less than 100 °C (Adesina et al. 2024). Further, based on the energy source and the working gas, cold plasma can be produced at atmospheric pressure or at low pressure (0–100 kPa) conditions. The commonly used discharge methods in food applications are dielectric barrier discharges, corona discharges, gliding arc discharges, plasma jets, and glow discharges using energy sources like AC, DC, pulsed DC power supply, microwave and radio frequency. Further, the plasma-activated medium such as plasma-activated water in which the water is activated by cold plasma first and then used for treating food

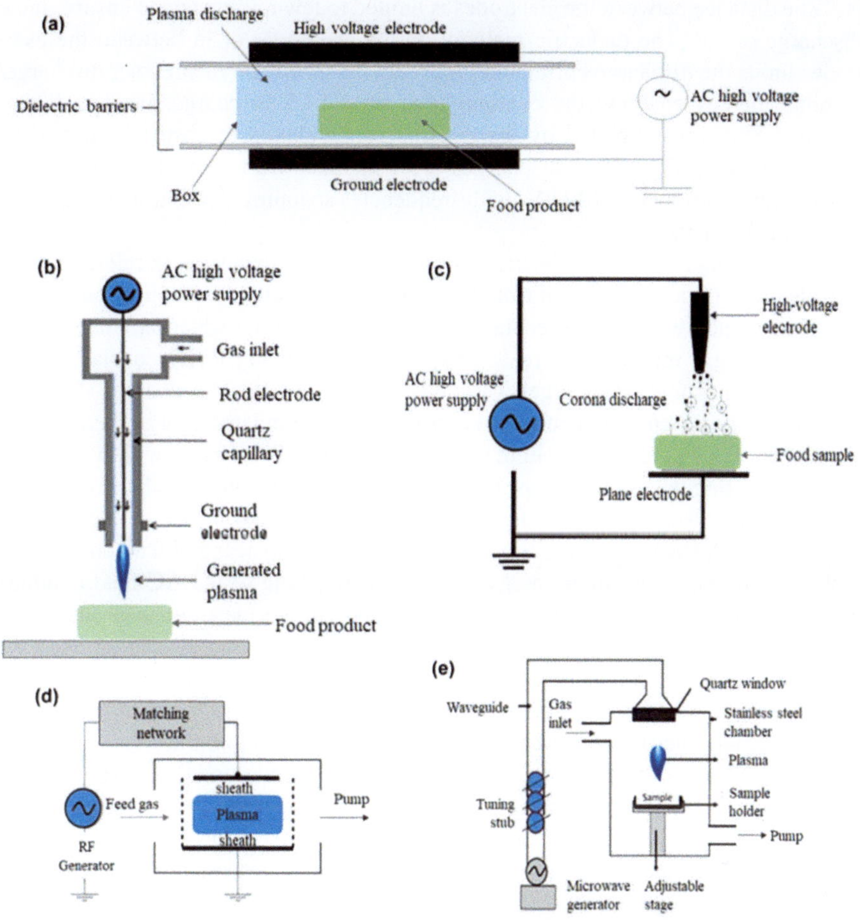

Fig. 15.2 Schematics of cold plasma generation systems (**a**) dielectric barrier discharge, (**b**) plasma jet, (**c**) corona discharge, (**d**) radiofrequency, and (**e**) microwave plasma system. (Nwabor et al. 2022)

materials. The major CP generation methods are discussed in detail below (Fig. 15.2).

15.3.1 Dielectric Barrier Discharge (DBD)

Dielectric barrier discharges (DBD) are also known as silent discharges, and it is one of the versatile and widely used methods of generating cold plasma at atmospheric pressure. This system consists of two electrodes with at least one electrode covered with the dielectric barrier material such as glass, quartz, acrylic, PET, polypropylene, etc. The distance between the electrodes is limited to few millimeters to ensure stable discharge of CP. The dielectric material on the electrodes or in between the electrodes limits the discharge current and prevents the formation of streamer discharge/arching, thereby enabling the continuous or pulsed CP discharge. Further, it also distributes streamers created by the electron accumulation to ensure homogenous discharge. Generally, DBD discharges are generated with ac voltages or pulsed power ranging from 1 to 100 kV, with frequencies spanning through line frequency to several kilohertz.

As the voltage rises, the electric field in the gas gap increases, accelerating any free electrons (naturally present from background radiation or pre-ionization). These electrons collide with gas molecules, leading to ionization, excitation, and dissociation, forming a multitude of reactive species and creating localized plasma filaments known as microdischarges. Crucially, when these microdischarges strike the dielectric surface, charge accumulates on it. This accumulated charge then rapidly creates an opposing electric field that quenches the microdischarge within nanoseconds, preventing the full breakdown into a hot arc. This self-limiting characteristic of the dielectric barrier is key to maintaining the non-thermal nature of the plasma, ensuring that the heavy particles (ions and neutrals) remain near ambient temperature while electrons possess high energy. The rapid switching of the AC field continuously re-ignites new microdischarges across the surface, leading to a sustained, spatially distributed cold plasma rich in reactive species.

Due to its simpler construction, various configurations of DBD discharges are available, such as symmetric, asymmetric, planar, co-planar, DBD brushes, cylindrical, coaxial, microarray DBDs, etc. Detailed information on the DBD discharge and the system configurations can be found at (Bardos and Barankova 2010; Brandenburg 2017; Conrads and Schmidt 2000; Flores-Fuentes et al. 2009).

While dielectric barrier discharge (DBD) cold plasma systems offer significant promise for various applications, their utility in food processing faces two key limitations:

1. DBD systems, particularly those using parallel plate or coaxial cylinder configurations, typically operate with a very narrow gap (often a few millimeters) between the dielectric surface and the opposing electrode. This narrow gap is crucial for maintaining the non-thermal nature of the plasma and preventing arcing. A narrow discharge gap makes it difficult to uniformly treat large or irregularly shaped products, leading to inefficient or incomplete decontamination.
2. DBD plasma characteristics are highly sensitive to even minor fluctuations in operating parameters. These parameters include the applied voltage and frequency, the type and flow rate of the process gas, the humidity and temperature of the ambient air (if using air), and even the surface properties of the dielectric material and the treated food product. A slight change in any of these can significantly alter the intensity, homogeneity, and reactive species composition of the plasma. If the plasma properties vary, the efficacy of microbial inactivation, enzyme deactivation, or surface modification can become inconsistent. For

example, a minor change in humidity could alter the generation of key reactive oxygen and nitrogen species (RONS), affecting decontamination effectiveness.

15.3.2 Corona Discharge

Corona discharges are another commonly used method for cold plasma (CP) generation, employed in both direct and indirect applications in food processing. This system typically employs an electrode configuration with a sharp point, wire, or edge (the high-voltage electrode) positioned at a significant distance from a larger, often grounded, counter-electrode or surface. When a high voltage (either DC or AC) is applied, the electric field becomes highly concentrated around the sharp features of the energized electrode. If this localized electric field exceeds the dielectric strength of the surrounding gas, a partial electrical breakdown occurs, forming a region of ionized gas around the pointed electrode. This creates a characteristic bluish or purplish glow, which is the visual indicator of the "corona." The mechanism involves the acceleration of initial free electrons in this strong field, leading to inelastic collisions with gas molecules that cause ionization and excitation, primarily at the point of highest field strength. While these collisions generate reactive species and energetic electrons, the discharge does not fully bridge the gap to form a hot arc, as the electric field rapidly weakens away from the sharp electrode, and the discharge extinguishes before it becomes too conductive at the end of each pulse. This localized and self-limiting nature allows the heavy gas particles to remain near ambient temperature. The main drawback of corona discharge is the thin plasma volume, posing limitations in the applications. Furthermore, as the cold plasma is directly formed on a metal electrode, concerns regarding metal etching and subsequent deposition onto food material also present a significant challenge (Lukes et al. 2012).

15.3.3 Gliding Arc Discharge

In gliding arc discharge is another type of cold plasma discharge in which the discharge "glides" between the electrodes under high voltage. Typically, this system consists of a set of two to four electrodes operating in the voltage range of 100 v to 20 kV. The core mechanism involves an arc forming at the point of shortest inter-electrode distance. This arc is then dynamically stretched and "blown" along the electrodes by the gas flow, often combined with magnetic fields, until it breaks into a plasma plume. Immediately, a new arc reforms, initiating a new cycle. This irregular arc development is influenced by several critical parameters, including the gas flow rate, the type of process gas, the electrode material and geometry, and even surface micro-corrosion (Pawłat et al. 2019).

15.3.4 Plasma Jets

Plasma jet devices offer a distinct advantage in plasma applications, particularly in food processing, due to their ability to project a stable plasma discharge onto a target surface without requiring the target to be an integral part of the electrical circuit. According to Lu et al. (2012), the noble gas plasma jets are classified into four types viz.

1. Dielectric barrier discharge (DBD) jets: Where at least one electrode is covered by a dielectric material, controlling the discharge and preventing arc formation.
2. Dielectric-free electrode jets: Lacking a dielectric barrier, often relying on specific gas flow dynamics or pulsed power to maintain non-thermal conditions.
3. DBD-like jets: Configurations that share characteristics with DBD systems but may have variations in electrode arrangement or dielectric placement.
4. Single-electrode jets: Utilizing only one active electrode, with the surrounding air or the treated material acting as the implicit ground.

The sustained, uniform discharge of noble gas plasma jets at higher frequencies are highly favored in biomedical applications as it avoids the possibility of arcing and enables the precise application of plasma. However, the air plasma jets are difficult to sustain at atmospheric pressures due to the presence of oxygen. (Hong et al. 2009) reported a device using a porous alumina dielectric to separate a high-voltage electrode from a ground electrode, producing a 2 cm plasma jet at 60 Hz, but with a relatively high gas temperature (60 °C at 10 mm). Additionally, "floating" electrode air plasma jets have been developed, which are notably safe and generate room-temperature plasma, suitable for both large surface and localized 3D treatments respectively (Lu et al. 2012). These plasma jets, including those that can also operate with N_2, vary in plasma plume length and specific operating parameters.

15.3.5 Radiofrequency Discharge

These discharges are made utilizing radio frequency energy to ionize the gas. RF discharges can be capacitively coupled (between parallel electrodes) or inductively coupled (using an electromagnetic field). In an RF discharge, an alternating voltage is applied to a gas at low pressure, typically in the range of 1–100 MHz. This oscillating electric field accelerates electrons, causing them to collide with neutral gas molecules, leading to ionization, and the formation of plasma.

15.3.6 Microwave Discharges

Microwave cold plasma discharges, also known as microwave-induced plasma, are a type of electrodeless discharge created by using electromagnetic waves, typically at a frequency of 2.45 GHz, to ionize a gas (Lebedev 2010). These discharges are characterized by a high electron density and relatively low electron temperatures, making them ideal for various applications, including food safety, materials science, and chemical processing.

15.3.7 Plasma Activated Media (PAM)

Another important, indirect method of utilizing cold plasma in food processing. As the optimization and control of reactive species exposure of the food material is difficult under direct cold plasma applications, the plasma-activated media, such as plasma-activated water (PAW), are being evaluated. In PAM, the liquid media is exposed to cold plasma so that the reactive species generated in the plasma penetrate the medium and remain stable for varying periods. This also enables the storage and transportation of PAM as opposed to on-the-spot treatment of cold plasma. Detailed information on plasma-activated liquid media can be found in (Kaushik et al. 2018; Perinban et al. 2019).

15.4 Applications in Food Processing

In food processing, cold plasma is being used for disinfection, inactivation of microorganisms and enzymes, cleaning of the food contact surfaces, to modify the functional properties of food materials etc. The interaction of cold plasma and the efficacy of the intended application, however, depends on many factors such as the CP generation method, working gas composition, treatment time, flow rate, operating conditions, type of food material, and the type of microorganism. Hence, it is challenging to comprehensively compare the efficacy of CP treatments.

As discussed earlier, cold plasma is a partially ionized gas, containing electrons, ions, neutral radicals, and UV radiation. Among these, ionized particles and neutral particles are not the significant components in determining the cold plasma chemistry because of the partial ionization of the gas. The electrons, however, are the major component responsible for the generation of reactive species by subsequent collisions (Whitehead 2016). These electrons collide with the discharge gas molecules through various physical processes like ionization, dissociation, and excitation, leading to the formation of reactive oxygen and nitrogen species. Reactive Oxygen Species (ROS) are the species formed from oxygen, and Reactive Nitrogen Species (RNS) are the species from nitrogen molecules in the discharge gas. If the feed gas contains both gases, then reactive oxygen and nitrogen species (RONS) will be formed in the cold plasma.

15.4.1 Applications in Microbial Disinfection

Cold plasma has shown significant promise in reducing microbial contamination on food surfaces, offering an effective non-thermal alternative to traditional decontamination methods. Its ability to inactivate a wide spectrum of foodborne pathogens, including bacteria, yeasts, molds, and spores, makes it especially attractive for ensuring food safety while preserving the sensory and nutritional qualities of food products. The antimicrobial efficacy of cold plasma arises from the complex interactions between reactive species generated during plasma discharge and the structural and functional components of microbial cells.

The inactivation of microorganisms by cold plasma is primarily attributed to the generation of reactive oxygen and nitrogen species (RONS), along with UV photons, charged particles, and electric fields. The effectiveness of microbial inactivation via cold plasma varies depending on several factors, including the type and physiology of the microorganism, treatment duration, input power, gas composition, and the characteristics of the food matrix. Gram-negative bacteria such as *Escherichia coli* and *Salmonella* are generally more susceptible to plasma treatment due to their thinner cell walls, while gram-positive bacteria and bacterial spores often require longer exposure times or higher energy input. Similarly, the physical surface properties of foods—such as roughness, porosity, and moisture content—can influence the distribution and penetration of reactive species, thereby affecting treatment outcomes (Kaur et al. 2024).

Cold plasma has been extensively studied for its microbial disinfection and inactivation properties against various food spoilage microorganisms regardless of food substances. In apple juice with in 30 s of cold plasma treatment, more than 4 log CFU/mL reduction in the *E.coli* was observed (Liao et al. 2018). Ziuzina et al. (2020) found that DBD-ACP treatment in static mode was effective in reducing the levels of *E. coli* and *L. innocua* by 2.0 and 1.3 log10 CFU/mL, respectively, whereas 3.8 log10 CFU/mL reduction of *L. innocua* was observed in continuous mode treatment of strawberries. (Choi et al. 2017) reported the decontamination of dried squid using corona discharge plasma generated at 20 kV (58 kHz) for 0–3 min. Upon termination, a 2.0, 1.6, and 0.9 log reduction was observed for aerobic bacteria, marine bacteria, and *Staphylococcus aureus*, respectively, with minimal changes to the sensorial profile and volatile basic nitrogen, when compared with the control. Similarly, the sporacidal effect of microwave-induced CP against onion powder stored at two distinct temperatures (4 and 25 °C) was examined. The investigation revealed that a 40 min duration of high microwave density CP could inhibit *B. cereus, A. brasiliensis, and E. coli* O157:H7 by 2.1, 1.6, and 1.9 log CFU/cm^2, respectively, after 21 days (Kim et al. 2017). Yong et al. (2017) employed a flexible thin-layer plasma technique to inactivate bacteria and mould on beef jerky in packaging. The populations of *E. coli O157:H7, Listeria monocytogenes, Salmonella typhimurium*, and *Aspergillus flavus* on beef jerky decreased by 2–3 log CFU/g after a 10 min plasma treatment.

15.4.1.1 Mechanism of Microbial Inactivation

The main mechanism of cold plasma inactivation of micro-organisms involves the development of oxidative stress inside the microbial cell and damaging the cellular structure (Fig. 15.3; Table 15.1). The reactive species, such as singlet oxygen, ozone, hydroxyl radicals, hydrogen peroxide, nitric oxide, and nitrogen dioxide, exert oxidative stress on microbial cells, attacking the cell membrane, DNA, proteins, and intracellular enzymes (Nasiru et al. 2021). Cold plasma has also demonstrated notable efficacy against microbial biofilms, which are typically more resistant to conventional sanitization techniques due to their protective extracellular matrix. The reactive species generated by plasma can penetrate and degrade this matrix, exposing the cells embedded within and allowing for more effective inactivation. This property is particularly beneficial in the decontamination of food contact surfaces and processing equipment, where biofilms can act as persistent sources of contamination (Zhu et al. 2020). The most important effects of cold plasma treatment are the disruption of the microbial cell membrane. Reactive species can cause lipid

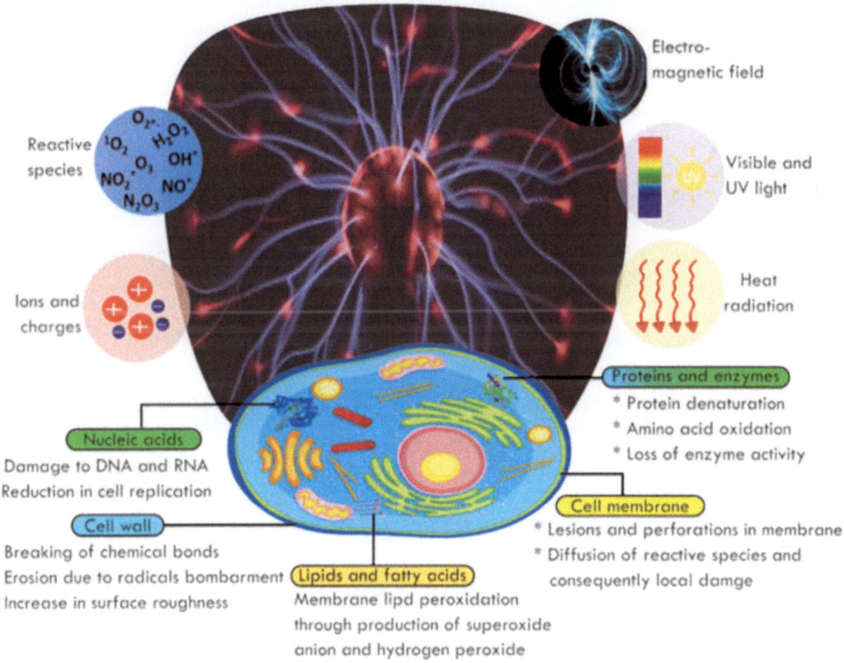

Fig. 15.3 Cold plasma microbial inactivation mechanism. (Kaur et al. 2024)

Table 15.1 Cold plasma disinfection of microorganisms in food products

Food	Microorganism	Treatment conditions	Findings	Reference
Apple juice	*Zygosaccharomyces rouxii*	Voltage: 21 kV, treatment time: 30 min, gas flow rate: 150 L/h	5.6 log CFU/g reduction	Wang et al. (2020)
Tender coconut water	*E. coli*, *L. monocyotogenes* and *S. enterica*	Voltage:90 kV; gas: Air and M65 (65% O_2, 30% CO_2, 5% N_2); treatment: 120 s	5 $\log_{10}$ reduction in population of *E. coli*, *L. monocyotogenes*, and *S. enterica* Structure collapse and shrinkage observed in gram-positive and gram-negative bacteria.	Mahnot et al. (2019)
Fresh-cut apple (granny smith) skin	*E. coli*, *L. innocua*	Low pressure cold plasma: treatment time: 20 min; gas: Ar, N_2, O_2, and Ar–O_2; power: 29.6 W; flowrate:40 L/min	20 min treatment effectively inactivated ($p < 0.05$) *E. coli* (1.68 log CFU/cm^2) with O_2, and *L. innocua* (0.88 log CFU/cm^2) with N_2.	Segura-Ponce et al. (2018)
Cherry tomatoes and strawberries	*Escherichia coli*, *S. enterica serovar Typhimurium* and *Listeria monocytogenes*	DBD, 70 kVRMS for 30 s—300 s in air and at atmospheric pressure	Reducing by 3.5, 3.8, and 4.2 log10 CFU/sample, respectively, within 300 s of treatment	Ziuzina et al. (2014)
Chicken breast	*Campylobacter jejuni*	Ar/5 lpm 2–3 kV 1 MHz, 3 min	2.5 log reduction	Rossow et al. (2018)
Egg	*Salmonella typhimurium*	N2/100 SCCM 600 W, for 2 min	2.6 log reduction	Lin et al. (2020)
Egg	*Salmonella enteritidis*	DBD air 100 kV 60 kHz 1 min	More than 5 log reduction	Illera et al. (2023)
Beef jerky	*Aspergillus flavus*	Flexible thin-layer plasma DBD, AIR, 15 kHZ, 10 min	3.18 log reduction	Yong et al. (2017)
Asian sea bass slice	Enterobacteriaceae	DBD, (Ar: O_2 (10: 90) 80 kVRMS 50 Hz 5 min	1.5 log reduction	Olatunde et al. (2020)
Fresh strawberries and spinach	*E. coli* and *L. innocua*	In-package DBD in PET containers	Strawberry: 2.0 and 1.3 log10 CFU/mL reductions of *E. coli* and *L. innocua*, respectively. Spinach: 2.2 and 1.7 log10 CFU/mL reductions for *E. coli* and *L. innocua*, respectively	Ziuzina et al. (2020)

peroxidation, compromise the integrity of the phospholipid bilayer and increase membrane permeability. This destabilization leads to leakage of intracellular contents and, in many cases, complete lysis of the cell. The disruption of membrane function not only affects nutrient uptake and waste elimination but also undermines the ability of the cell to maintain osmotic balance, which is essential for survival.

Further, reactive oxygen species (ROS) such as hydroxyl radicals can modify amino acid side chains, resulting in protein misfolding or denaturation. These structural changes impair the function of enzymes and transport proteins critical for cellular metabolism. Inhibition of enzymatic activity further weakens the cell's ability to respond to stress and repair damage, accelerating the progression to cell death. Another major target of cold plasma treatment is microbial DNA. Plasma-generated UV photons can directly induce DNA strand breaks and form pyrimidine dimers, particularly thymine dimers, which interfere with DNA replication and transcription. Additionally, indirect damage occurs through the interaction of RONS with nucleic acids, leading to base modifications, crosslinking, and fragmentation. DNA damage is especially critical in preventing cell recovery and reproduction, contributing to the irreversible inactivation of microbial cells (Perinban et al. 2019).

Reactive nitrogen species (RNS) also play a significant role by diffusing through damaged membranes and contributing to intracellular acidification (Bourke et al. 2017). The accumulation of nitrites, nitrates, and peroxynitrite within the cytoplasm can lower pH levels and disrupt the redox balance. Acidification impairs enzyme activity and further accelerates cellular dysfunction. The combined effect of oxidative and nitrosative stress can overwhelm microbial defense mechanisms, even in relatively resistant organisms such as spores or gram-positive bacteria with thicker cell walls.

15.4.2 Removal of Allergens and Enzymes

Cold plasma technology shows significant promise in reducing the allergenicity of various substances, particularly in food and airborne allergens. The reactive species in cold plasma interact with the allergen molecules. Cold plasma can cause changes in the conformational and linear epitopes of allergen proteins. This means it can modify the specific parts of the protein that are recognized by the immune system (IgE antibodies), thereby reducing their ability to trigger an allergic reaction. Cold plasma reactive species can cleave peptide bonds and modify amino acids, leading to the fragmentation of allergen proteins into smaller, less allergenic components. Oxidation, nitration, hydroxylation, and dehydrogenation of amino acids may also occur due to exposure to cold plasma, further altering the protein structure and reducing its allergenicity. Furthermore, cold plasma can induce protein crosslinking, forming insoluble aggregates that may be less accessible to immune recognition.

Venkataratnam et al. (2020) studied the effectiveness of cold plasma against the two main peanut allergens (Ara h 1 and Ara h 2). Using a pin to plate cold plasma reactor, dry, whole peanut (WP) and defatted peanut flour (DPF) were exposed to an atmospheric air discharge. SDS–PAGE study showed that the major peanut allergenic protein solubility decreased with increasing plasma exposure. Reduced antigenicity for both Ara h 1 and Ara h 2 was found by competitive ELISA using proteins isolated from plasma-treated WP or DPF. The greatest decrease in antigenicity was 66% for Ara h 2 and 65% for Ara h 1 from DPF. CD spectroscopy results indicate that changes in the allergen's secondary structure are the cause of the decrease in antigenicity. Venkataratnam et al. (2019) also evaluated the effect of cold atmospheric plasma at a voltage of 80 kV for varying treatment times (0, 15, 30, 45, and 60 min) on dry, defatted peanut flour (DPF), and whole peanut (WP). The SDS–PAGE, immunoblot, and competitive ELISA methods were used to examine the allergen samples. They also used circular dichroism to study the secondary structure. For both DPF and WP, the protein intensity bands corresponding to Ara h 1 did not alter, according to the SDS-PAGE data. The antigenicity was reduced by up to 43% for DPF and 9.3% for WP, according to competitive ELISA. Changes in secondary structure brought on by plasma reactive species were discovered by circular dichroism experiments.

Glow discharge plasma (80 W and 50 kHz power supply) was used to treat the cashew nuts under low pressure (Alves Filho et al. 2019). The total anacardic acid content of the nut rose from 0.7 to 1.2 $\mu g \cdot mg^{-1}$. Anacardic acid (C15:3) showed the biggest change, rising from 0.2 to 0.55 μg/mg of nut for samples treated with a flow of 10 $mL \cdot min^{-1}$ and 30 min of processing. The rabbit anti-cashew and human cashew allergy IgE binding were unaffected by plasma processing of cashew nuts. The treatment that had the least impact on the overall composition of the nuts was 10 min of plasma processing at a flow rate of 30 $mL \cdot min^{-1}$ of synthetic air, followed by 20 min at a flow rate of 5.8 $mL \cdot min^{-1}$. Ng et al. (2021) studied the effect of cold plasma using SDS–PAGE, FTIR, UPLC–MS/MS, and ELISA to assess casein, β-lactoglobulin, and α-lactalbumin both before and after plasma therapy. Following treatments, the casein and α-lactalbumin intensity bands decreased, according to SDS–PAGE data; however, the β-lactoglobulin intensity band stayed constant. According to FTIR experiments, plasma-induced changes in protein secondary structure were found, specifically in the amounts of β-sheet and β-turn. The antigenicity of casein and α-lactalbumin decreased, whereas that of β-lactoglobulin increased, according to ELISA. According to the study, ambient cold plasma can be modified to reduce the possibility of bovine milk allergies in dairy ingredients.

Wang et al. (2025) showed that ultrasound-assisted cold plasma treatment for 25 min (UC25) might help the secondary structure's transition from α-helix to β-sheet when compared to untreated peanut protein. According to in vitro allergenicity data, the high oxidizing characteristic of UC25 could destroy both native and exposed allergenic epitopes, resulting in a 74% reduction in IgG binding capacity when compared to the sole ultrasonic treatment (UL) group. Further supporting the substantial reduction of allergy symptoms by UC25, animal tests showed that UC25 administration could restore the Th1/Th2 balance by increasing IFN-γ secretion (by 2.11 times) and reducing IL-13 secretion (by 46%). The purpose of this study was to offer fresh perspectives for the creation of innovative hypoallergenic peanut products. Cheng et al. (2023) examined the effects of cold plasma (CP) from dielectric barrier discharge (DBD) on the allergenic qualities of tropomyosin (TM) in shrimp. After 20 min of treatment, the α-helix concentration dropped by 69% and the surface hydrophobicity rose by 578.8%. According to allergenicity analysis, following a 20-min treatment, the IgE binding capacity dropped by 96%, and the KU812 cells' degranulation indexes, including the β-HEX release rate, intracellular calcium ion intensity, histamine release, and inflammatory cytokines (IL-4, TNF-α), dropped by 32.5%, 31.0%, 37.3%, 51.7%, and 70.2%, respectively. The present investigation verified that DBD CP might decrease the allergenicity of TM by altering its structure.

These studies corroborate that cold plasma presents a novel and promising approach for mitigating the problem of food allergens by effectively modifying their protein structures. Its non-thermal nature and broad applicability make it an exciting area of research for developing hypoallergenic food products.

15.4.3 Modification of Food Functional Properties

Cold plasma technology presents a powerful and versatile tool for tailoring the functional properties of proteins, offering significant potential for innovation in the food, pharmaceutical, and biomaterial industries. Protein chemistry is impacted by the cold plasma treatment, which results in variations in how the protein behaves in solution and at surfaces. Moosavi et al. (2020) examined the effects of cold plasma treatments to alter the characteristics of protein films. After employing the solvent casting process to create the films from whey and gluten proteins, they were subjected to vacuum low-pressure glow plasma using argon and air gases at 50 KW for 5, 10, and 15 min. Atomic force microscopy (AFM) photos showed that after plasma treatment, the whey protein film roughness significantly increased, whereas the treated gluten films roughness significantly decreased. Additionally, after 10 min of processing, the tensile strength of the films increased dramatically, going from 6.902 to 10.772 MPa for the whey protein film and from 1.854 to 2.571 MPa for the gluten film. FTIR detected the introduction of functional groups, including C–O and O=C bonds, along with the formation of cross-links, which may result in modifications to several film characteristics. Other characteristics, such as the solubility and water vapor permeability of control and plasma-treated films, did not exhibit significant changes, despite the fact that the gas permeability of both edible polymers fell dramatically. Thus, protein films may be modified by the ideal low-pressure glow plasma conditions and timing.

Whey protein isolate (WPI) was treated with vacuum cold plasma (VCP) by Mohammadi et al. (2023). The VCP treatment was applied at low pressure using several gases (air, argon, and air combination, and sulfur hexafluoride (SF_6) in order to reduce the unwanted effects of excessive oxidation and to boost the efficiency of reactive species. As shown by oxidation processes, increased disulfide bond formation, and protein molecule cross-linking, the treatment resulted in a decrease in sulfur concentration and an increase in carbonyl content. According to fluorescence-based indicators, the proteins' hydrophobicity and aggregation appear to rise following VCP treatment with an argon–air gas mixture, while they decrease following VCP treatments with air and SF_6. The SF_6 plasma treatment causes the pH to significantly drop and the ζ-potential to rise, which in turn causes the aggregate size to significantly grow. Following SF_6VCP treatment, both the dynamic surface tension and the adsorption rates rise; whereas, following air–VCP and argon–air–VCP treatments, they fall. The type of VCP treatment determines the thickness of the strong viscoelastic interfacial layers formed by the adsorbed WPI particles.

Basak and Annapure (2022) reviewed the action of cold plasma on proteins and concluded that with little need for chemicals and solvents, cold plasma can be a sustainable and "green" way to modify proteins. Cold plasma treatment can result in the enhancement of the mechanical characteristics and biodegradability of plant protein-based films. In addition to the process of inactivating enzymes, cold plasma can also immobilize enzymes and create plant protein nanoparticles. Plant proteins' solubility, hydrophilicity, emulsifying, and foaming qualities can all be enhanced by plasma treatment. They also reported that protein functional qualities were negatively impacted by extended exposure to cold plasma treatments. Nonetheless, cold plasma is a promising technology to inactivate protein allergens and antinutritional components.

Wang et al. (2024) examined how the functional properties of chickpea protein isolate (CPI) were affected by the combination of pH-shifting and cold plasma (CP) treatment. The functional characteristics of CPI were considerably enhanced when the pH was changed to an acidic (pH = 2) or alkaline (pH = 12) state, followed by a 30-s CP treatment before neutralization at pH 7, as opposed to when the pH was changed alone. The most effective approach, which greatly increased solubility, was specifically carried out at pH 12, followed by a 30 s CP treatment. These functional enhancements were ascribed to improved surface hydrophobicity, decreased protein aggregation, and greater solubility. Additionally, after these treatments, changes in the secondary and tertiary protein structures had a favorable effect on the functional characteristics of CPI, providing significant implications for protein modification with enhanced functional profiles.

15.4.4 Applications in Food Packaging

Cold plasma technology is a new, environmentally friendly method with a wide range of possible uses in food packaging. Although it was initially created to boost polymer surface energy, improving adherence and printability, it has since become a potent instrument for surface sterilization of food items and food packaging materials. By providing non-thermal post-packaging food treatment, new trends seek to advance in-package decontamination. Furthermore, by bringing polar functional groups onto the bio-composite surface, cold plasma enhances the film surface properties, especially in protein-based films. This results in increased roughness, improved printability, increased adhesion, and a decreased contact angle; however, it is ineffective in enhancing the water vapor permeability. When subjected to microbial contaminations, cold plasma-treated edible packaging films showed notable improvement, which was primarily because cold plasma technology is non-thermal, which protects the antibacterial potency of bioactive chemicals and antimicrobial constituents. As a result, it can be said that cold plasma treatment is a novel way to improve the qualities of edible films and a viable substitute for the chemical and physical modification techniques that are now in use.

In order to lessen the sensitivity of films made with fish myofibrillar proteins to water, cold plasma, and a carnauba wax covering were used (Romani et al. 2020). Protein solution casting, plasma treatment (glow discharge), and carnauba wax coating were used to create the bi-layer films. The packaging-related characteristics of protein films (thickness, mechanical and barrier performance, water solubility, color, and opacity), microstructure, glass transition, and melting temperatures were assessed in relation to the impacts of plasma and wax coating. The application of glow discharge plasma and subsequent coating with carnauba wax produced fish protein films with a 175% increase in tensile strength and a 65% decrease in water vapor permeability when compared to control films. Thermal transitions, X-ray diffraction patterns, and scanning electron microscopy pictures all verified the changes in the characteristics of fish protein films. Thus, the combination of cold plasma and carnauba wax coating is a promising way to enhance the packaging properties of fish protein films.

15.4.5 Application for Removal of Pesticide Residues

In recent years, there has been a widespread awareness among consumers about the residual effects of pesticides in food. Cold plasma has been proven to be effective against various commonly used pesticides. Further, with a broader spectrum of reactive species, it is the most potent and versatile technology when compared to pulsed electric field and ozonation (Table 15.2). NTP treatment reduced the pesticide imidacloprid up to 91.4% by reactive oxygen species chemisorption degradation mechanism (Ni et al. 2024). Sarangapani et al. (2020) reported more than 70% reduction in pesticides chlorpyrifos and carbaryl in grapes and strawberries using plasma-activated water treatment. Similarly, upto 90% reduction in abamectin and fenpyroximate residues on Barhi dates was achieved after 120 s treatment under gliding-arch NTP. Further, in the case of methiocarb and aminocarb, more than 99% of these pesticides in water were degraded after 5 min of DBD plasma. Mahbubeh Mousavi et al. (2017) reported that DBD cold plasma was effective in degrading pesticides diazinon and chlorpyrifos, and the degrading efficiency was reduced when the concentration of these pesticides was increased from 500 ppm to 1000 ppm in apples and cucumbers.

Table 15.2 Applications of cold plasma on food pesticide residues

Food material	Type of pesticide	Processing conditions	Percentage reduction	Reference
Tomato	Chlorpyrifos	5 W for 6 min.	89	Ranjitha Gracy et al. (2019)
Mango	Chorpyrifos	8 kV for 5 min.	74	Phan et al. (2017)
	Cypermetrin		63	
Lycium barbarum	Dichlorovos	10 kV for 30 min.	96.8	Zhou et al. (2018)
	Omethoate		99.5	
Blueberry	Boscalid	80 kV for 5 min.	75	Sarangapani et al. (2017)
	Imidacloprid		80	
Apple	Chlorpyrifos	13 Kv for 10 min	86	Mahbubeh Mousavi et al. (2017)
	Diazinon		87	
Cucumber	Chlorpyrifos		58	
	Diazinon		82	
Cucumber	Diazinon	0.75 W for 15 min.	88	Dorraki et al. (2016)
Strawberry	Azoxystrobin	80 kV for 5 min.	69	Misra et al. (2014)
	Cyprodinil		45	
	Fludioxonil		71	
	Pyriproxyfen		46	
Black Grapes	Chlorpyrifos	120 V for 15 min	65.25	Rakesh et al. (2024)
Lettuce	Malathion Chlorpyrifos	80 kV for 120 s	53.1 51.4	Cong et al. (2021)
Corn	Chlorpyrifos Carbaryl	20 W for 60 s	86.2 66.6	Liu et al. (2021)

15.5 Conclusion

The growing demand for minimally processed, nutritious, and preservative-free foods has propelled research into non-thermal preservation methods like cold plasma (CP). This innovative technology offers a promising alternative to conventional thermal treatments, capable of preserving the nutritional integrity and extending the shelf life of various food products. CP demonstrates significant potential for commercial applications in ensuring microbial safety of foods, inactivation of enzymes, modification of functional properties of foods, eliminating, or reducing chemical and biological contaminants such as pesticides and allergens, thereby enhancing consumer safety while maintaining the desired organoleptic and nutritional qualities. Furthermore, its integration into packaging materials could revolutionize shelf-life extension, delivering substantial benefits across the food industry by ensuring prolonged product freshness and minimizing waste.

However, optimizing CP's performance, economic viability, and environmental sustainability hinges on precise control of key parameters, including gas composition, pressure, electric field strength, frequency, and temperature. Achieving optimal outcomes also necessitates a profound understanding of microbial diversity, substrate characteristics, and the synergistic effects of UV radiation, reactive species, and free radicals. While CP represents a notable advancement in food safety and quality assurance, it is crucial to acknowledge its limitations. It is not a universal solution, and its judicious application often requires a hybrid approach, combining CP with other preservation techniques, to maximize food safety and nutritional integrity effectively within complex food systems.

References

Adesina K, Lin TC, Huang YW, Locmelis M, Han D (2024) A review of dielectric barrier discharge cold atmospheric plasma for surface sterilization and decontamination. IEEE Trans Radiat Plasma Med Sci 8(3):295–306. https://doi.org/10.1109/TRPMS.2024.3349571

Afshari R, Hosseini H (2012) Atmospheric pressure plasma technology: a new tool for food preservation

Alves Filho EG, Silva LMA, Oiram Filho F, Rodrigues S, Fernandes FAN, Gallão MI, Mattison CP, de Brito ES (2019) Cold plasma processing effect on cashew nuts composition and allergenicity. Food Res Int 125:108621. https://doi.org/10.1016/j.foodres.2019.108621

Bardos L, Barankova H (2010) Cold atmospheric plasma: sources, processes, and applications [Article]. Thin Solid Films 518(23):6705–6713. https://doi.org/10.1016/j.tsf.2010.07.044

Basak S, Annapure US (2022) Recent trends in the application of cold plasma for the modification of plant proteins – a review. Future Foods 5:100119. https://doi.org/10.1016/j.fufo.2022.100119

Bourke P, Ziuzina D, Han L, Cullen PJ, Gilmore BF (2017) Microbiological interactions with cold plasma. J Appl Microbiol 123(2):308–324. https://doi.org/10.1111/jam.13429

Brandenburg R (2017) Dielectric barrier discharges: progress on plasma sources and on the understanding of regimes and single filaments. Plasma Sources Sci Technol 26(5):053001. https://doi.org/10.1088/1361-6595/aa6426

Cheng J-H, Li J, Sun D-W (2023) Effects of dielectric barrier discharge cold plasma on structure, surface hydrophobicity and allergenic properties of shrimp tropomyosin. Food Chem 409: 135316. https://doi.org/10.1016/j.foodchem.2022.135316

Choi S, Puligundla P, Mok C (2017) Effect of corona discharge plasma on microbial decontamination of dried squid shreds including physico-chemical and sensory evaluation. LWT-Food Sci Technol 75:323–328. https://doi.org/10.1016/j.lwt.2016.08.063

Cong L, Huang M, Zhang J, Yan W (2021) Effect of dielectric barrier discharge plasma on the degradation of malathion and chlorpyrifos on lettuce. J Sci Food Agric 101(2):424–432

Conrads H, Schmidt M (2000) Plasma generation and plasma sources. Plasma Sources Sci Technol 9(4):441–454. https://doi.org/10.1088/0963-0252/9/4/301

Dorraki N, Mahdavi V, Ghomi H, Ghasempour A (2016) Elimination of diazinon insecticide from cucumber surface by atmospheric pressure air-dielectric barrier discharge plasma. Biointerphases 11(4):041007

Flores-Fuentes A, Pena-Eguiluz R, Lopez-Callejas R, Mercado-Cabrera A, Valencia-Alvarado R, Barocio-Delgado S, Piedad-Beneitez A d l (2009) Electrical model of an atmospheric pressure dielectric barrier discharge cell. IEEE Trans Plasma Sci 37(1):128–134. https://doi.org/10.1109/TPS.2008.2006844

Fridman A (2008) Plasma chemistry. Cambridge University Press

Hong YC, Kang WS, Hong YB, Yi WJ, Uhm HS (2009) Atmospheric pressure air-plasma jet evolved from microdischarges: eradication of E. coli with the jet. Phys Plasmas 16(12). https://doi.org/10.1063/1.3272089

Huang H, Tang L (2007) Treatment of organic waste using thermal plasma pyrolysis technology. Energy Convers Manag 48(4):1331–1337

Illera AE, Souza VR, Tang L, Nikmaram N, Keener KM (2023) Effect of high voltage atmospheric cold plasma on chicken eggs quality during refrigerated storage. Food Biosci 53:102754

Kaur S, Kumar Y, Singh V, Kaur J, Panesar PS (2024) Cold plasma technology: reshaping food preservation and safety. Food Control 163:110537. https://doi.org/10.1016/j.foodcont.2024.110537

Kaushik NK, Ghimire B, Li Y, Adhikari M, Veerana M, Kaushik N, Jha N, Adhikari B, Lee SJ, Masur K, von Woedtke T, Weltmann KD, Choi EH (2018) Biological and medical applications of plasma-activated media, water and solutions. Biol Chem 400(1):39–62. https://doi.org/10.1515/hsz-2018-0226

Kim JE, Oh YJ, Won MY, Lee K-S, Min SC (2017) Microbial decontamination of onion powder using microwave-powered cold plasma treatments. Food Microbiol 62:112–123

Lebedev YA (2010) Microwave discharges: generation and diagnostics. J Phys Conf Ser 257(1): 012016. https://doi.org/10.1088/1742-6596/257/1/012016

Liao X, Li J, Muhammad AI, Suo Y, Chen S, Ye X, Liu D, Ding T (2018) Application of a dielectric barrier discharge atmospheric cold plasma (Dbd-Acp) for Eshcerichia coli inactivation in apple juice. J Food Sci 83(2):401–408. https://doi.org/10.1111/1750-3841.14045

Lin L, Liao X, Li CZ, Abdel-Samie MA, Cui HY (2020) Inhibitory effect of cold nitrogen plasma on Salmonella typhimurium biofilm and its application on poultry egg preservation. LWT-Food Sci Technol 126:109340. ARTN 109340

Liu H, Guo D, Feng X (2021) Plasma degradation of pesticides on the surface of corn and evaluation of its quality changes. Sustainability 13(16):8830

Lu X, Laroussi M, Puech V (2012) On atmospheric-pressure non-equilibrium plasma jets and plasma bullets [Article]. Plasma Sources Sci Technol 21(3):17. https://doi.org/10.1088/0963-0252/21/3/034005

Lukes P, Locke BR, Brisset JL (2012) Aqueous-phase chemistry of electrical discharge plasma in water and in gas–liquid environments. In: Plasma chemistry and catalysis in gases and liquids, pp 243–308

Mahbubeh Mousavi S, Imani S, Dorranian D, Larijani K, Shojaee M (2017) Effect of cold plasma on degradation of organophosphorus pesticides used on some agricultural products. J Plant Prot Res 57(1)

Mahnot NK, Mahanta CL, Keener KM, Misra N (2019) Strategy to achieve a 5-log Salmonella inactivation in tender coconut water using high voltage atmospheric cold plasma (HVACP). Food Chem 284:303–311

Misra N, Pankaj S, Walsh T, O'Regan F, Bourke P, Cullen P (2014) In-package nonthermal plasma degradation of pesticides on fresh produce. J Hazard Mater 271:33–40

Mohammadi OE, Yeganehzad S, Hesarinejad MA, Dabestani M, Schneck E, Miller R (2023) Effects of various types of vacuum cold plasma treatment on the chemical and functional properties of whey protein isolate with a focus on interfacial properties. Colloids and Interfaces, 7(3):54. https://doi.org/10.3390/colloids7030054

Moosavi MH, Khani MR, Shokri B, Hosseini SM, Shojaee-Aliabadi S, Mirmoghtadaie L (2020) Modifications of protein-based films using cold plasma. Int J Biol Macromol 142:769–777. https://doi.org/10.1016/j.ijbiomac.2019.10.017

Nasiru MM, Frimpong EB, Muhammad U, Qian J, Mustapha AT, Yan W, Zhuang H, Zhang J (2021) Dielectric barrier discharge cold atmospheric plasma: influence of processing parameters on microbial inactivation in meat and meat products. Compr Rev Food Sci Food Saf 20(3): 2626–2659

Ng SW, Lu P, Rulikowska A, Boehm D, O'Neill G, Bourke P (2021) The effect of atmospheric cold plasma treatment on the antigenic properties of bovine milk casein and whey proteins. Food Chem 342:128283. https://doi.org/10.1016/j.foodchem.2020.128283

Ni JB, Jia XF, Zhang JY, Ding CJ, Tian WL, Peng WJ, Fang XM (2024) Efficient degradation of imidacloprid by surface discharge cold plasma: Mechanism of interaction between ROS and molecular structure and evaluation of residual toxicity. Journal of Hazardous Materials, 465, 133066. https://ssrn.com/abstract=4492734

Nwabor OF, Onyeaka H, Miri T, Obileke K, Anumudu C, Hart A (2022) A cold plasma technology for ensuring the microbiological safety and quality of foods. Food Eng Rev 14(4):535–554. https://doi.org/10.1007/s12393-022-09316-0

Olatunde OO, Benjakul S, Vongkamjan K (2020) Cold plasma combined with liposomal ethanolic coconut husk extract: a potential hurdle technology for shelf-life extension of Asian sea bass slices packaged under modified atmosphere. Innov Food Sci Emerg Technol 65:102448

Pawłat J, Terebun P, Kwiatkowski M, Tarabová B, Kovaľová Z, Kučerová K, Machala Z, Janda M, Hensel K (2019) Evaluation of oxidative species in gaseous and liquid phase generated by mini-gliding arc discharge. Plasma Chem Plasma Process 39:627–642

Perinban S, Orsat V, Raghavan V (2019) Nonthermal plasma–liquid interactions in food processing: a review. Compr Rev Food Sci Food Saf 18(6):1985–2008. https://doi.org/10.1111/1541-4337.12503

Phan KTK, Phan HT, Brennan CS, Phimolsiripol Y (2017) Nonthermal plasma for pesticide and microbial elimination on fruits and vegetables: an overview. Int J Food Sci Technol 52(10): 2127–2137

Rakesh B, Anbarasan R, Kamalapreetha B, Purushothaman R, Ashutosh BR, Mahendran R (2024) Chlorpyrifos pesticide removal from black grapes using plasma-activated water produced by plasma bubbling technology. J Food Process Preserv 2024(1):7856706

Ranjitha Gracy TK, Gupta V, Mahendran R (2019) Influence of low-pressure nonthermal dielectric barrier discharge plasma on chlorpyrifos reduction in tomatoes. J Food Process Eng 42(6): e13242. https://doi.org/10.1111/jfpe.13242

Romani VP, Olsen B, Collares MP, Oliveira JRM, Prentice C, Martins VG (2020) Cold plasma and carnauba wax as strategies to produce improved bilayer films for sustainable food packaging. Food Hydrocolloids, 108:106087. https://doi.org/10.1016/j.foodhyd.2020.106087

Rossow M, Ludewig M, Braun PG (2018) Effect of cold atmospheric pressure plasma treatment on inactivation of campylobacter jejuni on chicken skin and breast fillet. LWT 91:265–270

Sarangapani C, O'Toole G, Cullen PJ, Bourke P (2017) Atmospheric cold plasma dissipation efficiency of agrochemicals on blueberries. Innovative Food Sci Emerg Technol 44(Supplement C):235–241. https://doi.org/10.1016/j.ifset.2017.02.012

Sarangapani C, Scally L, Gulan M, Cullen PJ (2020) Dissipation of pesticide residues on grapes and strawberries using plasma-activated water. Food Bioprocess Technol 13(10):1728–1741. https://doi.org/10.1007/s11947-020-02515-9

Segura-Ponce LA, Reyes JE, Troncoso-Contreras G, Valenzuela-Tapia G (2018) Effect of low-pressure cold plasma (LPCP) on the wettability and the inactivation of Escherichia coli and Listeria innocua on fresh-cut apple (Granny Smith) skin. Food Bioprocess Technol 11: 1075–1086

Subrahmanyam K, Gul K, Sehrawat R, Tiwari BK, Sahoo S (2024) Cold plasma-mediated inactivation of microorganisms for the shelf-life extension of animal-based foods: efficiency, mechanism of inactivation, and impact on quality attributes. Food Control 162:110464. https://doi.org/10.1016/j.foodcont.2024.110464

Thakur M, Modi VK (2020) Emerging technologies in food science – focus on the developing world. Springer Nature

Thakur M, Belwal T (2023) Bioactive components. Springer, Singapore. https://doi.org/10.1007/978-981-19-2366-1_1

Venkataratnam H, Cahill O, Sarangapani C, Cullen PJ, Barry-Ryan C (2020) Impact of cold plasma processing on major peanut allergens. Sci Rep 10(1):17038. https://doi.org/10.1038/s41598-020-72636-w

Venkataratnam H, Sarangapani C, Cahill O, Ryan CB (2019) Effect of cold plasma treatment on the antigenicity of peanut allergen Ara h 1. Innovative Food Sci Emerg Technol 52:368–375. https://doi.org/10.1016/j.ifset.2019.02.001

Wang Y, Wang Z, Zhu X, Yuan Y, Gao Z, Yue T (2020) Application of electrical discharge plasma on the inactivation of Zygosaccharomyces rouxii in apple juice. LWT 121:108974

Whitehead JC (2016) Chapter 3 – The chemistry of cold plasma. In: Misra NN, Schlüter O, Cullen PJ (eds) Cold plasma in food and agriculture. Academic, pp 53–81. https://doi.org/10.1016/B978-0-12-801365-6.00003-2

Yong HI, Lee H, Park S, Park J, Choe W, Jung S, Jo C (2017) Flexible thin-layer plasma inactivation of bacteria and mold survival in beef jerky packaging and its effects on the meat's physicochemical properties. Meat Sci 123:151–156. https://doi.org/10.1016/j.meatsci.2016.09.016

Zhou R, Zhou R, Yu F, Xi D, Wang P, Li J, Wang X, Zhang X, Bazaka K, Ostrikov KK (2018) Removal of organophosphorus pesticide residues from Lycium barbarum by gas phase surface discharge plasma. Chem Eng J 342:401–409

Zhu Y, Li C, Cui H, Lin L (2020) Feasibility of cold plasma for the control of biofilms in food industry. Trends Food Sci Technol 99:142–151

Ziuzina D, Misra N, Han L, Cullen P, Moiseev T, Mosnier J-P, Keener K, Gaston E, Vilaró I, Bourke P (2020) Investigation of a large gap cold plasma reactor for continuous in-package decontamination of fresh strawberries and spinach. Innov Food Sci Emerg Technol 59:102229

Ziuzina D, Patil S, Cullen PJ, Keener K, Bourke P (2014) Atmospheric cold plasma inactivation of Escherichia coli, salmonella enterica serovar Typhimurium and listeria monocytogenes inoculated on fresh produce. Food Microbiol 42:109–116

Wang Y, Zhang L, Zhao J, Raghavan V, Qian J, Wang J (2025) Insight into the mechanism of ultrasound-assisted cold plasma alleviated the allergenicity of peanut protein with improved functional properties. Food Hydrocoll 166:111295. https://doi.org/10.1016/j.foodhyd.2025.111295

Wang J, Zhou X, Li J, Pan D, Du L (2024) Enhancing the functionalities of chickpea protein isolate through a combined strategy with pH-shifting and cold plasma treatment. Innov Food Sci Emerg Technol 93:103607. https://doi.org/10.1016/j.ifset.2024.103607

Chapter 16
Fruit-Flavored Kombucha: Crafting a Healthier, Functional Beverage to Replace Sugary Drinks

C. Rajna, R. Jenisha Priyadharshini, and S. Madumitha

Abstract Kombucha is a fermented beverage produced through the activity of a symbiotic culture of bacteria and yeast (SCOBY) on sweetened tea, and it is increasingly viewed as a healthier substitute for sugar-laden drinks due to its functional properties. The objective of this endeavor was to formulate a nutritionally enhanced kombucha fruit drink by incorporating grape juice, basil extract, and star anise, thereby creating a beverage with improved health attributes. Three formulations were prepared using different proportions of kombucha tea base, grape juice, basil extract, and star anise: V1 (80 ml, 15 ml, 5 ml, 0.5 g), V2 (70 ml, 20 ml, 10 ml, 0.5 g), and V3 (60 ml, 25 ml, 15 ml, 0.5 g). Sensory evaluation revealed that the V3 variant received the highest acceptance, scoring 4.8 ± 0.3 for appearance and 4.8 ± 0.5 for flavor. Nutritional analysis of V3 indicated that each 100 ml of the drink contained 31.17 kcal of energy, 179 mg of phenolic compounds, 91.3% DPPH antioxidant activity, and 4.87 g of alcohol. Shelf-life assessment over 45 days showed minimal changes in color and taste at ambient temperature, while refrigerated samples exhibited no notable alterations. The developed kombucha fruit beverage offers a refreshing, probiotic-rich option that may support digestive health, possess antioxidant properties, and contribute to overall well-being.

Keywords Kombucha · SCOBY · Fermentation · Probiotics · Immune booster

C. Rajna · R. Jenisha Priyadharshini · S. Madumitha (✉)
Department of Food Science and Nutrition, Dr. N.G.P Arts and Science College, Coimbatore, Tamil Nadu, India

M. Thakur, T. Belwal (eds.), *Food Biodiversity and Consumer Engagements*, Sustainable Development and Biodiversity 46,
https://doi.org/10.1007/978-981-95-7777-4_16

16.1 Introduction

Asian-inspired kombucha is a fermented beverage. But because of its medicinal properties, which include antimicrobial, antioxidant, antidiabetic, anticarcinogenic, and treatment for high cholesterol and stomach ulcers, it has become more and more well-liked in the West. Additionally, it has demonstrated effects on liver detoxification and the immune response (Chakravorty et al. 2016).

The traditional beverage is produced by fermenting black tea (*Camellia sinensis*) that is then sweetened. But various teas can also be used in its production (Rodrigues et al. 2018). Tea fermentation results from a symbiotic colony of yeasts and bacteria enclosed in a cellulose film (De Filippis et al. 2018). The name symbiotic colony of bacteria and yeasts (SCOBY) refers to this type of cellulose film. But it is also referred to as kombucha mother or tea fungus (Santos et al. 2020).

Yeasts that thrive in high-sugar environments break down the sugars in the tea to produce ethanol, and then acetic acid-producing bacteria convert that ethanol into acetic acid (Teoh et al. 2004). Acetic acid is not the only organic acid that is produced; other acids like gluconic, lactic, malic, citric, and tartaric also have antibacterial properties and keep harmful bacteria from spoiling kombucha (Neffe Skocinska et al. 2017).

The presence of probiotic microorganisms (lactic and acetic bacteria), antibiotics, amino acids, polyphenols from tea, sugars, organic acids, ethanol, water-soluble vitamins, and a range of micronutrients produced during fermentation are all believed to be beneficial to the health benefits of kombucha (Fu et al. 2017). Years later, Swiss researchers found that the acids in kombucha stimulated the growth of beneficial microbes in the intestine, which renders it just as healthy to consume as yogurt (Frank 1991). Since, kombucha is mildly carbonated and acidic, customers are more likely to accept it. Because of its high degree of carbonation, it can serve as a healthy, low-alcoholic alternative to sparkling wines or soft drinks.

The work aimed to formulate a kombucha beverage obtained by fermentation of kombucha tea and fruit juice under different conditions, develop a value-added shelf-stable product, evaluate the influence of the storage conditions on shelf life, bioactive compounds, and finally evaluate its acceptance.

16.2 Materials and Methodology

16.2.1 Selection of Raw Material

The ingredients required to produce kombucha were purchased from the local markets of Coimbatore and the local village areas. Grapes, sugar, basil leaves, star anise, water. Grapes of good quality, without bruises and dark colored grapes, because they have more antioxidants. Basil leaves are selected fresh without the stems. Potable water is selected, clean, and free from dust particles.

16.2.2 Processing of Product

16.2.2.1 Production of Kombucha Tea

Water (1000 ml)

↓

Addition of black tea powder (10g) and sugar (250g)

↓

Boiling for 10 minutes

↓

Cooling & filtering

↓

Inoculation of SCOBY

↓

Fermentation at room temperature for 7-15 days

↓

Kombucha

16.2.2.2 Preparation of Grape Flavored Kombucha

Addition of grape juice to the prepared kombucha tea

↓

Addition of basil leaf extract and spice star anise

↓

Re-Fermentation (7 days)

↓

Storage

16.2.3 Production of Grape Flavored Kombucha

Dark red-colored grapes (*Vitis vinifera*) were selected and blanched for 10 min. The juice was prepared by adding the required amount of water and sugar. And then it was filtered and transferred into a clean glass bottle with the fermented tea, basil extract, and star anise. Then the bottles were labelled and kept for fermentation at room temperature for 45 days.

16.2.4 Standardization of the Developed Product

The product is formulated under four categories of control: V-1, V-2, and V-3. The standardization of the developed product is given in Table 16.1.

16.2.5 Sensory Evaluation of the Developed Product

The kombucha prepared using all the variations was subjected to organoleptic evaluation for its quality attributes such as appearance, color, texture, taste, flavor, and overall acceptability using 5-point hedonic rating scale. A total of 25 semi-trained panel members were involved. Among the selected Variations, they were kept for 45 days at room temperature and cold temperature, and sensory evaluation

Table 16.1 Quantity of ingredients for formulation of the product

Ingredients	V1	V2	V3
Kombucha tea	80 ml	70 ml	60 ml
Grape juice	15 ml	20 ml	25 ml
Basil extract	5 ml	10 ml	15 ml
Star anise	0.5 g	0.5 g	0.5 g

was conducted to determine any changes in their quality parameters. The scores obtained from the sensory evaluation were calculated for mean and SD using MS Excel 2007 to find the most acceptable product.

16.3 Results and Discussion

In this section, the key findings are discussed with a combined analysis that aims to provide a comprehensive understanding of the research outcomes and patterns observed in the data.

16.3.1 Organoleptic Evaluation of the Developed Product

The organoleptic evaluation involves the assessment of various attributes of the product such as appearance, color, texture, aroma, flavor, and mouth feel. These attributes are evaluated by panel members, who are trained to detect and quantify differences in the sensory properties of products. The grape-flavored kombucha was prepared and analyzed by 30 semi-trained panel members using a five-point hedonic rating scale. In this evaluation, the quality of the product was asked to be judged by the panel members on appearance, color, flavor, taste, consistency, and overall acceptability.

Figure 16.1 indicates the average sensory score of the grape-flavored kombucha. Among the three variations, V3 was highly accepted in all sensory characteristics.

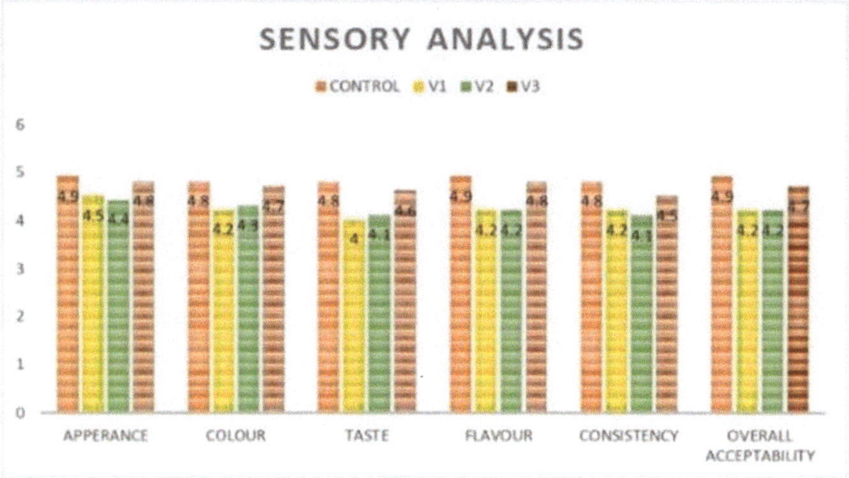

Fig. 16.1 Sensory analysis of Variation 3 (*expressed in means ± SD)

16.3.2 Nutrient Analysis of the Grape Flavor Kombucha

The proximate principle of analysis includes the determination of nutrients like energy, carbohydrates, protein, fat, calcium, iron, sodium, potassium, vitamin C, and the physicochemical properties like moisture, ash, pH, alcohol content, total soluble solids, phenolic compounds, antioxidant activity, and total plate count (Table 16.2). The nutrient analysis is done for the selected variations from the developed product.

The selected variation, V-3, contains 7.23 g of carbohydrates, 0.54 g of protein,, 31.17kcal of energy, 12.3 mg of calcium, 0.36 mg of iron, 6.11mg of sodium, 107mg of potassium, and 2.32mg of vitamin C per 100 ml.

16.3.3 Shelf Life of the Developed Kombucha

The shelf life of the product is tested by storing the product at room temperature and in the refrigerator. The shelf life can be determined by keeping the product at room temperature and in the refrigerator for 45 days. The developed grape-flavored kombucha showed no changes in its color, flavor, taste, and consistency after the study period under cold temperature. Whereas kombucha stored at room temperature after the study period showed a slight decline in its flavor and consistency. Therefore, kombucha stored in the refrigerator for weeks or even a few months can be consumed with a satisfactory sensory property, depending on the initial level of acidity and the conditions of storage and handling practices. Over time, the acidity level of the kombucha may increase, which can make the beverage taste more sour or acidic.

Table 16.2 Nutrient content of the grape-flavored kombucha

S. no.	Parameters	Nutritive value/100 ml
01	Energy	31.17 kcal
02	Carbohydrate	7.23 g
03	Protein	0.54 g
04	PH	2.13
05	Sodium	6.11 mg
06	Calcium	12.3 mg
07	Potassium	107 mg
08	Vitamin C	2.32 g
09	Iron	0.36 g
10	Phenolic compounds	179 mg
11	Moisture	81.7 g
12	Ash	6.73 g
13	Alcohol	4.87 g
14	Total plate count	1×10^6 CFU/ml
15	Total soluble solids	13.2 g
16	Antioxidant activity	91.3 DPPH

16.4 Conclusion

Fermentation is an ancient form of bio-preservation that is common to all regions of the world. Functional beverages have become more and more popular from a scientific, consumer, and commercial aspect as traditional fermented products continue to thrive in numerous fields. Kombucha is a good source of bioactive compounds, which make it a functional beverage. Kombucha tea is now regarded as a "health drink". Also, fruit juices are a good source of nutrients with a lot of health benefits as they strengthen the immune system, aid digestion, prevent allergies, detoxifies our body, and enhance weight loss. The nutritive value and health benefits of fruit juices are improved by flavoring the fruit juices with kombucha and other natural flavors such as basil, Indian borage, cardamom, and star anise.

References

Chakravorty S, Bhattacharya S, Chatzinotas A, Chakraborty W, Bhattacharya D, Gachhui R (2016) Kombucha tea fermentation: microbial and biochemical dynamics. Int J Food Microbiology 220:63–72

De Filippis F, Troise AD, Vitaglione P, Ercolini D (2018) Different temperatures select distinctive acetic acid bacteria species and promote organic acid production during kombucha tea fermentation. Food Microbiol 73:11–16

Frank GW (1991) Kombucha: healthy beverage and natural remedy from the Far East; its correct preparation and use. Ennsthaler

Fu QY, Li QS, Lin XM, Qiao RY, Yang R, Li XM, Liang YR (2017) Antidiabetic effects of tea. Molecules 22(5):849

Neffe-Skocinska, K., Sionek, B., Scibisz, ´ I., Kolozyn-krajewska, D., 2017. Acid contents and the effect of fermentation condition of Kombucha tea beverages on physicochemical, microbiological and sensory properties. CyTA-J Food 15 (4), 601–607

Paludo Rodrigues RDS, Machado MRG, Barboza G, Soares LS, Heberle T, Leivas YM (2018) Características Físicas e químicas de Kombucha à base de chá de Hibisco (Hibiscus sabdariffa, L.). 6° Simpósio de Segurança Alimentar, Rio Grande do Sul

Santos JA, Fraga H, Malheiro AC, Moutinho-Pereira J, Dinis LT, Correia C, Schultz HR (2020) A review of the potential climate change impacts and adaptation options for European viticulture. Appl Sci 10(9):3092

Teoh AL, Heard G, Cox J (2004) Yeast ecology of Kombucha fermentation. Int J Food Microbiol 95(2):119–126

Chapter 17
Innovative Food Solutions for Health: Integrating Nutrition, Safety, and Sustainability in Modern Food Systems

Karuna Singh, Monika Thakur, and Corinne Malpuech-Bruger

Abstract The field of food innovation for health is rapidly advancing, focusing on the creation of new food products, processes, and technologies to boost nutritional value, enhance food safety, and encourage sustainable practices within the food industry. This review delves into the essential elements of food innovation for health, such as nutritional enhancement, functional foods, alternative proteins, clean label and natural ingredients, personalized nutrition, food safety and traceability, and sustainable practices. Nutritional enhancement involves fortifying foods with essential nutrients, biofortification, and targeted supplementation to address dietary deficiencies and improve public health. Functional foods are enriched with ingredients beyond basic nutrition that are utilized by the body for specific health benefits, like omega-3 fatty acids for heart health. The alternative protein market, driven by environmental and ethical concerns, encompasses plant-based proteins, cultured meats, and insect-based foods. Consumers increasingly demand clean label and natural ingredients, spurring innovations in minimally processed foods. Personalized nutrition utilizes technological advancements to offer individualized dietary recommendations based on genetic, health, and lifestyle factors. For the improvement of food safety and traceability, blockchain technology is being employed. Sustainable practices in the food industry aim to reduce food waste, utilize sustainable packaging materials, and promote environmentally friendly farming practices. The food innovation for the health sector spans various industries, propelled by consumer demand, regulatory changes, and advancements in food science and technology, presenting opportunities for growth and collaboration among health

K. Singh (✉)
Sharda school of Allied Health Sciences, Sharda University, Greater Noida, Uttar Pradesh, India
e-mail: karuna.singh@sharda.ac.in

M. Thakur
Amity Institute of Food Technology, Amity University Uttar Pradesh, Noida, Uttar Pradesh, India

C. Malpuech-Bruger
Clermont Auvergne University (UCA), Clermont-Ferrand, France

M. Thakur, T. Belwal (eds.), *Food Biodiversity and Consumer Engagements*, Sustainable Development and Biodiversity 46,
https://doi.org/10.1007/978-981-95-7777-4_17

professionals, food scientists, and entrepreneurs to enhance public health through innovative food solutions.

Keywords Enhancement · Functional food · Personalized nutrition · Alternative protein · Sustainable practices

17.1 Introduction

Food innovation for health encompasses the creation of new food products, processes, and technologies that emphasize health benefits. This field investigates methods to boost nutritional value, enhance food safety, and establish sustainable practices in food production (Thakur 2024; Thakur et al. 2024a, b). The rising demand for nutritious and sustainable food options has led to the creation of new alternatives, even for traditional products. Enhanced foods can be developed by reducing certain ingredients, incorporating new ones, or substituting traditional ingredients with others. By altering their products, manufacturers can provide healthier options to a growing number of consumers who are keen on maintaining a balanced diet (Hsieh and Ofori 2007). Foods now go beyond fulfilling basic physical needs and are expected to support health and well-being. In today's society, people seek both good health and longevity, thus requiring nutritious and functional foods that enhance their well-being, enjoyment, and active lifestyle. Convenient health foods or those offering additional health benefits have become a top priority in the food industry. Similarly, the increasing range of dietary supplements available in health stores is also gradually finding its way into new food formulation. (Salifou et al. 2023).

The following are some key aspects related to food innovation for health and its markets:

17.1.1 Nutritional Enhancement

Nutritional Enhancement plays an essential part in enhancing global nutrition and public health. As research progresses, new technologies and methods are continually being developed, providing innovative solutions to tackle nutritional challenges worldwide. These innovations aim to enrich foods with vital nutrients, such as vitamins, minerals, and proteins, to combat dietary deficiencies and improve health outcomes. Nutritional enhancement involves the process of increasing the nutritional value of food products through various techniques. This practice aims to address nutrient deficiencies, promote public health, and enhance overall food quality.

Key aspects of nutritional enhancement include the following:

17.1.1.1 Fortification

Incorporating essential nutrients into foods to boost their nutritional value is a widespread practice. Examples include salt fortified by iodine, vitamin D-fortified milk, and flour with folate (Allen et al. 2006). Food fortification is a strategic approach to mitigate nutrient deficiencies within populations over medium to long-term periods. This process involves adding a nutrient-rich 'premix' of essential vitamins and minerals into widely consumed foods during manufacturing (Mannar and Wesley 2016). Depending on the food processing techniques used, as shown in Fig. 17.1, various methods can be applied to ensure these micronutrients are retained as much as possible.

A well-implemented food fortification program can sustainably and consistently provide large populations with significant amounts of essential micronutrients (Mannar and Wesley 2016).

17.1.2 *Biofortification*

Developing crops with higher nutrient contents through selective breeding or genetic modification. Examples include golden rice (enhanced with β-carotene) and iron-rich beans. (Ofori et al. 2022).

Biofortification enhances the concentration and bioavailability of essential nutrients in crops during growth (Malik and Maqbool 2020). This approach develops crops with elevated micronutrients, alongside traits such as increased yield and disease resistance. Unlike food fortification, which adds nutrients during post-harvest processing, biofortification incorporates nutrients before harvest (Umar et al. 2019). Current strategies for biofortifying crops include agronomic biofortification, plant breeding, and transgenic methods. These methods have been applied to cereals, legumes, oilseeds, vegetables, and fruits (Malik and Maqbool

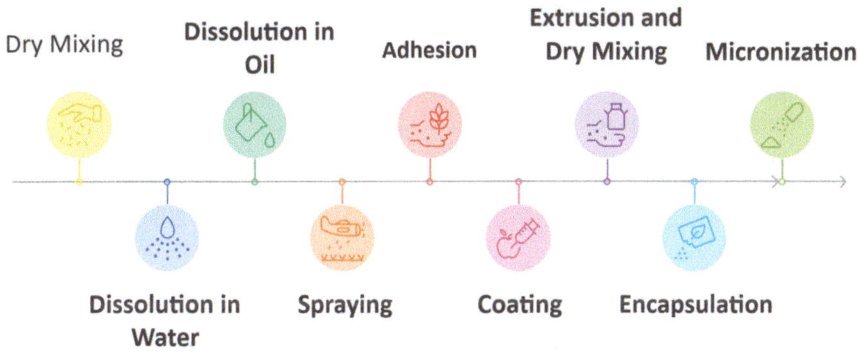

Fig. 17.1 Food fortification methods

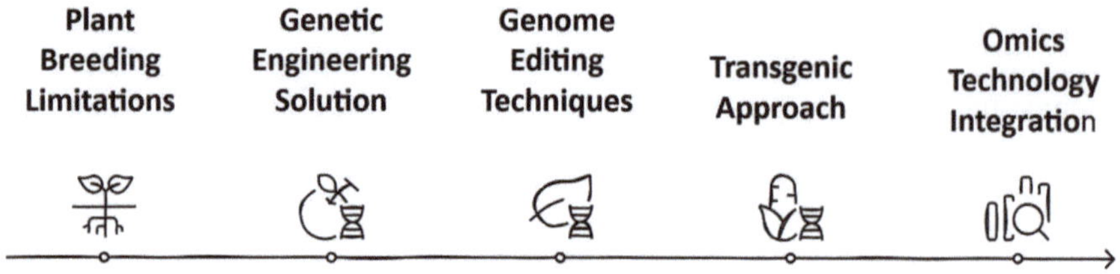

Fig. 17.2 Biofortification strategies in plant breeding

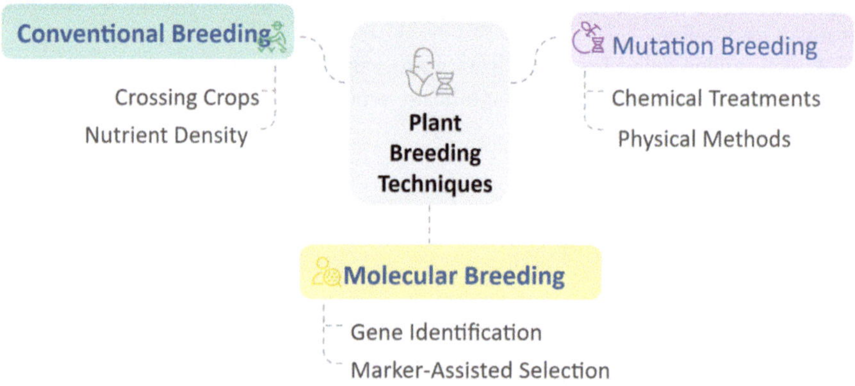

Fig. 17.3 Plant breeding techniques for biofortification

2020). Other common plant breeding technology used are depicted in Figs. 17.2 and 17.3.

Traditional plant breeding techniques have been used to improve micronutrient content, developing new varieties across countries where farmers and consumers now cultivate and consume them (Saltzman et al. 2016).

17.1.3 Nutrient Preservation

Implementing processing and storage techniques to minimize nutrient loss during food production and distribution. Food preservation is crucial for maintaining the quality and safety of food products (Szczepańska et al. 2024). It highlights the need for effective methods that can prolong shelf life while retaining essential nutrients. The food industry is increasingly focusing on alternative and innovative preservation techniques. Example: the use of acetic acid treatment for fresh-cut water chestnuts not only prolongs shelf life but also improves storage quality, indicating its effectiveness in maintaining nutrient levels during storage. Another innovation, like transcriptome analysis, is done to understand the molecular mechanisms behind nutrient retention. This approach can lead to the development of targeted preservation technologies that enhance the quality of food products (Takehisa and Sato 2021).

17.1.4 Targeted Supplementation

Developing specialized products to address the specific nutritional needs of different population groups, such as infant formulas or sports nutrition products. These specialized products cater to distinct nutritional needs, leveraging bioactive compounds and tailored formulations to support optimal health outcomes for their respective user groups (Pinstrup-Andersen 1991).

Enhancing nutrition is essential for addressing micronutrient deficiencies in communities, improving public health, and reducing diet-related diseases. This endeavor advances food security and nutrition in developing countries by meeting the dietary needs of specific populations (e.g., elderly, athletes, and pregnant women):

Challenges and considerations:

- Ensuring the bioavailability and stability of added nutrients.
- Maintaining sensory qualities and consumer acceptance of enhanced foods.
- Addressing regulatory and safety concerns related to fortification and biofortification.
- Balancing cost-effectiveness with nutritional benefits.
- Educating consumers about the importance and benefits of nutritionally enhanced foods.

17.2 Functional Foods

Incorporating specific ingredients to provide health benefits beyond basic nutrition, such as probiotics in yogurt or omega-3 fatty acids in fortified eggs (Jendyose 2024). Functional foods offer the benefit of "using food as medicine" and have emerged as a preferred option. The effectiveness of these foods is derived from the bioactive compounds present in their raw materials. These products with added ingredients that offer specific health benefits beyond basic nutrition, such as probiotics for gut health or antioxidants for heart health. Functional foods are food products modified or enriched with ingredients to provide health benefits beyond basic nutrition. These foods promote optimal health and reduce disease risk when part of a balanced diet (Jones and Jew 2007; Thakur and Belwal 2023; Thakur et al. 2024a, b; Fig. 17.4).

Articulating specific health benefit claims is essential in transitioning functional foods from concept to market product. The development of functional foods requires multiple phases and collaboration among commercial, academic, and regulatory sectors, with consumer acceptance as a key factor. India, with its tradition of

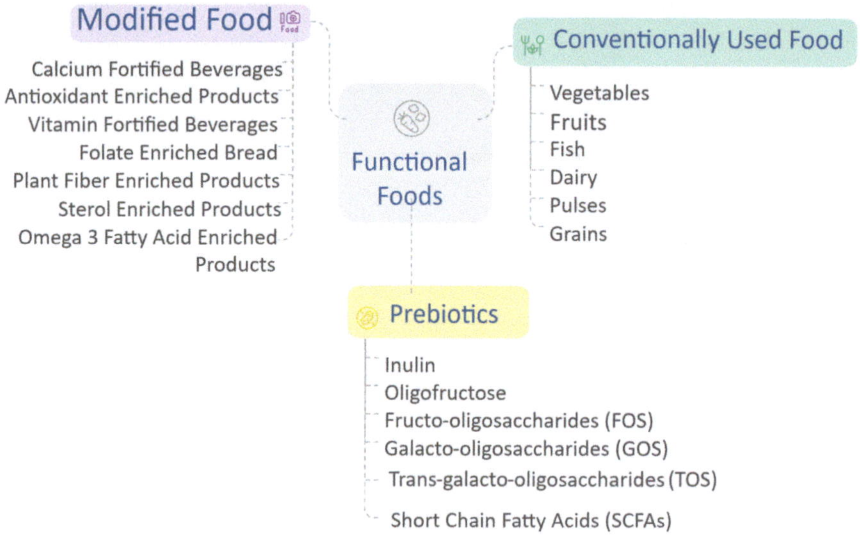

Fig. 17.4 Classification and benefits of functional foods

health-promoting foods, ranks among the top ten countries in functional food purchases, and the market is expected to nearly double in 5 years. By leveraging biodiversity and traditional knowledge, these nations can create products appealing to local and global market (Essa et al. 2023).

Key characteristics of functional foods include the following:

1. Added ingredients: Functional foods are fortified with specific ingredients that are incorporated or enhanced to confer distinct health benefits.
2. Scientific basis: Research findings typically validate the health advantages associated with functional foods.
3. Regulatory oversight: Many countries have implemented regulations concerning the advertising and labelling of functional foods to protect consumers.

Examples of some functional foods available in market include the following:

- Probiotics: Fermented products such as yogurt and kefir, which contain beneficial bacteria, contribute positively to the maintenance of gut health and the enhancement of immune function.
- Omega-3 fortified products: Added to eggs, milk, or bread to support heart and brain health.
- Antioxidant-rich foods: Consumables such as beverages or snacks fortified with antioxidants, including vitamins C and E, are designed to mitigate oxidative stress.
- Fiber-enriched products: Grains or baked products enriched with fiber to enhance digestive wellness and provide a feeling of fullness.

- Plant sterol-fortified foods: Products such as margarine and yogurt that contain plant sterols have the potential to contribute to the reduction of cholesterol levels.
- Vitamin D-fortified foods: Milk or orange juice fortified with vitamin D to enhance bone health.

The functional food industry has experienced substantial growth, driven by consumer awareness of diet-health relationships. However, it is crucial to note that while functional foods can complement a nutritious diet, they should not replace a balanced diet and healthy lifestyle (Hasler 2002).

Challenges in functional food sector:

- High upfront costs.
- Lack of scientific validation.
- Regulatory hurdles.
- Consumer awareness and acceptance.

17.3 Alternative Proteins

Amid growing concerns about sustainability and animal welfare, there has been a surge in innovations such as plant-based proteins, lab-cultured meats, and insect-derived foods, which serve as substitutes for conventional animal protein sources (Fig. 17.5). The alternative protein market is expanding rapidly in response to environmental, ethical, and health concerns associated with traditional animal-based proteins. (Hefferon et al. 2023).

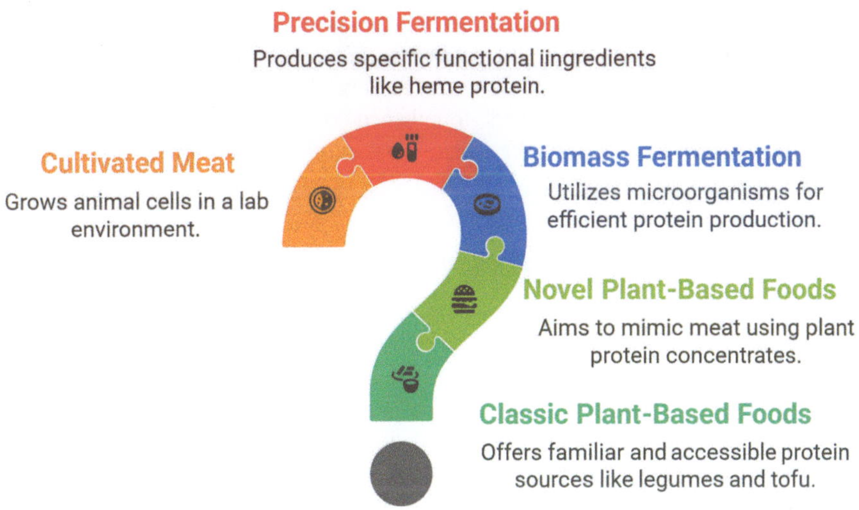

Fig. 17.5 Alternative protein category: Which one should be pursued for sustainable food production?

This innovation encompasses three main categories.

17.3.1 Plant-Based Proteins

Plant-based proteins, such as those derived from soy and pea, are currently the most widely accepted alternatives. The industry is focusing on improving taste, texture, and nutritional profiles to better mimic traditional meat products. Plant-based proteins can be used in various forms, including seeds, powders, and protein isolates. The characteristics and composition of these protein sources determine their applications in food products. Pea protein is frequently used in baked goods like cookies, bread, pizza bases, and cakes, and in extruded snacks. Soybeans are utilized as meat extenders for beef, pork, and poultry, and in meat substitutes and sports nutrition, replacing dairy proteins in protein bars. Rice protein is widely applied in various contexts (Fact Sheet: Alternative Protein 2019; Sharma et al. 2024).

17.3.2 Insect-Based Foods

Incorporating insects into human diets or as animal feed offers advantages similar to plant-based proteins. Although more expensive due to limited supply, insects provide superior nutrition, being rich in protein, unsaturated fatty acids, minerals, and Vitamin B. The protein content ranges from 60% to 77% (dry matter), depending on species. Common insects for protein include crickets, mealworms, and grasshoppers. Insect protein comes in forms like raw insects, coated products like bars, candy, chocolate, cookies, and snack packs, as well as powder/flour and paste. In Asia, insects are cleaned, seasoned, and cooked or fried before serving whole. In the EU/UK, insects are processed into flour for food products, such as Eat Grub's fruit-flavored insect-based chocolate bars, to enhance consumer acceptance. Insects are frequently used in bakery items, chocolate bars, muffins, and biscuits.

17.3.3 Microbial-Based Protein

Microbial protein comes from desiccated cells of microorganisms, including yeasts, microalgae, and fungi, cultivated in large-scale systems. The benefits align with those mentioned before, expanding protein source diversity for consumers. The protein content comprises 40–60% of dry matter. Microalgae and mycoprotein from fungi are the main sources for the food industry. A key example of mycoprotein's success is the Quorn brand, which uses it in products like sausages. Certain microalgae are also used to enhance beverages like shakes and smoothies.

17.3.4 Cultured Meats (Lab-Grown Meats)

Lab-grown meat, also called in vitro or cultured meat, is produced in laboratories through tissue engineering. This process involves incubating a muscle cell from beef or pork in a nutrient-rich solution. Cultured meat offers several advantages, including environmental and health benefits, reduced contamination risk, and greater ethical acceptability. However, scaling up production remains costly due to technical difficulties. Despite these challenges, cultured meat may contribute to meat product development within 10–20 years. The goal is to provide genuine meat without animal slaughter, addressing ethical concerns and environmental impact.

The alternative protein market is poised for significant growth and diversification, driven by several key factors:

- Technological advancements: Ongoing advancements in food science and biotechnology are improving the flavor, texture, and nutritional quality of alternative proteins.
- Consumer awareness: The increasing concerns regarding environmental sustainability, animal welfare, and individual health are driving the demand for plant-based and lab-grown protein alternatives.
- Investment and innovation: Substantial investment from venture capitalists and established food corporations is propelling advancements in the development of alternative protein products and their associated production techniques.
- Regulatory support: In an effort to advance sustainability, certain governments implement policies that facilitate the development and utilization of alternative proteins.
- Product diversification: The market is expanding its focus beyond traditional plant-based products to include innovative sources such as algae, insects, fungi, and cultured meat.
- Improved scalability: Recent advancements in manufacturing techniques have rendered the large-scale production of alternative proteins more economically viable and practical.
- Mainstream adoption: The increasing availability of alternative proteins in dining establishments, quick-service restaurants, and grocery stores is enhancing their familiarity among a broader spectrum of consumers.

These modifications are set to revolutionize global food systems by changing agricultural practices and land use patterns, restructuring supply chains and distribution systems and updating food processing and production techniques. Also, influencing dietary guidelines and nutrition policies can help reshaping the food systems.

As alternative proteins become increasingly prevalent, consumer preferences may evolve, leading to several potential outcomes. Enhanced acceptance and demand for plant-based and lab-cultivated options and modifications in culinary practices and meal preparation. As well as shifts in cultural food traditions and norms, and the development of novel marketing and branding strategies within the food industry.

The continuous growth and diversification of the alternative protein industry have the potential to significantly impact global food production, consumption patterns, and environmental sustainability.

17.4 Clean Label and Natural Ingredients

Consumers are increasingly demanding foods made with natural, minimally processed ingredients, leading to innovations in clean-label products free from artificial additives. Clean label products are distinguished by attributes prioritizing transparency, simplicity, and natural ingredients (Nabeshima et al. 2024).

Although there is no formal regulatory definition for clean label foods, they are characterized by specific attributes. These include simplicity and familiarity, with labels listing minimal, easily recognizable ingredients. Clean label foods should contain only natural ingredients, excluding synthetic flavors, colors, preservatives, or artificial additives. Transparency is fundamental to clean labeling, as information about ingredient sources, sourcing methods, and manufacturing processes assures consumers of food safety. Additionally, foods with clean labels should undergo minimal processing using traditional methods. These characteristics define clean label products, emphasizing natural, simple ingredients while highlighting transparency and ethical practices. The Clean Label Ingredient market is influenced primarily by consumer demand. Health awareness and preference for natural, minimally processed foods drive the market, as consumers seek clean-label ingredients with familiar names and few additives. Transparency is crucial, with consumers favoring clear ingredient lists free of artificial substances, building trust. Skepticism towards synthetic ingredients and safety concerns guides consumers toward clean label products, reflecting a trend towards natural alternatives (Clean Label Ingredient Market-Global Industry Analysis and Forecast (2025–2032) 2024).

The Clean Label Ingredient Market faces several challenges. Substituting synthetic ingredients with clean label alternatives creates technical difficulties in preserving functionality, as alternatives may not provide equivalent texture or shelf life. Standardizing consistent sourcing of dry ingredients poses challenges. Achieving desired flavor with clean-label ingredients requires innovation. The absence of clear regulations for "clean label" creates consumer confusion and potential greenwashing. Supply chain complexity and limited ingredient availability compound these issues.

17.5 Personalized Nutrition

Advances in technology allow for personalized diets based on genetic, health, or lifestyle factors, leading to tailored food products and services. Personalized nutrition is an emerging field that leverages technological advancements to create individualized dietary recommendations based on a person's unique genetic makeup, health status, and lifestyle. This approach moves beyond one-size-fits-all nutrition guidelines to offer more targeted and effective dietary interventions (Boland 2008).

Key aspects of personalized nutrition include the following:

1. Genetic analysis: DNA testing can reveal genetic variations that influence nutrient metabolism, food sensitivity, and disease risk. This information can help tailor dietary recommendations according to an individual's genetic profile.
2. Health status assessment: Comprehensive health evaluations, including blood tests, body composition analysis, and medical history, provide insights into current health conditions and nutritional needs.
3. Lifestyle factors: Consideration of physical activity levels, sleep patterns, stress levels, and personal preferences helps create realistic and sustainable dietary plans.
4. Microbiome analysis: Gut microbiome testing can provide dietary recommendations to optimize digestive health and overall well-being.
5. Wearable technology: Devices that track physical activity, heart rate, and sleep patterns provide crucial data for developing personalized nutritional plans.
6. Artificial intelligence and machine learning: These technologies analyze complex datasets to generate personalized nutrition recommendations and predict individual responses to specific diets.

In the contemporary era of personalized medicine, nutrition, and neurotherapeutic strategies, significant transformations are also occurring in the fields of food engineering and processing technologies (Donovan et al. 2025). Thermal and non-thermal food processing methods present distinct advantages and disadvantages. To address global demands, 3D food printing offers a cost-effective alternative to manual customization for small-scale production, as it does not require expensive setups. The complexity of culinary creations depends more on process and planning than on individual skills. While food printers in domestic settings remain impractical, the technology has attracted global interest and may become feasible soon. Currently, 3D food printing serves as a complementary technology in food processing. With automated, customized food production, this innovation that has transformed the food industry is poised to impact home kitchens, enabling personalized diets through domestic 3D food printers (GT 2019).

Personalized nutrition has enabled bespoke food products and services, including meal delivery services tailored to individual dietary needs, supplement plans based on genetic and health data, and applications that provide dietary advice and track progress. Additionally, functional foods address specific health issues or genetic predispositions, and nutrition coaching services combine technology with expert guidance. As personalized nutrition evolves, it can revolutionize dietary interventions, improve health outcomes, and enhance well-being through science-based nutritional strategies.

17.6 Food Safety and Traceability

Technologies such as blockchain are increasingly used to ensure food safety, ingredient traceability, and transparency within supply chains. This decentralized digital ledger system offers key advantages. Blockchain creates an immutable record of every stage from farm to consumer, enhancing traceability. This capability enables swift tracking of ingredients and products, facilitating rapid identification of contamination sources or quality issues. Supply chain participants can access and verify information, fostering trust among producers, distributors, retailers, and consumers, improving transparency. Real-time data access provides immediate information on origin, processing methods, and storage conditions, supporting faster decision-making. The tamper-proof nature of blockchains reduces food fraud risk and aids in fraud prevention. During contamination or quality issues, blockchain enables quicker recalls, mitigating health risks and economic losses. Blockchain can automate compliance checks and documentation, ensuring adherence to food safety regulations. Consumers can access detailed product information to make informed decisions about their food purchases, empowering them with information. The technology improves inventory management and reduces waste by providing real-time data on product movement and conditions, optimizing the supply chain.

In India, food product recalls due to safety concerns remain infrequent. However, as the Food Safety and Standards Authority of India (FSSAI) strengthens its regulatory capabilities and branded foods evolve, recalls are increasingly used to protect consumers when food production issues arise. India has advanced significantly in food traceability, implementing tracking systems like the Grape Net software, which ensures traceability for table grapes exported to the European Union. The application of blockchain technology in food safety and traceability remains developmental, facing challenges in standardization, scalability, and system integration. Yet, its potential to transform food safety practices and enhance consumer confidence is substantial.

17.7 Sustainable Practices

Innovations focus on reducing food waste, using sustainable packaging materials, and promoting environmentally friendly farming practices. The current global food systems face a range of interconnected challenges that threaten their sustainability and impact various aspects of society, the environment, and the economy (Moshood et al. 2022). Details of some challenges are depicted in Fig. 17.6.

Technology plays a crucial role in transforming and enhancing the sustainability of food systems. Innovations in various technological domains are reshaping the way we produce, distribute, and consume food, leading to more efficient, environmentally friendly, and resilient practices.(León-Bravo et al. 2019; Fig. 17.7). Some key technology innovations in sustainable food systems include:

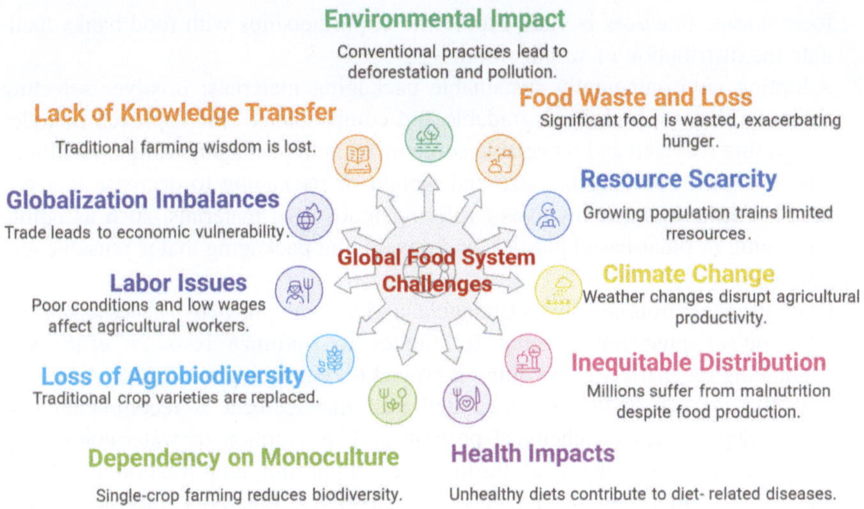

Fig. 17.6 Global food system challenges

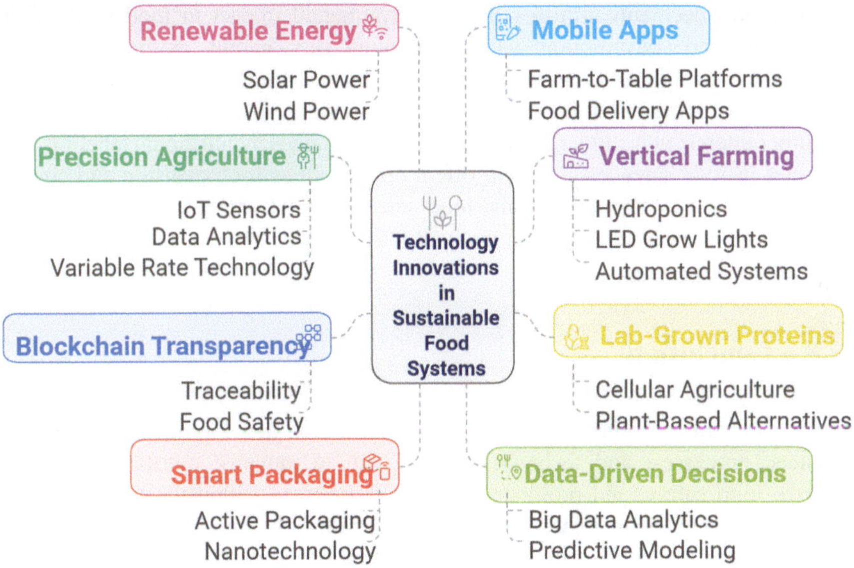

Fig. 17.7 Technology innovations in sustainable food systems

Sustainable practices in the food industry encompass three key areas:

1. Strategies for reducing food waste: Implementing inventory management systems to optimize stock levels is essential. Using predictive analytics to forecast demand can reduce overproduction. Developing preservation techniques is crucial for extending food products' shelf life. Enhancing consumer awareness of proper

food storage practices is vital. Furthermore, partnerships with food banks facilitate the distribution of surplus food.

2. Adopting environmentally sustainable packaging materials: involves selecting solutions that are both biodegradable and compostable. This approach includes integrating recycled and recyclable components into packaging design. Additionally, it entails reducing the size and weight of packaging to decrease material usage. Furthermore, it involves exploring innovative materials, such as edible packaging or plant-based plastics, and developing packaging that is reusable and refillable.
3. Promoting environmentally sustainable agricultural practices: necessitates the adoption of precision farming techniques to optimize resource efficiency. Embracing regenerative agriculture is crucial for enhancing soil health. Additionally, the implementation of integrated pest management is recommended to reduce dependency on chemical pesticides. The adoption of water-conserving irrigation systems is also vital. Furthermore, supporting crop rotation and diversification is important for increasing biodiversity. Finally, exploring vertical farming and urban agriculture can significantly reduce transportation-related emissions.

These advancements aim to reduce environmental impact in food production, distribution, and consumption, while promoting sustainability within the sector. Innovations in food health span multiple industries, including food manufacturing, biotechnology, pharmaceuticals, and consumer goods. This movement is driven by consumer interest in healthier options, regulatory changes in nutritional labeling, and advances in food science. This field offers opportunities for expansion and collaboration among health professionals, food scientists, and entrepreneurs working to enhance public health through innovative food solutions.

17.8 Conclusion

Health innovation is fundamentally concerned with the development of novel food products, processes, and technologies that aim to enhance nutritional content, improve food safety, and promote sustainable practices within the food industry. Key components include the enhancement of nutrition through fortification and biofortification, the creation of functional foods with additional health benefits, the exploration of alternative proteins such as plant-based and lab-grown meats, the emphasis on clean labels and natural ingredients, the customization of nutrition to meet individual needs, the assurance of food safety and traceability through blockchain technology, and the adoption of sustainable methods to reduce waste and support environmentally friendly agriculture. Innovation within the health-related food industry spans multiple sectors and is driven by consumer preferences, regulatory changes, and advancements in food science and technology. This landscape presents opportunities for collaboration among health experts, food scientists, and entrepreneurs to advance public health through innovative food solutions.

References

Allen L, De Benoist B, Dary O, Hurrell R (2006) Guidelines on food fortification with micronutrients. World Health Organization, pp 3–17. Retrieved from https://iris.who.int/bitstream/handle/10665/43412/9241594012_eng.pdf

Boland M (2008) Innovation in the food industry: personalized nutrition and mass customization. Innovation 10(1):53–60

Clean Label Ingredient Market-Global Industry Analysis and Forecast (2025–2032) (2024) Global Market insight. Retrieved from: https://www.gminsights.com/industry-analysis/clean-label-ingredients-market

Donovan SM, Abrahams M, Anthony JC, Bao Y, Barragan M, Bermingham KM, Blander G, Keck AS, Lee BY, Nieman KM, Ordovas JM, Penev V, Reinders MJ, Sollid K, Thosar S, Winters B (2025) Personalized nutrition: perspectives on challenges, opportunities, and guiding principles for data use and fusion. Crit Rev Food Sci Nutr:1–18

Essa MM, Bishir M, Bhat A, Chidambaram SB, Al-Balushi B, Hamdan H, Govindarajan N, Friedland RP, Qoronfleh MW (2023) Functional foods and their impact on health. Food Sci Technol 60(3):820–834

Fact Sheet: Alternative Protein (2019). Retrieved from: https://www.foodinnovationcentre.co.uk

GT (2019) Innovative technologies in tailoring designer food and personalized nutrition. Novel Tech Nutr Food Sci 4(2)

Hasler CM (2002) Issues and opinions functional foods: benefits, concerns and challenges – a position. Am Counc Sci Health 1:3772–3881

Hefferon KL, De Steur H, Perez-Cueto FJA, Herring R (2023) Alternative protein innovations and challenges for industry and consumer: an initial overview. Front Sustain Food Syst 7

Hsieh YHP, Ofori JA (2007) Innovations in food technology for health. Asia Pac J Clin Nutr 16(1): 73

Jendyose M (2024) Development of functional foods with enhanced health benefits. Int J Food Sci 5(2):29–42

Jones PJ, Jew S (2007) Functional food development: concept to reality. Trends Food Sci Technol 18(7):387–390

León-Bravo V, Moretto A, Cagliano R, Caniato F (2019) Innovation for sustainable development in the food industry: retro and forward-looking innovation approaches to improve quality and healthiness. Corp Soc Responsib Env Manage 26(5):1049–1062

Malik KA, Maqbool A (2020) Transgenic crops for biofortification. Front Sustain Food Syst 4

Mannar MGV, Wesley AS (2016) Food fortification. In: International encyclopedia of public health, vol 1, pp 143–152

Moshood TD, Nawanir G, Mahmud F, Mohamad F, Ahmad MH, Abdul Ghani A, Kumar S (2022) Green product innovation: a means towards achieving global sustainable product within biodegradable plastic industry. J Clean Prod 1:363

Nabeshima EH, da Rocha Tavares PE, da Silva Corrêa Lemos AL, de Moura SCSR (2024) Emerging ingredients for clean label products and food safety. Braz J Food Technol 1(27):1–29

Ofori KF, Antoniello S, English MM, Aryee ANA (2022) Improving nutrition through biofortification – a systematic review. Front Nutr 9

Pinstrup-Andersen P (1991) Targeted nutrition interventions. Food Nutr Bull 13(3):161–169

Salifou A, Christian Tétédé RK, Alexandrine B, Nicodème WC, Fidèle PT, Mohamed MS (2023) Innovative approaches in food processing: enhancing quality, preservation, and safety through advanced technologies: a review. World J Adv Res Rev 20(2):637–648

Saltzman A, Andersson M S, Asare-Marfo D, Lividini K, De Moura F F, Moursi M, Oparinde A, Taleon V (2016) Biofortification techniques to improve food security reference module in food science Elsevier

Szczepańska E, Liedtka A, Czech E (2024) Knowledge and eating habits regarding functional food among adults. Ann Nat Inst Hyg 75(1):35–44

Takehisa H, Sato Y (2021) Transcriptome-based approaches for clarification of nutritional responses and improvement of crop production. Breed Sci 71(1):76–88

Thakur M (2024) Sustainable food systems (volume II). World sustainability series. Springer, Cham. https://doi.org/10.1007/978-3-031-46046-3_1

Thakur M, Belwal T (2023) Bioactive components. Springer, Singapore. https://doi.org/10.1007/978-981-19-2366-1_34

Thakur M, Modi VK, Feng S (2024a) Editorial: food sustainability: challenges and opportunities for the future. Front Nutr 10:1341427. https://doi.org/10.3389/fnut.2023.1341427

Thakur M, Ferrante C, Sanches Silva A (2024b) Editorial: sustainable development in the agro-food and nutrition sector. Front Nutr 11:1419937. https://doi.org/10.3389/fnut.2024.1419937

Umar M, Nawaz R, Sher A, Ali A, Hussain R, Khalid MW (2019) Current status and future perspectives of biofortification in wheat. Asian J Res Crop Sci:1–14

Chapter 18
From Pulse Waste to Functional Food: Enhancing Biodiversity-Driven Nutrition Through Resource Valorization

Parul Chauhan, Ashish M. Mohite, Pankaj B. Pathare, and Neha Sharma

Abstract Pulses offer essential, eco-friendly nutrition, yet processing generates by-products like husks, stems, and pod coats often discarded or used as basic animal feed. These pod coats, rich in dietary nutrients and bioactive compounds, hold promise for health benefits and added commercial worth, advancing sustainable farming, and healthier food chains. They serve as bases for edible films, packaging, coatings, or functional additives with antioxidant and antimicrobial effects. Pulse pod coat fibres also act as prebiotics, deliver anti-inflammatory and anti-cancer benefits, and help regulate blood sugar and cholesterol. For instance, pigeon pea pod extract boosts oxidative stability in soybean oil, while bean and lupin coats yield GRAS-approved dietary fibre products. This chapter explores repurposing pulse by-products from low-value waste into innovative functional foods.

Keywords By-products · Pod coat · Antioxidant · Probiotics · Dietary fibre

18.1 Introduction

Pulses represent a cornerstone of global food and nutritional security due to their rich composition of proteins, complex carbohydrates, dietary fibre, minerals, and bioactive compounds. As the second most widely consumed plant-based food group after cereals, pulses are cultivated across diverse agro-ecological regions worldwide, making them an integral component of sustainable food systems (Langyan et al. 2022). Their affordability and nutritional density have positioned them as staple foods, particularly in developing economies where they play a crucial role in addressing protein and micronutrient deficiencies (Jha and Warkentin 2020).

P. Chauhan · A. M. Mohite · N. Sharma (✉)
Amity Institute of Food Technology, Amity University Uttar Pradesh, Noida, India
e-mail: parul.chauhan3@s.amity.edu; ammohite@amity.edu; nsharma9@amity.edu

P. B. Pathare
Sultan Qaboos University, Al Seeb, Oman

M. Thakur, T. Belwal (eds.), *Food Biodiversity and Consumer Engagements*, Sustainable Development and Biodiversity 46,
https://doi.org/10.1007/978-981-95-7777-4_18

In many regions of Asia, Africa, and Latin America, pulses are deeply embedded in traditional cropping systems and agricultural practices (Joshi and Rao 2017). These crops are frequently grown under rain-fed conditions, on marginal soils, or as intercrops alongside cereals and oilseeds, contributing to both farm resilience and soil fertility. Despite occupying a relatively modest proportion of global arable land—approximately 5.8% during the 2011–2013 period—pulses make a disproportionately high contribution to human nutrition and agricultural sustainability (Kumar et al. 2023).

Over recent decades, global pulse production has shown a steady upward trend, increasing from approximately 45 million tonnes in the early 1980s to more than 70 million tonnes by 2013. This growth has been accompanied by an expansion in cultivated area from nearly 63 million hectares to around 80 million hectares, reflecting rising demand driven by population growth and shifting dietary preferences (Martineau-Cote et al. 2022). At the national level, pulses account for a substantial share of arable land in several countries, including Myanmar, Niger, Kenya, Mozambique, and India, underscoring their agronomic and socio-economic importance (Herrington et al. 2019).

According to the Food and Agriculture Organization, pulses encompass a diverse group of leguminous crops, including dry beans, peas, lentils, chickpeas, cowpeas, pigeon peas, lupins, and Bambara beans, among others (Tiwari et al. 2021). While pulses contribute approximately 5% of total global protein intake, their role is considerably more pronounced in low- and middle-income countries, where they may supply between 10% and 40% of dietary protein (Bhat et al. 2022). Projections indicate that demand for pulses is likely to increase further in developing regions due to rising incomes, urbanization, and growing recognition of plant-based diets (Magrini et al. 2018).

Beyond direct human consumption, pulses generate substantial quantities of agricultural by-products that serve as valuable resources within integrated food–feed systems. Crop residues remaining after seed harvest possess considerable potential as livestock feed, thereby strengthening the feed–food nexus and indirectly enhancing food security (Wodajo et al. 2020; Shinde et al. 2022). The utilization of pulse residues in animal nutrition can reduce dependence on cereal grains and soybean meal, while maintaining dietary quality for both animals and humans (Simitzis and Deligeorgis 2018).

The nutritional value of pulse residues is influenced by several factors, including plant species, cultivar, maturity stage at harvest, and environmental conditions such as soil fertility and climate (Singh 2017; Zhong et al. 2018). Compared to cereal straws, pulse straws generally exhibit higher metabolizable energy, improved dry matter digestibility, and lower neutral detergent fibre content. These attributes are largely attributed to their higher proportion of digestible cell contents and pectin-rich carbohydrates, which are readily fermented by rumen microorganisms (Patras et al. 2011; Verma 2018). Additionally, pulse residues often contain higher levels of digestible protein, further enhancing their suitability as animal feed (Kumar et al. 2020).

18.1.1 Sustainable Approach Towards Valorization of Pulses By-Products

Recent global assessments, including the IPCC report on Climate Change and Land, emphasize the role of plant-based dietary transitions in mitigating climate change while delivering co-benefits for human health and environmental sustainability (Mersha 2016). In this context, pulses have gained increasing attention as sustainable sources of nutrition due to their low environmental footprint and high nutrient density. Preserving pulse genetic diversity and promoting its utilization in food systems can enhance agricultural resilience and support the development of healthier diets.

However, the expansion of pulse production and processing has also resulted in the generation of significant quantities of by-products along the value chain, including seed coats, husks, pods, and processing wastewater (Hawu 2022). In many cases, these materials constitute more than 25% of the total biomass during pulse splitting and dehulling operations (Slominski 2018). Effective valorization of these by-products presents an opportunity to reduce food waste while creating economic and nutritional value (Saha and Pathak 2021).

Pulse processing residues are increasingly recognized as rich sources of polyphenols, dietary fibre, proteins, and other bioactive compounds. Studies have demonstrated strong associations between the phenolic content of pulse seed coats and antioxidant activity, highlighting their potential health-promoting properties (Cohen et al. 2021). Consequently, pulse by-products can serve as functional ingredients within circular economy frameworks, contributing to sustainable food innovation (Aschemann-Witzel et al. 2021).

A variety of pulse-derived residues—including pomace, hulls, and peels—have been successfully incorporated into value-added food products such as bakery items, extruded snacks, noodles, beverages, and frozen desserts (Pugliese et al. 2024). The incorporation of these functional ingredients can enhance dietary fibre and protein content while influencing techno-functional properties such as water absorption, emulsification, gel formation, and textural attributes (Sharma et al. 2022). Comprehensive techno-functional characterization, combined with sensory evaluation, is therefore essential to ensure both product performance and consumer acceptance (Acharya et al. 2019; Bhat 2021) (Fig. 18.1).

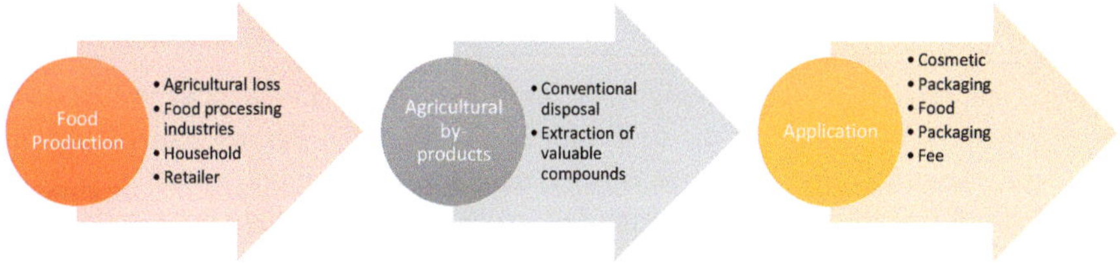

Fig. 18.1 Process of sustainable approach towards the valorization of pulses by-products

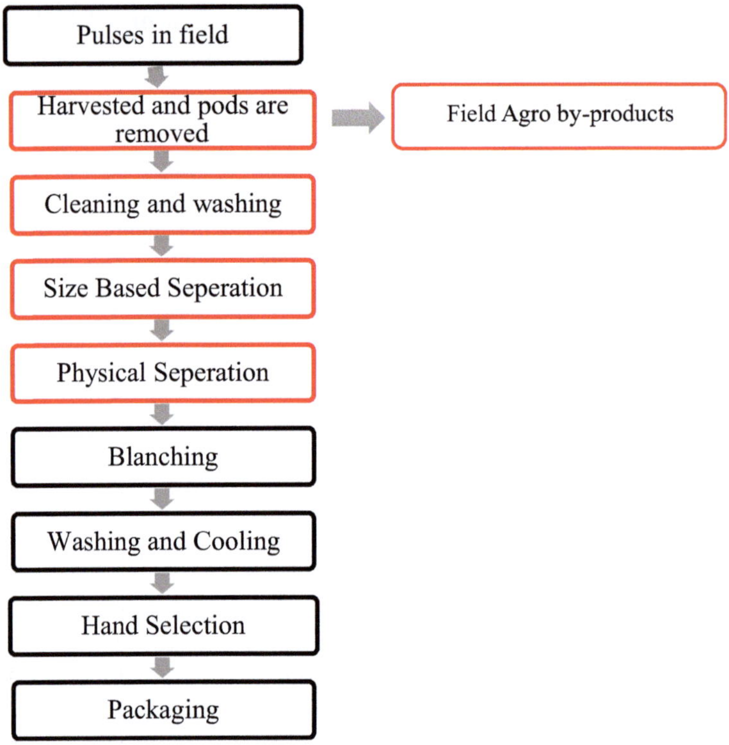

Fig. 18.2 The process scheme of pulse production, the box in red represents the by-product generation part of the chain

The processing of pulses generates substantial quantities of residues and by-products at multiple stages, beginning at harvest and continuing through industrial transformation. During field operations, a proportion of the harvested crop is rejected due to physical damage, immaturity, or visual defects, resulting in early-stage waste generation (Nino-Medina et al. 2017). Additional residual materials are produced during post-harvest handling, including cleaning, de-podding, and splitting operations, where non-edible components such as pods and seed fragments are separated from marketable grains (Fig. 18.2).

In the case of fresh pulse processing, particularly for crops such as beans and lentils, agro-industrial residues typically consist of a heterogeneous mixture of leaves, stems, and empty pods removed prior to further processing (De Aguiar Saldanha Pinheiro et al. 2021). Estimates from major European pulse processors indicate that residues generated along the agro-industrial value chain may account for approximately 5–25% of the total biomass obtained at harvest, depending on crop type, processing intensity, and quality standards (Selvamuthukumaran 2024).

Following harvest, pulses are commonly collected and shelled either in the field or at centralized facilities, producing large volumes of empty pods, stalks, and foliage that are often classified as agricultural waste. The extracted seeds are then transported for industrial processing, where further by-products arise during sequential sorting and quality control steps (Garcia-Amezquita et al. 2018). Initial separation typically involves density-based techniques designed to isolate fully matured grains, as pulse density tends to increase with physiological maturity (Siddiqui et al. 2024).

Subsequent operations, including size grading, optical sorting, and manual inspection, generate additional residues composed of undersized, misshapen, or visually defective seeds. Final processing stages, particularly those preceding preservation or packaging, further contribute to by-product formation. Commonly generated materials during these phases include rejected seeds, seed coats, hulls, and pod fragments (Sharif et al. 2017). Although traditionally considered waste, these by-products represent valuable biomass streams with considerable potential for recovery and valorization within sustainable food and bioresource systems.

18.2 Pulses By-Products Extraction Technologies

A range of processing technologies has been developed for the recovery of proteins and other bioactive compounds, including dietary fibres and phenolic constituents, from pulse seeds. Recent studies have demonstrated that many of these extraction strategies are equally applicable to pulse-derived by-products, thereby expanding their potential for value-added utilization (Liu et al. 2023).

Despite the significant volume of post-harvest residues generated during pulse processing, only a limited number of investigations have explored their nutritional and functional potential. For example, protein-rich fractions with promising nutritional value have been successfully recovered from the pod coats of moth bean (*Vigna aconitifolia*), highlighting the feasibility of valorizing pulse processing waste streams (Boroomand 2018).

At the industrial scale, solvent-based extraction methods employing organic solvents such as ethanol are commonly used to separate lipids, proteins, and other valuable components through solid–liquid extraction systems. These processes yield ingredients with favourable nutritional profiles suitable for food and feed applications (Aboye 2022). However, solvent-based extraction may adversely affect the functional properties of proteins, as reported for soybean pod-derived proteins, thereby restricting their technological applicability. Organic solvents have also been widely used to isolate phenolic compounds from pulse by-products, offering an effective route for recovering antioxidant-rich fractions (Essien et al. 2021).

To enhance extraction efficiency while reducing solvent consumption, ultrasound-assisted extraction (UAE) has emerged as a cost-effective and environmentally friendly alternative to conventional techniques. UAE enhances solvent penetration into plant tissues and promotes mass transfer through acoustic cavitation, facilitating the release of intracellular compounds. The reduced reliance on hazardous solvents positions UAE as a green extraction approach aligned with U.-S. Environmental Protection Agency (EPA) sustainability guidelines (de Koff 2008).

Recent findings indicate that ultrasound application can significantly improve the recovery of valuable constituents from pulse residues. For instance, ultrasound treatment of soybean pod material resulted in approximately 10% higher yields of proteins, oils, and total solids. Additionally, UAE has been successfully applied to extract natural antioxidants from mung bean seed coats. Lafarga et al. reported improved protein solubilization from Ganxet bean pods using an acid–alkaline extraction process assisted by acoustic energy, with maximum yields observed at elevated sodium hydroxide concentrations (0.3–0.4 M). From an industrial perspective, UAE is particularly attractive because it can be integrated into existing processing lines with minimal infrastructural modifications while delivering enhanced extraction efficiency (Eder 2021).

Microwave-assisted extraction (MAE) has also been investigated as an advanced technique for improving the recovery of bioactive compounds from pulse matrices. This method utilizes electromagnetic radiation within the frequency range of 300 MHz to 300 GHz, generating rapid internal heating through molecular vibration and thereby accelerating mass transfer processes. Laboratory-scale studies have demonstrated the successful extraction of soluble proteins from soybean seeds using MAE. However, limited research has addressed the application of microwave-assisted methods for protein extraction from pulse by-products, indicating a clear need for further investigation in this area (Chandran et al. 2023).

18.3 Nutritional and Bio-active Properties of Pulses By-Products

The incorporation of pulse by-products into food formulations offers an environmentally sustainable strategy for valorizing processing residues while enhancing nutritional quality. Pulse by-products are nutritionally diverse, containing appreciable levels of protein (1.5–24.9%), dietary fibre (3.1–21.7%), and lipids (0.03–33.7%). Current trends in pulse-based foods emphasize increasing protein and fibre content while reducing lipid levels; however, achieving this balance remains challenging, particularly in products formulated with pulse processing residues (Goyal et al. 2005). Protein and dietary fibre are key macronutrients that not only improve the nutritional profile of foods but also contribute desirable functional and technological properties. In this context, pulse by-products present significant potential as both nutritional enhancers and functional ingredients in food systems (Fierascu et al. 2019).

18.3.1 Protein: Ingredients in Food Applications

Pulses are recognized for their high protein content and well-balanced amino acid composition, with sulphur-containing amino acids and tryptophan generally being limiting. Pulse by-products—including seed coats, hulls, and husks—contain lower protein levels than whole seeds; however, certain fractions, such as broken or damaged seeds, often discarded during processing, retain relatively high protein concentrations. Average protein contents reported include approximately 7% in seed coats, 12% in pods, 10% in bagasse and chickpea okara, and over 20% in broken seeds, while negligible amounts are present in soaking and cooking water (Deng et al. 2022).

In most food applications, protein-rich extracts from pulse by-products are seldom used directly. Instead, minimally processed forms such as flours or powders derived from hulls and pods are incorporated into formulations. Combining pulse-derived ingredients with cereals is a common strategy to improve protein quality by achieving a more complete amino acid profile. Consequently, pulse–cereal blends are widely applied in bakery and snack products, with protein contents typically ranging from 10% to 20% depending on the product category (Escobedo and Mojica 2021).

18.3.2 Dietary Fibre as an Ingredient in Food Applications

Dietary fibre is one of the most important bioactive components of pulses, with more than half of commercial functional foods containing fibre-enriched ingredients. Regular intake is particularly beneficial for metabolic health, and a daily consumption of at least 25 g is recommended by the Food and Agriculture Organization and the European Food Safety Authority. Pulse-derived dietary fibre comprises a diverse group of compounds, including cellulose, hemicelluloses, pectin, gums, β-glucans, lignin, resistant starch, and non-digestible oligosaccharides (Ciudad-Mulero et al. 2020).

Based on solubility, dietary fibre is classified into soluble (SDF) and insoluble fractions (IDF). A balanced SDF:IDF ratio of approximately 30–50% SDF and 50–70% IDF is considered optimal for both health and technological functionality, although processing treatments such as heating may modify this profile. Pulse by-products, particularly seed coats and pods, are rich in total dietary fibre, with seed coats containing up to ~80% fibre, predominantly in the insoluble form, while broken seeds generally show lower fibre levels (Campos et al. 2010).

Concentrated fibre fractions (up to 90%) can be produced and incorporated into foods as fortifying ingredients. As a result, fibre-enriched products formulated with pulse by-products typically contain 3–22% dietary fibre. High-fibre bread, extruded snacks, gluten-free pasta, beverages, and meat products developed using pulse residues demonstrate substantial fibre enhancement, with product-dependent variations in IDF/SDF ratios (Domínguez Díaz et al. 2020).

Beyond their nutritional role, pulse fibres also act as carriers of bioactive compounds. Insoluble fibre fractions, in particular, are associated with phenolic compounds such as hydroxybenzoic and hydroxycinnamic acids, flavanols, and procyanidins, contributing to the overall bioactive potential of fibre-rich foods (Grasso 2020).

18.3.3 Total Phenolic Content: Ingredients and Foods

Pulses are rich sources of bioactive phytochemicals, including phenolic acids, flavonoids, tannins, anthocyanins, and isoflavones, which are predominantly concentrated in the seed coat. Most phenolic compounds occur in insoluble, bound forms associated with structural polysaccharides such as cellulose, hemicellulose, lignin, and pectin. Consequently, the phenolic profile of pulse by-products varies with genetic, agronomic, and processing factors, which must be considered when designing standardized functional food ingredients.

Total phenolic content (TPC) is commonly measured to evaluate the suitability of pulse by-product extracts for use as natural antioxidants, preservatives, or nutraceutical ingredients. These extracts are typically obtained using aqueous or hydroalcoholic solvents. Reported TPC values are high for seed coats and hulls, reaching over 4000 mg GAE/100 g in pigmented bean and broad bean by-products, with chickpea seed coats showing the highest concentrations (>12,000 mg GAE/100 g extract). In contrast, soaking water contains minimal phenolic content (Ciudad et al. 2018).

Incorporation of pulse by-products into food formulations can significantly enhance antioxidant potential. Fibre- or hull-enriched breads, beverages, and hydrogel-based products demonstrate elevated TPC levels, confirming the feasibility of using pulse residues as functional antioxidant ingredients. However, further studies on diverse food matrices are required to better understand phenolic stability and shelf-life effects in formulated products.

18.3.4 Tocopherol: Ingredients and Foods

Tocopherols and carotenoids are lipid-soluble micronutrients with antioxidant activity, with tocopherols representing biologically active forms of vitamin E (α, β, γ, and δ isomers). Among pulses, γ-tocopherol is the predominant form; however, both pulses and their by-products generally contain low tocopherol levels. Although tocopherols have been detected in certain by-products, such as lentil hulls and pods of cowpea, mung bean, and moth bean, their limited concentrations restrict broader functional food applications (Ciudad et al. 2018).

Carotenoids, which also function as vitamin A precursors, are more abundant in chlorophyll-rich tissues such as peas and green beans. Overall, pulses are relatively

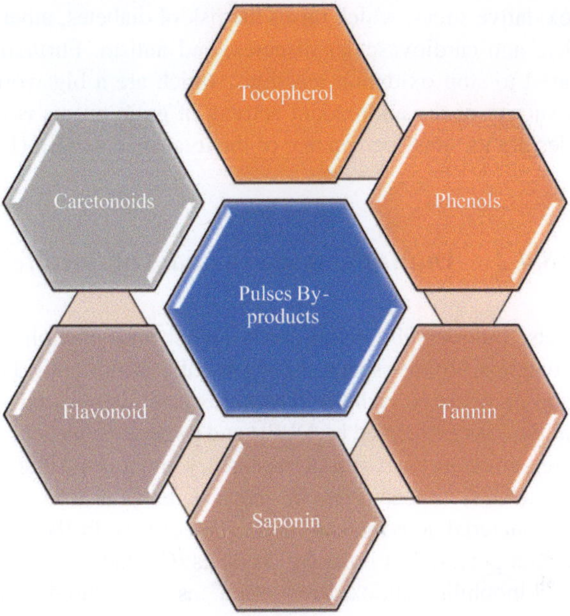

Fig. 18.3 Bioactive composition of pulses by-products

poor sources of carotenoids compared with leafy vegetables and fruits. Within pulse fractions, seed coats—particularly in black gram—exhibit higher carotenoid levels than cotyledons or germ tissues; however, these fractions are commonly removed during milling, reducing their contribution to final food products (Ciudad et al. 2018) (Fig. 18.3).

18.4 Health Properties

Pulse by-products are rich sources of bioactive compounds, including phenolics, carotenoids, tocopherols, and tocotrienols, which collectively contribute to their antioxidant potential. These compounds play an important role in neutralizing free radicals and may help reduce the risk of several chronic diseases.

18.4.1 Cardiovascular Health

During cellular metabolism, free radicals are produced from oxygen, nitrogen, and sulphur. These molecules react with other molecules through their unpaired electrons, and these reactions are crucial for gene expression, apoptosis, and cell signalling. However, excessive levels of free radicals can damage double bonds in fatty acids, nucleic acid bases, and amino acid side chains in proteins. This results in oxidative stress, which raises the risk of diabetes, atherosclerosis, cancer, neurological and cardiovascular diseases, and autism. Furthermore, antioxidants are often used to stop oxidation reactions, which are a big worry in the food industry. The majority of the antioxidant activity in grain pulses is contributed by the seed coat, despite its small percentage of the total seed weight (Bessada et al. 2019).

18.4.2 Anti-inflammatory and Anti-proliferative Properties

Pulse hulls exhibit significantly higher total phenolic content than cotyledons and contain a broader spectrum of phenolic compounds. For example, lentil seed coats are particularly rich in flavonoids and demonstrate strong antioxidant activity in vitro, as evidenced by DPPH and β-carotene bleaching assays. These polyphenols are known to exert anti-inflammatory and anti-proliferative effects, including activity against cancer cells. In addition, extracts from pigeon pea hulls have shown antibacterial activity against *Bacillus cereus*, further supporting the multifunctional health potential of pulse by-products (Ciudad et al. 2018).

Lipophilic antioxidants such as carotenoids and tocopherols, especially γ-tocopherol, are present at higher concentrations in hulls than in whole seeds. Although their levels in pulses are relatively low, compounds such as lutein and zeaxanthin contribute to cellular protection and eye health.

18.4.3 Anti-cancer Properties

Pulse pods and husks contain polyphenols, saponins, and dietary fibre with demonstrated antioxidant and antimicrobial activities. Extracts from common bean, broad bean, and pea pods show strong in vitro antioxidant capacity, while pectin derived from chickpea husks exhibits effective free-radical scavenging activity. These bioactive compounds are increasingly recognized for their role in cancer prevention and cardiovascular protection, partly through antioxidant and prebiotic mechanisms (Ciudad et al. 2018).

Overall, detailed characterization of bioactive compounds in pulse by-products provides a strong foundation for their use in value-added functional foods. Future research should focus on product development and in vivo validation to confirm health benefits, consumer acceptance, and disease risk reduction.

18.5 Applications of Pulse By-Products

18.5.1 Feed Applications

The growing demand for animal-derived foods has increased the use of grain pulses in livestock diets, resulting in large quantities of pulse biomass and by-products requiring effective utilization. Pulse residues have gained attention as feed ingredients due to their relatively high protein content, good palatability, and digestibility. In intensive livestock systems, their inclusion can partially replace cereals and soybean meal, with pulses contributing approximately 10–20% of livestock diets and, in some cases, up to 50% (faostat 2020).

FAO reports highlight that the suitability of pulse by-products as feed is mainly determined by their nutritional value and relatively low levels of anti-nutritional factors. These by-products provide amino acids, carbohydrates, fibre, minerals, vitamins, phenolics, and essential fatty acids, while generally exhibiting higher dry matter digestibility and metabolizable energy than cereal residues. However, the presence of compounds such as lectins and saponins may limit direct use, particularly in monogastric animals. Appropriate processing methods—including soaking, heat treatment, fermentation, and de-hulling—can significantly reduce these constraints.

Specific residues such as pea pods and chickpea straw have shown high potential for ruminant feeding. Pea pods are rich in minerals, fibre, soluble carbohydrates, phenolics, and crude protein, while chickpea straw demonstrates greater palatability and nutritional value than cereal straws, making it a cost-effective alternative forage (Ciudad-Mulero et al. 2020; Bhat 2021).

18.5.2 Food Applications

Pulse by-products are increasingly explored for food applications due to their high protein and dietary fibre content, aligning with the growing demand for plant-based, vegetarian, and vegan foods. Pulse-derived proteins contribute to satiety, improved lipid metabolism, and blood pressure regulation, making them attractive ingredients for functional foods.

Protein-rich flours, concentrates, and isolates from pulse residues have been incorporated into pasta, snacks, baked goods, meat analogues, beverages, and baby foods to enhance nutritional and technological properties. Additionally, enzymatic or chemical hydrolysis of pulse proteins yields bioactive peptides with antioxidant and antihypertensive potential, expanding their use in functional food formulations.

Pulse by-products are also excellent sources of dietary fibre and phenolic compounds. Ingredients derived from chickpea husks, pea pods, and broad bean residues have been successfully used to fortify bakery and snack products, improving fibre content and antioxidant capacity. Natural phenolics from pulses are of particular interest as clean-label antioxidants and antimicrobial agents, especially in meat products, where they help limit lipid oxidation and off-flavour development.

18.5.3 Packaging

Recent advances focus on valorizing pulse by-products and agro-industrial waste for the development of sustainable packaging materials. The packaging sector is increasingly exploring bio-based and biodegradable polymers that meet functional requirements for food and cosmetic preservation while reducing environmental impact. A major objective is to lower reliance on costly biopolymers such as polylactic acid (PLA), polyhydroxyalkanoates (PHA), polybutylene succinate (PBS), and polycaprolactone (PCL), while achieving performance comparable to fossil-based plastics.

In this context, residues and bioactive compounds recovered from pulse by-products are used as fillers or reinforcing agents in biopolymer matrices. Fibre-rich biomass generated after protein or bioactive compound extraction can be incorporated into bio-composites, improving mechanical strength and reducing material costs. Natural fibres derived from pulse residues are particularly attractive due to their renewability, biodegradability, low cost, and ability to enhance polymer performance (Cuj-Laines et al. 2018).

Pulse-derived residues have been successfully combined with biodegradable polymers such as PHA and PLA to produce rigid packaging materials with favorable mechanical properties. Beyond polymer composites, bean pod ash has also shown promise as a low-cost reinforcement material in composite systems due to its availability and low density. Emerging innovations include the development of biodegradable bioplastics from chickpea aquafaba and fully recyclable packaging paper produced from bean-processing waste. These materials reduce dependence on virgin cellulose, lower greenhouse gas emissions, and align with clean-label, vegan, and circular economy principles.

18.6 Uses of Pulse By-Products in the Food Industry

Pulses are staple foods in many regions, with lentil, broad bean, chickpea, lupin, pea, and common beans accounting for nearly 80% of global pulse production (FAOSTAT 2020). However, substantial material losses occur along the production and processing chain, with up to one-quarter of harvested pulses discarded due to broken grains, quality non-compliance, and removal of pods, leaves, and stems. Field residues primarily include pods, husks, leaves, and bagasse, while processing waste consists of seed coats, hulls, broken seeds, and cooking or soaking water generated during drying, milling, dehulling, and sorting operations (Fathonah et al. 2020).

The dehulling process, commonly applied because pulses are largely consumed as split seeds, generates considerable by-products such as husks, broken fractions, protein-rich flours, and fibre residues. Although broken grains constitute a small fraction of total output, they are often discarded despite retaining high nutritional value. Seed coat proportions vary across pulse types, accounting for approximately 7–13% in beans, peas, and lentils, up to 25% in lupin, and nearly 28% in chickpea (Deng et al. 2022).

Pods represent a major source of agricultural residue, comprising up to two-thirds of the pod mass in peas and about 70% in broad beans. Additional underutilized by-products include okara, generated during pulse milk extraction, and viscous cooking and soaking water, which contain soluble proteins and phytochemicals leached from seeds during processing. Collectively, these by-products present significant opportunities for valorization as functional ingredients in the food industry.

18.7 Future Perspective

Pulse by-products derived from crops such as beans, peas, lentils, chickpeas, and soybeans possess considerable nutritional value due to their rich content of macronutrients, micronutrients, and bioactive compounds. Although pulses are staple foods in many regions, conventional processing often removes bioactive-rich fractions, particularly hulls and seed coats, which are abundant in polyphenols, carotenoids, tocopherols, and tocotrienols. Recognizing the functional potential of these discarded materials offers an effective strategy to reduce agri-food waste while supporting the growth of functional food and nutraceutical sectors.

Pulse residues represent promising sources of bioactive molecules, including proteins that can be converted into health-promoting peptides. Growing evidence from in vitro and in vivo studies suggests that compounds derived from pulse by-products may contribute to the prevention or management of chronic conditions such as cardiovascular diseases, inflammation, obesity, type 2 diabetes, and certain cancers. However, their practical application is often limited by low bioaccessibility, which can be addressed through advanced processing approaches such as steam explosion, high-pressure treatment, and other emerging technologies.

In the context of climate change and resource scarcity, the valorization of pulse by-products aligns with sustainable production models and circular economy principles. Full utilization of pulse biomass across the food chain—from farm to fork—can significantly reduce environmental impacts, enhance economic returns, and unlock the therapeutic and functional potential of these underutilized resources.

18.8 Conclusions

Pulse by-products derived from crops such as lentils and moth beans offer significant opportunities for value creation across food, feed, and packaging industries. These residues are rich sources of bioactive and nutritive compounds, including proteins and dietary fibres, which can be recovered using environmentally sustainable extraction technologies. In parallel, the packaging sector is increasingly exploring pulse-derived biomass as a renewable feedstock for bio-based materials, supporting reduced dependence on petroleum-derived polymers. Efficient utilization of pulse agro-industrial residues contributes to lowering both environmental and economic burdens associated with food processing. Adoption of green technologies and circular economy principles enables the transformation of pulse by-products into functional ingredients such as flours, protein–fibre fractions, extracts, and cooking or soaking water, while minimizing waste generation. Their techno-functional properties support diverse food applications, including bakery products, beverages, emulsified systems, and plant-based formulations.

Although current studies demonstrate promising techno-functional performance, challenges remain in standardizing analytical methods and improving sensory evaluation practices. Addressing these gaps will support the development of high-quality, consumer-acceptable products and accelerate the transition toward a sustainable, zero-waste bio-based food system.

References

Aboye AD (2022) Pilot review study on the importance of treated agro-industrial wastewater to reuse for irrigating agriculture. Earth Env Sci Res Rev 5(1):01, 26

Acharya B, Mukherjee B, Lalramhlimi B, Dutta S, Chakraborty I, Nath R (2019) Value addition to pulses through post harvest technology and processing. In: Recent trends & advances in food science & post harvest technology. Satish Serial Publishing House, New Delhi

Aschemann-Witzel J, Gantriis RF, Fraga P, Perez-Cueto FJ (2021) Plant-based food and protein trend from a business perspective: markets, consumers, and the challenges and opportunities in the future. Critic Rev Food Sci Nutr 61(18):3119–3128

Bessada SMF, Barreira JCM, Oliveira MBPP (2019) Pulses and food security: dietary protein, digestibility, bioactive and functional properties. Trends Food Sci Technol 93(228):53–68

Bhat R (2021) Sustainability challenges in the valorization of agri-food wastes and by-products. In: Valorization of agri-food wastes and by-products. Academic, pp 1–27

Bhat S, Aditya KS, Kumari B, Acharya KK, Sendhil R (2022) Pulses production, trade and policy imperatives: A global perspective. In: Advances in legumes for sustainable intensification. Academic, pp 639–656

Boroomand, A. (2018) Subcritical water extraction of tannins, phenolics and fibre from Canadian faba bean hull and green pea pod

Campos-Vega R, Loarca-Piña G, Oomah BD (2010) Minor components of pulses and their potential impact on human health. Food Res Int 43(2):461–482

Chandran AS, Suri S, Choudhary P (2023) Sustainable plant protein: an up-to-date overview of sources, extraction techniques and utilization. Sustain Food Technol 1(4):466–483

Ciudad-Mulero M, Barros L, Fernandes Â, Berrios JDJ, Cámara M, Morales P, Fernández-Ruiz V, Ferreira ICFR (2018) Bioactive compounds and antioxidant capacity of extruded snack-type

products developed from novel formulations of lentil and nutritional yeast flours. Food Funct 9(2):819–829
Ciudad-Mulero M, Fernández-Ruiz V, Cuadrado C, Arribas C, Pedrosa MM, Berrios JDJ, Pan J, Morales P (2020) Novel gluten-free formulations from lentil flours and nutritional yeast: evaluation of extrusion effect on phytochemicals and nonnutritional factors. Food Chem 315(126175):5
Cohen B, Cowie A, Babiker M, Leip A, Smith P (2021) Co-benefits and trade-offs of climate change mitigation actions and the Sustainable Development Goals. Sustainable Prod Consumption 26:805–813
Cuj-laines R, Hernández-santos B, Herman-lara E, Martínezsánchez CE, Juárez-barrientos JM, Torruco-uco JG, Rodríguez-miranda J (2018) Relevant aspects of the development an alternative to reduce global undernourishment. In: Holban AM, Grumezescu AM (eds) Alternative and replacement foods. Elsevier, pp 141–166
De Aguiar Saldanha Pinheiro AC, Martí-Quijal FJ, Barba FJ, Tappi S, Rocculi P (2021) Innovative non-thermal technologies for recovery and valorization of value-added products from crustacean processing by-products—an opportunity for a circular economy approach. Foods 10(9): 2030
de Koff JP (2008) Sustainable waste management: the beneficial reuse of industrial, municipal, and agricultural waste products as soil amendments (Doctoral dissertation, Purdue University)
Deng X, Liao J, Zhao Z, Qin Y, Liu X (2022) Distribution and speciation of selenium in soybean proteins and its effect on protein structure and functionality. Food Chem 370:130982
Domínguez Díaz L, Fernández-Ruiz V, Cámara M (2020) An international regulatory review of food health-related claims in functional food products labeling. Jour of Func Foods 68 (March):103896
Eder S (2021) Adsorption and ultrafiltration as techniques for value addition to plant-based by-products (Doctoral dissertation, ETH Zurich)
Escobedo A, Mojica L (2021) Pulse-based snacks as functional foods: processing challenges and biological potential. Compr Rev Food Sci Food Saf 20(5):4678–4702
Essien BA, Okereke PO, Oyeleye AD (2021) Industrial crops production. In: Agricultural Technology for Colleges, p 122
FAO (2020) FAOSTAT
Fathonah S, Rosidah Amalia B, Humaizah S (2020) The formulation of alternative gluten-free mung bean biscuits. J Phys Conf Ser 1444(1):012004
Fierascu RC, Fierascu I, Avramescu SM, Sieniawska E (2019) Recovery of natural antioxidants from agro-industrial side streams through advanced extraction techniques. Molecules 24(23): 4212
Garcia-Amezquita LE, Tejada-Ortigoza V, Serna-Saldivar SO, Welti-Chanes J (2018) Dietary fibre concentrates from fruit and vegetable by-products: processing, modification, and application as functional ingredients. Food Bio Tech 11:1439–1463
Goyal RK, Wanjari OD, Ilyas SM, Vishwakarma RK, Manikantan MR, Mridula D (2005) Pulse milling technologies. Technical bulletin No CIPHET/Pub/05
Grasso S (2020) Extruded snacks from industrial by-products: a review. Trends Food Sci Technol 99:284–294
Hawu O (2022) Legumes in maize straws: nutritive value, palatability indices and silage fermentation parameters (Doctoral dissertation, North-West University (South Africa))
Herrington C, Lividini K, Angel MD, & Birol E (2019) Prioritizing countries for biofortification interventions: biofortification priority index second edition (BPI 2.0) (Vol. 40). Harvest Plus Working Paper
Jha AB, Warkentin TD (2020) Biofortification of pulse crops: status and future perspectives. Plants 9(1):73
Joshi PK, Rao PP (2017) Global pulses scenario: status and outlook. Ann N Y Acad Sci 1392(1): 6–17
Kumar S, Meena RS, Datta R, Verma SK, Yadav GS, Pradhan G et al (2020) Legumes for carbon and nitrogen cycling: an organic approach. In: Carbon and nitrogen cycling in soil, pp 337–375

Kumar S, Gopinath KA, Sheoran S, Meena RS, Srinivasarao C, Bedwal S et al (2023) Pulse-based cropping systems for soil health restoration, resources conservation, and nutritional and environmental security in rainfed agroecosystems. Front Microbiol 13:1041124

Langyan S, Yadava P, Khan FN, Bhardwaj R, Tripathi K, Bhardwaj V et al (2022) Nutritional and food composition survey of major pulses toward healthy, sustainable, and biofortified diets. Front Sustain Food Syst 6:878269

Liu Z, de Souza TS, Holland B, Dunshea F, Barrow C, Suleria HA (2023) Valorization of food waste to produce value-added products based on its bioactive compounds. Processes 11(3):840

Magrini MB, Anton M, Chardigny JM, Duc G, Duru M, Jeuffroy MH et al (2018) Pulses for sustainability: breaking agriculture and food sectors out of lock-in. Front Sustain Food Syst 2:64

Martineau-Côté D, Achouri A, Karboune S, L'Hocine L (2022) Faba bean: an untapped source of quality plant proteins and bioactives. Nutrients 14(8):1541

Mersha A (2016) Determination of cultivar-dependent variation in food-feed traits in lentil (Lens culinaris) (Doctoral dissertation, Hawassa University)

Niño-Medina G, Muy-Rangel D, Urías-Orona V (2017) Chickpea (Cicer arietinum) and soybean (Glycine max) hulls: byproducts with potential use as a source of high value-added food products. Waste Biomass Valor 8:1199–1203

Patras A, Oomah BD, Gallagher E (2011) By-product utilization. In: Tiwari BK, Gowen A, McKenna BM (eds) Pulse foods processing, quality and nutraceutical applications, pp 325–362

Pugliese G, Losacco C, Roselli V, Laudadio V, Piemontese L, Naz S et al (2024) Lentil (Lens culinaria) and its by-products inclusion in livestock nutrition: present insights and emerging trends in rabbit and poultry system. J Appl Anim Res 52(1):2362254

Saha SK, Pathak NN (2021) Fundamentals of animal nutrition. Springer, Singapore, pp 219–246

Selvamuthukumaran M (2024) Wealth out of food processing waste: ingredient recovery and valorization. CRC Press

Sharif MK, Butt MS, Sharif HR, Nasir M (2017) Sensory evaluation and consumer acceptability. In: Handbook of food science and technology, vol 10, pp 362–386

Sharma P, Gaur VK, Gupta S, Varjani S, Pandey A, Gnansounou E et al (2022) Trends in mitigation of industrial waste: global health hazards, environmental implications and waste derived economy for environmental sustainability. Sci Total Environ 811:152357

Shinde R, Shahi DK, Mahapatra P, Singh CS, Naik SK, Thombare N, Singh AK (2022) Management of crop residues with special reference to the on-farm utilization methods: A review. Ind Crops Prod 181:114772

Siddiqui SA, Kamran F, Stankov S, Rathod NB, Teixeira-Costa BE, Fidan H et al (2024) Unveiling the diversity of non conventional proteins-from sources, extraction, technofunctionality, nutraceutical potential to advancement in food applications – a systematic review. Waste Biomass Valor:1–23

Simitzis PE, Deligeorgis SG (2018) Agroindustrial by-products and animal products: a great alternative for improving food-quality characteristics and preserving human health. In: Food quality: balancing health and disease. Academic, pp 253–290

Singh N (2017) Pulses: an overview. J Food Sci Technol 54(4):853–857

Slominski, B. A. (2018) Advances in the understanding of dietary fibre and its components in relation to the use of alternative feed ingredients in modern poultry and livestock production

Tiwari BK, Gowen A, McKenna B (2021) Advances in pulse foods. In: Pulse foods. Academic, pp 1–7

Verma P (2018) Processing and value addition of pulses. AISECT Univ J 06(13):17–20

Wodajo HD, Gemeda BA, Kinati W, Mulem AA, van Eerdewijk A, Wieland B (2020) Contribution of small ruminants to food security for Ethiopian smallholder farmers. Small Rumi Res 184: 106064

Zhong L, Fang Z, Wahlqvist ML et al (2018) Seed coats of pulses as a food ingredient: characterization processing and applications. Trends Food Sci Technol 80:35–42

Chapter 19
Harnessing Biodiversity: Fortifying Functional Hard-Boiled Candies with Fruit-Based Nutrients for Safe and Sustainable Confectionery

Reeba Iqbal, Charlotte Peraut, and Monika Thakur

Abstract With the rise in awareness about healthy and nutritious food products, the food industry is under persistent pressure to cater to such demands and deliver the desired food products to consumers. Confectionery products, such as hard-boiled candies, are products of interest for people of all age groups, ranging from children to adults. Due to the presence of insubstantial calories in these hard-boiled candies, they are constantly being catechized for their nutritional and health benefits. To address this rising demand for such candies, this research work aimed to incorporate micronutrients, namely, vitamin C and iron, into the candy formulations. This was done to address and target common micronutrient deficiencies and to confer a health benefit on the consumers. The aim to develop the products with apt sensory properties that are shelf-stable was achieved using kiwi extract to address vitamin-C deficiency and date extract to address iron deficiency. The hard-boiled candies were formulated using chocolate extract as the flavouring agent with both the respective ingredients to develop two different types of hard-boiled candies.

Keywords Hard-boiled candies · Healthful candies · Nutritional candies · Deficiency

---

R. Iqbal · M. Thakur (✉)
Amity Institute of Food Technology, Amity University Uttar Pradesh, Noida, Uttar Pradesh, India
e-mail: mthakur1@amity.edu

C. Peraut
Université Clermont Auvergne, Clermont-Ferrand, France
e-mail: charlotte.peraut@etu.uca.fr

M. Thakur, T. Belwal (eds.), *Food Biodiversity and Consumer Engagements*, Sustainable Development and Biodiversity 46,
https://doi.org/10.1007/978-981-95-7777-4_19

19.1 Introduction

Candies are sugar-based confectionery products that are amorphous, hard, chewy, and homogeneous in nature. Candies are loved by kids as well as adults due to their sweet taste. Owing to their compact size and high calorific content, candies are one of the most preferred forms of convenience foods (Mauliza and Dalimunthe 2014). The basic principle behind the technology of making candy is the art and science of attaining different textural conformities by the alteration of sugar, which is the major and primary ingredient in the manufacture of candies.

The main objectives of the study include addressing the vitamin-C and iron deficiencies by developing candies with the required ingredients. Vitamin C, a water-soluble vitamin, is one of the vitamins that plays a significant role in the growth and nourishment of the human body. Humans cannot synthesize vitamin C from glucose and galactose as they are deficient in the enzyme named gulonolactone oxidase; therefore, it is important to incorporate it in the human diet through various sources. Encompassing a vitamin-C source in hard-boiled candies is an extremely attractive idea to include the vitamin-C in a commodity that is appealing to consumers of all ages.

The deficiency of iron is one of the major nutrient deficiencies that is observed in people of all age groups. Anaemia is one of the major problems associated with the deficiency of iron in the human body. It occurs due to the below-par bioavailability of iron, which affects its absorption and utilization by the human body (Singh and Thakur 2025). In India, 70–80% of the children suffer from anaemia, while the figures are approximately 70% in the case of pregnant women (NFHS 2007; Kohli et al. 2025). Fortification, supplementation, and diversification of diets with iron are some methods that can be used to combat iron deficiencies among consumers (Prakash and Singh 2016).

Candies can prove to be a great food product to serve as an iron-fortified food item, as it is consumed by large sections of the population and by consumers of all age brackets.

The various ingredients that are used in the production of candies are discussed in detail as follows:

1. Sugar (Sucrose): The principal ingredient of all sugar-based candies is sugar or sucrose obtained from cane sugar or beet sugar. The sugar concentration is used to control the water content of the candies, as the sugar concentration is directly proportional to the boiling point of the overall solution. The different concentrations of solutions of sugar and water yield different types of candies. For example, highly concentrated solutions may yield amorphous glass, fully crystalline mass, or partially crystalline mass, which may further be made into different types of candies.
2. Invert Sugar: In the bakery industry, glucose is referred to as dextrose, whereas fructose is referred to as levulose. When the sucrose is hydrolysed using acids or enzymes, it results in the production of invert sugar. Invert sugar is used in the confectionery industry in order to control the degree of crystallization of sucrose

and to facilitate the formation of smaller crystals, which in turn affects the texture of the candies obtained. Moreover, it helps in the over-drying of the candies due to its hygroscopic nature. It is also sweeter than sucrose alone (Potter and Hotchkiss 1995).

3. Glucose Syrup: It is derived from starch by hydrolysis. The glucose syrup is used to slow down crystallization of sugar and to control the viscosity of candies, and also to impart sweetness to the products (Potter and Hotchkiss 1995).
4. Corn Syrup: Corn syrup is a viscous liquid that is obtained by the hydrolysis of corn starch. It helps in retarding crystallization of sucrose and also adds viscosity to the candies and slows the dissolution of candies in the mouth, thus enhancing the mouth feel (Potter and Hotchkiss 1995).
5. Colourants: Colourants are used to impart vibrant colours to the candies to attract consumers, especially young children. The colourants used can be natural, like caramel, or they may be synthetic in nature, that is, the ones that are made by humans, like FD&C Yellow No. 6 or aluminium lake, etc. (Hodgson 2002).
6. Flavourings: Flavours enhance the taste of the candies and are thus used to increase the variety of candies available in the market to the consumers. They can be natural flavours like peppermint or laboratory-made synthetic flavours like ethyl vanillin (artificial vanilla flavour) (Hodgson 2002).
7. Humectants: Humectants are substances that are used to hold moisture within the candies. These are the products that have low moisture content and thus are susceptible to drying out easily. The most common examples of humectants that are used in candies are invert sugar, glycerol, sorbitol, pectins, and gums (Potter and Hotchkiss 1995).
8. Preservatives: Preservatives are the substances used to prevent the spoilage of candies by inhibiting the growth of bacteria, yeast, and moulds in the candies.

   There are 2 main classes of preservatives-

   (i) Class-I Preservatives: These are the preservatives that are procured from nature, that is, these preservatives are derived from sources of natural origin. Some examples include sugar, salt, honey, spices, etc. (Dwivedi et al. 2017).

   (ii) Class-II Preservatives: These are the preservatives that are produced by humans in factories or laboratories and can be chemical, synthetic, or semi-synthetic in nature. Examples of class-II preservatives include sorbates, benzoates, glutamates, glycerides, nitrates, and nitrites of potassium, among others (Anand and Sati 2013).

9. Stabilizers and Thickeners: They are used to stabilize the mixtures and to impart them a smooth and homogeneous consistency. Pectin and corn starch are commonly used in the candy industry to serve the purpose (Hodgson 2002).
10. Emulsifiers: Emulsifiers are used to help in the dispersion of ingredients, to reduce viscosity, to prevent crystallization of sugar, and to stabilize the structure of the candies (Godshall 2016). Glycerol monostearate and gum Arabica are the commonly used emulsifiers in the candy industry (Hodgson 2002).

11. Acidulants: The substances called acidulants can also be used in candies to maintain the pH of the candy, to provide tart flavour, and to act as a preservative. The most widely used acidulant in candies is maleic acid, which imparts a sour flavour to the candies (Hodgson 2002).

The sugar-based candies can be classified into the following types based on their texture:

1. Hard-boiled candies.
2. Soft-boiled candies.

Hard-boiled candies are basically the candies that have a low moisture content ranging from 1% to 2% and have a total solids content (TSS content) of 97% (Lees and Jackson 1973; Alikonis 1979). Hard-boiled candies are recognized by their hard texture and shiny appearance (Mauliza and Dalimunthe 2014).

A natural phenomenon called graining is associated with hard-boiled candies, and it classifies these hard-boiled candies into grained candies and non-grained candies. Graining is basically the physical phenomenon of recrystallization of sugar in hard candies upon storage. The grainy hard-boiled candies are opaque in appearance, while the ungrained hard-boiled candies are clear in appearance (Efe and Dawson 2022). One of the major characteristics of the hard-boiled candies is the glassy state of the product. The glassy state of a product is achieved when the liquid is supercooled, thereby restricting the production of a crystalline structure (Lees and Jackson 1973). The glassy state limits the molecular mobility of the components of the candies and thus enhances the stability of the end product (Lungade et al. 2012).

The hard-boiled candies come in various shapes, sizes, and colours. All of the physical attributes are dependent on the ingredients used in the hard-boiled candies and the specific end product requirements. The two major components of the hard-boiled candies are sugar (cane sugar or beet sugar) and liquid glucose (Efe and Dawson 2022). Sugar forms the basis of the hard-boiled candies and imparts sweetness to the product. However, glucose syrup not only provides sweetness to the product but also affects the functional properties of the candies, like viscosity, as well as the nutritional value of the candies. It also helps in preventing graining in the hard-boiled candies. The glucose syrup in hard-boiled candies is utilized as per its DE value or the dextrose equivalent value, which is related to the amount of reducing sugars present in the glucose syrup. The glucose syrup with a high DE value is known to impart higher sweetness to the candies. 42 DE and 63 DE are the two commonly used glucose syrups in the candy industry (Efe and Dawson 2022). Apart from these two major ingredients, the other ingredients that constitute the hard candies are water, colorants, flavourings, etc. (Jackson 1995).

There are various types of processes for the manufacture of hard-boiled candies. However, the basic procedure involves mixing of the two major ingredients, that is, sugar and glucose syrup, followed by their dissolution in water and then heating up the sugar solution to high boiling temperatures of around 150 °C. This is followed by the mixing of the additional ingredients, like colourants and flavourings, into the solution with continuous stirring. Once the desired consistency has been achieved, the ingredients are transferred into moulds and allowed to cool and harden into candies. The set candies are then removed from the moulds and packed (Efe and Dawson 2022; Mauliza and Dalimunthe 2014).

Soft-boiled candies are a type of sugar-based confectionery product that are soft in nature and are made with the help of gelling agents such as gelatine, pectin, or starch, individually or in combination (Lees and Jackson 1973). Soft-boiled candies are widely preferred due to their gelatinized morphology (Gunes et al. 2022). Soft-boiled candies are soft and can easily be chewed and swallowed by people of all age groups. A good soft-boiled candy has moderate softness and does not have the tendency to cohere to the teeth (Kimura and Ohno 2004). The soft-boiled candies have a moisture content ranging from 8% to 22%, which accounts for their soft texture (Efe et al. 2019; Lees and Jackson 1973).

One of the distinct features of the soft-boiled candies that distinguishes them from the hard-boiled candies is their ability to remain in a rubbery state. Hence, it is extremely essential to monitor the glass transition temperature in the case of soft-boiled candies, to prevent them from attaining the glassy state by crystallization (Kalogeras and Hagg Lobland 2012). Maintaining the glass-transition temperature in the soft-boiled candies manufacturing process is one of the most crucial steps in order to achieve the desired soft texture of the soft-boiled candy by making it retain the rubbery state (Kalogeras and Hagg Lobland 2012).

The basic steps involved in the production of soft-boiled candies involve the mixing of all the ingredients, followed by their dissolution in water, concentrating them to the desired consistency, and then transferring them to moulds. Once the mixture has been transferred to the moulds, it is dried to remove the water vapour in order to get the desired end product. The soft-boiled candies are then packed in the desired packaging (Lees and Jackson 1973).

Some important factors affecting the properties of both hard-boiled candies and soft-boiled candies are discussed as follows:

1. Glass-Transition Temperature: It is the temperature at which the substances change from a glassy and hard state to an amorphous or rubbery state. The glass transition temperature acts as the governing parameter for the texture of candies. It depends on a number of factors. However, in the case of hard-boiled candies, maintaining the cooking temperature under the glass transition temperature results in better sensory properties and longer shelf-life of the hard-boiled candies (Lungade et al. 2022).
2. Solubility: Sweeteners are the most important component of the candies, and their solubility in the mixture is of utmost importance. It is important to take into consideration the solubility percentages and corresponding temperatures of the sweeteners used, as they affect the texture of the ultimate candies. Preparing candies without crystal formation is vital to have good quality attributes in candies (Lungade et al. 2022).
3. Moisture Content: Water used in the candy-making process is generally incorporated to mix the different ingredients in the candy mixture. Hard-boiled candies usually should have a moisture content of 1–2%. Therefore, understanding the role and concentration of the ingredients used in the overall moisture content of

the candies is vastly crucial. It is vital to remember that ingredients like inverted sugar syrup, glucose, corn syrup, etc., act as humectants and reduce the moisture content of the end product, and hence their concentrations must be regulated in order to get the candies with desirable moisture content (Lungade et al. 2022). It must also be noted that candies tend to lose moisture if they have a moisture content higher than the relative humidity of the environment they are kept in. On the contrary, if candies have less moisture content than the relative humidity of the environment, they tend to uptake moisture. This leads to undesirable changes like loss of flavour, reduction of glass transition temperature, and graining in candies (Lungade et al. 2022).

The consumption of sugar-based candies accounts for about 39% of all the confectionery products that are consumed all over the world (Godshall 2016). Candy has been one of the most preferred products of consumption by consumers of all age groups. This can be attributed to the rapid urbanization that is occurring across the globe and to the ceaseless increase in the innovative and gourmet food product demands. Various studies that have been conducted to study the candy consumption patterns across the globe have found that in developing countries, which are showing a rise in the middle class, there is also a rise in the consumption of candies. In Asian countries like India, China, etc., candies serve the purpose of gourmet gift alternatives, whereas in the case of Western countries like America, candies are a part of the cultural and traditional processions and festivals like Halloween, Easter, etc. Therefore, the sugar-based candies are known to have a potential growing market all across the globe (Godshall 2016). Hard-boiled candies are known to provide empty calories without imparting any beneficial nutritional impacts on human health and the body. This has led to hard-boiled candies being questioned for their health-associated risks and their abandonment by the health-conscious people who want to have healthy and nourishing diets. This has led to the need to address the demands of consumers and to shift to healthier and sustainable options for candies. This means that the candy manufacturers not only have to come up with innovative flavour options, but they also need to incorporate ingredients in their candies that confer a health benefit to the consumer and are devoid of any ingredients that can negatively affect the health of the consumers. This demand can be catered to through the fortification and supplementation of these candies with additional beneficial nutrients like vitamins, minerals, bio-colorants, etc. These nutrients can be added through various sources in different forms to the hard-boiled candies.

19.2 Materials and Methods

1. Incorporation of Vitamin-C in Candy Development:
   As a part of this research, kiwi fruit was selected to serve as the source of Vitamin C. For this, water-soluble liquid kiwi fruit extract was procured from the market, which had an herb-to-extract strength of 4:1, which means 4 kg of

kiwi fruit were used to yield 1 kg of kiwi fruit extract. Kiwi fruit, also known as *Actinidia deliciosa,* is a native fruit of China and is available in two varieties, namely the green cultivar or Hayward cultivar (*A. Deliciosa*) and the golden cultivar (*A. Chinensis*). However, currently, New Zealand is the largest producer of kiwi fruit globally. The most typical feature of the kiwi fruit is its vitamin-C content that sets it apart from all other fruits (Mike 2013). The green cultivar of kiwi fruit contains 92.7 mg of vitamin C per 100 g, whereas the golden cultivar contains 161.3 mg of vitamin C per 100 g, which is almost thrice more than the vitamin-C content of fruits like oranges and strawberries (on edible flesh weight basis).

2. Incorporation of Iron in Candy Development:

   As a part of this research, date fruit was selected to serve the purpose of acting as the source of iron. For this, water-soluble liquid date extract was procured from the market.

   Dates, also known as *Phoenix dactylifera*, are a member of the palm family Palmae (Zohary and Hopf 1993). Nearly 200 different varieties of dates are prevalent across the globe (Amer 1994). Dates are known to be a majorly cultivated crop in the Middle East countries, Saudi Arabia, and Egypt (Rahmani et al. 2014). Dates are considered nutritionally superior foods owing to their excellent nutritional composition. Dates are abundant in carbohydrates like soluble sugars and are also rich in dietary fibre content, have a moderate amount of moisture, and are fairly low in levels of ash, protein, and lipids (Ghnimi et al. 2017). Dates are a good source of iron, as 100 g of dates furnish approximately 0.3–10.4 mg of iron. Henceforth, consuming even a little amount of dates each day will result in fulfilling the body's daily iron demands (Irandegani et al. 2019).

3. Incorporation of Chocolate for Flavour in Candies:

   From times unknown, chocolate has been one of the major attractions of the confectionery industry. The use of chocolate is widespread, whether it is candies, toffees, gums, etc., chocolate flavour is one of the most chosen flavours among all the flavours. Therefore, chocolate is an ideal choice for the addition of flavour to hard candies owing to its nutritional benefits, and in order to improve the sensory characteristics of the product being developed and to enhance the appeal of the candies to consumers of all groups.

4. Development of Hard-Boiled Candies:

   As a part of this research study, the open-pan method was used to make the candies. The procedure that was adopted for the development of candies is discussed below:

   - Weighing of raw materials in a clean pan
   - Addition of sugar & liquid glucose to the pan
   - Mixing of all ingredients with potable water
   - Addition of extract and flavour to the mixture with constant stirring
   - Heating the mixture with constant stirring
   - Transferring to moulds

- Cooling and hardening of candies
- Packaging of candies

As a part of this research study, two different kinds of hard-boiled candies were developed:

1. Kiwi candies with chocolate flavour, to provide candies rich with vitamin C (as a result of using kiwi extract in candies).
2. Date candies with chocolate flavour, to provide candies rich with iron (as a result of using date extract in candies).

In order to develop the kiwi candies, 1 ml of kiwi extract was added per 100 g of the candy. To provide flavour to the candies, 1 ml of chocolate flavour was added per 100 g of the candy. In order to develop the date candies, 1 ml of date extract was added per 100 g of the candy. To provide flavour to the candies, 1 ml of chocolate flavour was added per 100 g of the candy.

Evaluation and assessment of various attributes of the candies is a crucial step in analysing the quality and consumer acceptability of the candies. Therefore, a number of tests were conducted in order to analyse the quality of the candies developed.

1. Organoleptic Evaluation of Candies:

   Organoleptic tests help in the development of products that are profoundly accepted by consumers. Organoleptic tests include the assessment and description of the shape, colour, texture, aroma, and flavour of a food product (Table 19.1; Mauliza and Dalimunthe 2014).
2. Moisture Content Evaluation of the Candies:

   The moisture content evaluation of the hard candies was done by the hot air oven method, wherein 5 g of each sample was dried in a hot air oven at 105 °C for 3 hours (Table 19.2).

   The moisture content of the hard candies usually ranges from 1% to 2% and both types of candies developed have moisture content within these limits, hence ensuring a shelf-stable product has been developed.
3. pH Evaluation of the Candies:

   The pH evaluation of the hard candies was done using a pH meter, wherein 1 g of sample was dissolved in 10 ml of distilled water, and the pH was checked by dipping the probe of the pH meter in the solution developed (Table 19.3).

Table 19.1 Organoleptic evaluation of candies

Organoleptic test	Kiwi candies	Date candies
Shape	Round	Round
Colour	Orangish brown	Yellow
Texture	Brittle	Brittle
Aroma	Pleasant	Sweet
Flavour	Acidic with undertones of chocolate	Caramel with undertones of chocolate

Table 19.2 Moisture content evaluation of candies developed

Sample name	Moisture content (%)
Kiwi candy with chocolate flavour	0.58%
Date candy with chocolate flavour	0.97%

Table 19.3 pH evaluation of candies developed

Sample name	pH
Kiwi candy with chocolate flavour	4.2
Date candy with chocolate flavour	4.6

Table 19.4 Sensory evaluation of kiwi candies with chocolate flavour

Attribute	Average feedback for kiwi candies	Average feedback for date candies
Colour and appearance	7.5	8.4
Aroma	7.0	8.2
Taste	7.2	8.5
Flavour	7.3	8.5
Sweetness	7.0	8.5
Texture	7.1	8.3
After taste	7.0	8.0
Overall acceptability	7.1	8.3

4. Sensory Evaluation of the Candies:

   In this study, the 9-point Hedonic Scale was used that ranged from "like extremely" to "dislike extremely," with the following ratings—(9: like extremely), (8: like very much), (7: like), (6: like slightly), (5: neither like nor dislike), (4: dislike slightly), (3: dislike moderately), (2: dislike), and (1: dislike extremely). Data from 70 random people were collected for this purpose (Table 19.4).

   The results of the Hedonic Scale Testing carried out for the kiwi candy with chocolate flavour are shown in Table 19.4, with the average values for each of the attributes.

   It can be concluded from the sensory evaluation tests that the date candies with chocolate flavour were preferred and liked by more consumers than the kiwi candies with chocolate flavour, which can be used for analysing the future of both the products in terms of consumer preference and acceptability.

19.3 Discussion

The date candies with chocolate flavour can help combat the problem of anaemia among children and other consumers. On the other hand, the kiwi candies with chocolate flavour can help in tackling the vitamin C deficiency among consumers. Moreover, it is a great proposition to provide nutrition to the consumers in the form of delectable and convenient food items like candies that are cheap, readily available, handy, and toothsome.

19.4 Future Prospects in Sustainable Confectionery

Looking ahead, the sustainable development of nutrient-fortified confectionery holds significant promise both scientifically and commercially. With increasing consumer demand for healthier food options, future research can expand on innovative formulations that incorporate diverse bioactive fruit compounds and alternative plant-based gelling agents to improve nutritional profiles without compromising taste or texture, as seen in broader trends within functional candy development. Collectively, these avenues not only align fortified candies with global trends toward health-oriented and sustainable food systems but also offer opportunities for novel product diversification, regulatory guidance, and broader public health impact. Hence, the new functional, sustainable food product development have the potential not only to meet contemporary health and wellness demands but also to foster product diversification, support regulatory frameworks for quality and safety, and contribute to improved biodiversity in public health outcomes on a broader scale (Thakur et al. 2024, 2025).

19.5 Conclusion

Due to the increase in awareness about nutrition and health, more and more individuals across the globe are shifting towards healthier and more sustainable options that are not only scrumptious but healthy as well. This has made it the need of the hour for the food industry to address these changes and to adapt itself according to the stipulations of the consumers. The debut of healthful and nutrient-rich candies in the market can give a major boost to the candy and confectionery industry all over the world.

References

Alikonis JJ (1979) Candy technology. AVI Publishing Co.

Amer WM (1994) Taxonomic and documentary study of food plants in ancient Egypt (Ph.D. thesis). Cairo University, Giza

Anand SP, Sati N (2013) Artificial preservatives and their harmful effects: looking toward nature for safer alternatives. Int J Pharm Sci Res 4(7):2496

Dwivedi S, Prajapati P, Vyas N, Malviya S, Kharia A (2017) A review on food preservation: methods, harmful effects and better alternatives. Asian J Pharm Pharmacol 3(6):193–199

Efe N, Dawson P (2022) A review: sugar-based confectionery and the importance of ingredients. Eur J Agric Food Sci 4:10.24018/ejfood.2022.4.5.552

Efe N, Bielejewski M, Tritt-Goc J, Mert B, Oztop MH (2019) NMR relaxometry study of gelatin-based low-calorie soft candies. Mol Phys. https://doi.org/10.1080/00268976.2018.1564392

Ghnimi S, Umer S, Karim A, Kamal-Eldin A (2017) Date fruit (*Phoenix dactylifera* L.): an underutilized food seeking industrial valorization. NFS J 6:1–10. https://doi.org/10.1016/j.nfs.2016.12.001

Godshall MA (2016) Candies and sweets: sugar and chocolate confectionery. In: Encyclopedia of food and health. Academic, pp 621–627. https://doi.org/10.1016/B978-0-12-384947-2.00679-6

Gunes R, Palabiyik I, Konar N, Toker OS (2022) Soft confectionery products: quality parameters, interactions with processing and ingredients. Food Chem 385:132735. https://doi.org/10.1016/j.foodchem.2022.132735

Hodgson AS (2002) Common food additives in candy. Food Safety and Technology (FST-9). Department of Tropical Plant and Soil Sciences, University of Hawai'i at Manoa

Irandegani F, Arbabisarjou A, Ghaljaei F, Navidian A, Karajibani M (2019) The effect of a date consumption-based nutritional program on iron deficiency anemia in primary school girls aged 8 to 10 years old in Zahedan (Iran). Pediatr Health Med Ther 10:183–188. https://doi.org/10.2147/PHMT.S225816

Jackson EB (1995) Sugar confectionery manufacture, 2nd edn. Blackie, New York

Kalogeras IM, Hagg Lobland HE (2012) The nature of the glassy state: structure and glass transitions. J Mater Educ

Kimura T, Ohno Y (2004) Soft candy and its production method. U.S. Patent No. 6805894B1

Kohli A, Singh A, Thakur M (2025) Fortifying confectionery food products to combat hidden hunger and micronutrient deficiencies. In: Thakur M, Brugère CM (eds) Revolutionizing agri-food systems. World sustainability series. Springer, Cham. https://doi.org/10.1007/978-3-031-98200-2_21

Lees R, Jackson BE (1973) Sugar confectionery and chocolate manufacture. Blackie Academic & Professional

Lungade P, Karadbhajne SV, Mande V (2012). Int J Food Nutr Sci. ISSN 2319–1775

Lungade P, Karadbhajne SV, Mande V (2022) A review of sugar-based confectionery: properties and types of products. IJFANS Int J Food Nutr Sci 11(3)

Mauliza S, Dalimunthe GI (2014) Formulation and evaluation of the preparation of hard candy source of sweet orange (*Citrus sinensis* (L.) Osbeck) as a nutraceutical. Int J Sci Technol Manage ISSN 2722-4015

Mike B (2013) Nutritional benefits of kiwifruit. In: Kiwifruit proteins and enzymes, pp 59–80

NFHS (2007) National Family Health Survey (NFHS-3) 2005–06. International Institute for Population Sciences, Mumbai

Prakash J, Singh MC (2016) Nutritional and sensory quality of iron fortified tamarind candies. Nutr Food Sci 1:1–7

Rahmani AH, Aly SM, Ali H, Babiker AY, Srikar S, Khan AA (2014) Therapeutic effects of date fruits (*Phoenix dactylifera*) in the prevention of diseases via modulation of anti-inflammatory, anti-oxidant and anti-tumour activity. Int J Clin Exp Med 7(3):483–491

Singh V, Thakur M (2025) Management of iron deficiency anemia among adolescent girls through the intervention of Fe-Vit C-rich powder mix. BIO Web Conf 178:02018. https://doi.org/10.1051/bioconf/202517802018

Thakur M, Sanches Silva A, Singh K, Modi V (2025) Editorial: recent advances and future prospects for sustaining a healthier food system. Front Nutr 12:1696614. https://doi.org/10.3389/fnut.2025.1696614

Thakur M, Modi VK, Feng S (2024) Editorial: food sustainability: challenges and opportunities for the future. Front Nutr 10:1341427. https://doi.org/10.3389/fnut.2023.1341427

Zohary D, Hopf M (1993) Domestication of plants in the Old World, 2nd edn. Clarendon Press, Oxford

Part III
Eating with Intent: Sustainability in Market and Consumer Choices

Chapter 20
Beverage Sector Transformation: Biodiverse Sourcing, Innovative Products & Consumer Engagement

Aadya, Mahua Bhattacharjee, and Shivani Mehta

Abstract The beverage industry, encompassing both alcoholic and non-alcoholic segments, plays a pivotal role in the global and Indian food processing sectors. Globally, the beverage market was valued at USD 1180 billion in 2020 and is projected to grow at a compound annual growth rate (CAGR) of 7.3%, reaching an 84.1% increase by 2026. India's beverage sector, driven by shifting consumer preferences, rising disposable incomes, and a youthful demographic, is expected to register a CAGR of 48.31% between 2024 and 2028, reaching an estimated market volume of USD 5971 million. This study conducts a comparative analysis of the alcoholic and non-alcoholic beverage markets in India to evaluate their growth trajectories. The non-alcoholic segment, comprising bottled water, carbonated drinks, juices, energy drinks, and ready-to-drink (RTD) tea, is dominated by carbonated beverages in terms of market share. Notably, states such as Bihar and Odisha have shown the highest growth in non-alcoholic beverage consumption with CAGRs of 17.69% and 17.67%, respectively. Meanwhile, in the alcoholic segment, revenue generation varies significantly across states. In Haryana and Assam, "country spirits" dominate, contributing over 80% of the total revenue. In contrast, states like Uttar Pradesh and Rajasthan demonstrate a balanced contribution from both "foreign liquors and spirits" and "country spirits." The findings underscore the sector's potential for economic contribution, job creation, and value addition, while also highlighting the need for sustainable practices and regulatory reforms to enhance growth and environmental responsibility.

Aadya (✉)
Amity School of Economics, Amity University, Noida, Uttar Pradesh, India

M. Bhattacharjee
Foundation for Development of Rural Value Chains (FDRVC), New Delhi, India

S. Mehta
Manav Rachna International Institute of Research and Studies, Faridabad, Haryana, India
e-mail: shivanimehta.sbss@mriu.edu.in

M. Thakur, T. Belwal (eds.), *Food Biodiversity and Consumer Engagements*,
Sustainable Development and Biodiversity 46,
https://doi.org/10.1007/978-981-95-7777-4_20

Keywords Compound annual growth rate · Alcoholic beverages · Non-alcoholic beverages

20.1 Introduction

The food processing sector in India has emerged as a pivotal contributor to the nation's economic growth and employment generation. In 2023, the Hon'ble Prime Minister of India referred to it as the "Sunrise Sector" during the inauguration of the World Food India meet, underscoring its increasing importance in national development (Ministry of Food Processing Industries 2024). Over the past nine years, the sector has attracted foreign direct investment (FDI) worth approximately ₹50,000 crores, reflecting investor confidence and the sector's untapped potential (Ministry of Food Processing Industries 2024). Within this expansive domain, the beverage industry constitutes a significant and dynamic segment, demonstrating substantial growth both domestically and globally. India holds the distinction of being one of the largest global producers of key raw materials essential to the beverage industry, providing a robust foundation for the sector's expansion. The country's rich culinary heritage and diverse cultural practices have historically fostered a strong affinity for food and beverages. According to Mukherjee and Mukherjee (2021), the beverage industry not only contributes to reducing post-harvest wastage through efficient value chains but also enhances employment opportunities and attracts capital investments, particularly in the manufacturing sector.

On the demand side, changing consumer preferences, especially among the younger demographic, have fuelled a surge in the consumption of innovative and diverse beverage products. The youth in India are increasingly inclined toward experimenting with newer varieties, including both alcoholic and non-alcoholic beverages, thus expanding the market base (Mukherjee and Mukherjee 2021). This trend is reinforced by the rise in disposable incomes, urbanization, and lifestyle changes, all of which have contributed to the sector's evolving dynamics. India's beverage industry is witnessing a growing influx of entrepreneurial activity. Startups across the country are exploring innovative beverage formulations ranging from traditional tea and coffee to energy drinks, mojitos, and various craft beverages, including wines and beers. This entrepreneurial momentum is catalyzing innovation, enhancing product diversity, and reshaping the competitive landscape of the sector (Tiwari and Kapoor 2019).

The sector's growth trajectory has positioned India as an increasingly attractive destination for domestic and international investments. However, alongside the promising outlook, the industry faces several structural and environmental challenges. These include supply chain inefficiencies, high taxation rates, and the need for sustainable practices across production and packaging (Sovacool et al. 2021). Given this context, the present study explores the growth prospects of India's beverage industry, particularly in relation to global trends. It examines the contributions of both alcoholic and non-alcoholic segments, analyzes market dynamics, and evaluates the sector's role in employment generation, value addition, and sustainable development. Drawing upon secondary data from reliable sources such as Statista, Euromonitor, MOFPI, and industry reports, this research aims to provide an informed understanding of the beverage industry's trajectory in India's food processing ecosystem.

20.2 Overview of Alcoholic and Non-alcoholic Beverages

The Indian beverage industry is bifurcated into two broad segments: alcoholic and non-alcoholic beverages, each demonstrating unique consumption trends and growth trajectories. Non-alcoholic beverages encompass a wide array of categories, including carbonated soft drinks, bottled water, tea, coffee, juices, and energy drinks. Recent shifts in consumer lifestyle preferences, particularly among health-conscious urban populations, have led to a growing demand for low-calorie and functional beverages such as natural juices, infused waters, and diet sodas, especially in restaurants and cafés (Mukherjee and Mukherjee 2021). Accurate estimation of market size and revenue in the non-alcoholic segment is challenging due to the predominance of informal players, which constitute nearly 80% of the sector (Mukherjee and Mukherjee 2020). Nevertheless, product-based analysis reveals that carbonated beverages continue to hold the highest consumption share, followed by tea, coffee, bottled water, and fruit juices. The Indian non-alcoholic beverage market, valued at INR 1376 billion in 2023, is projected to reach INR 1,986 billion by 2028, reflecting a Compound Annual Growth Rate (CAGR) of 7.29%. In contrast, the alcoholic beverage sector in India has seen significant expansion, driven by factors such as rapid urbanization, a rising middle class, changing social norms, and increased disposable income. According to the Ministry of Food Processing Industries (MOFPI), the alcoholic beverage manufacturing sector experienced a growth rate of 23.8% between 2015–16 and 2021–22 (MOFPI 2024). In 2019 alone, the sector generated approximately USD 48.8 billion in revenue and supported over 1.5 million jobs nationally. Furthermore, state-level excise revenues from alcoholic products constitute one of the top three revenue streams across most Indian states (Mukherjee and Mukherjee 2020). Policy incentives for the alcoholic beverage sector, such as those offered in Maharashtra and Karnataka to promote domestic wine production, have catalyzed both domestic and foreign investments. Additionally, globalization and liberalization have encouraged greater international collaboration, contributing to product diversification and quality enhancement (Tiwari and Kapoor 2019).

The subsequent sections assess the growth potential of India's beverage industry, analyzing both alcoholic and non-alcoholic categories through secondary data sources such as Statista, Euromonitor, MOSPI, and Kantar Worldpanel India. These data provide insights into product-wise consumption patterns, sales volumes, and emerging consumer preferences. Furthermore, state budgetary receipts are examined to evaluate the fiscal contribution of the alcoholic beverage sector to the Indian economy. The findings indicate that both segments are poised for continued growth, albeit along different trajectories influenced by socio-economic, regulatory, and cultural factors. While the non-alcoholic beverage segment is being reshaped by health and wellness trends, the alcoholic sector is expanding through increased acceptance and evolving lifestyle choices (Munjal and Sharma 2017). A nuanced understanding of these dynamics is critical for policymakers and industry stakeholders to foster sustainable development within the broader food and beverage processing ecosystem.

20.3 Growth of the Beverage Sector

The Indian beverage sector, comprising both alcoholic and non-alcoholic segments, has demonstrated robust growth in recent years, driven by evolving consumer preferences and economic expansion. Increasing health consciousness among consumers has significantly contributed to the surge in non-alcoholic beverage consumption. According to, this sector experienced a compound annual growth rate (CAGR) of 14.7% between 2010 and 2019, underlining its growing importance in employment generation and value addition within the Indian economy. Rao and Guha (2022) highlights that the alcoholic beverage segment remains one of the top three revenue-generating consumer goods sectors in India. Both Rao and Mukherjee underscore the beverage sector's critical role in contributing to the national economy. As per a study by the Indian Council for Research on International Economic Relations (ICRIER 2022), the non-alcoholic segment includes eight primary categories, with carbonated beverages holding the largest market share, followed by bottled water and packaged juices.

The Ministry of Food Processing Industries reported that between 2013 and 2018, the consumption of beer and wine increased significantly, with projections indicating that by 2030, approximately 50% of consumers are expected to maintain consistent preferences for specific alcoholic beverages. Additionally, a primary survey noted a 70% increase in the alcoholic beverage sector during that period. Furthermore, employment impact is notable, with Jenson (2018) estimating that the beverage industry directly employs 25,000 individuals and indirectly supports another 15,000. He further noted gender-inclusive training programs delivered via mobile vans and initiatives to train farmers in sustainable cultivation practices of fruits like mango, orange, and grape. Sustainability is increasingly becoming integral to the sector's growth strategy. Menon and Mukherjee (2021) discuss the adoption of eco-friendly packaging, with several companies transitioning from plastic to paper. Furthermore, a global initiative toward "green supply chains" has emerged. However, India remains a relatively minor player in the global beverage market and levies some of the highest Goods and Services Tax (GST) rates on carbonated beverages—28% plus a 12% cess—regardless of sugar content. This contrasts with global practices where low-sugar or alternative sweetener beverages are taxed more favorably. Inconsistent taxation—such as 18% GST on both mineral and aerated waters—raises concerns regarding the health rationale behind tax structures. Despite these challenges, the sector remains integral to India's food processing industry, aptly termed the "Sunrise Sector," with significant prospects for job creation, innovation, and sustainable growth (Rao 2016).

20.3.1 Global Non-alcoholic Market

The beverage sector plays an important role in the environmental subsystem, and so the sector is important to discuss for a sustainable food economy (Kee et al. 2023). Benjamin Savacool et al. (2021) examine the benefits of sector decarbonization, including energy and carbon savings, cost savings, and other co-benefits related to sustainability or health, as well as barriers across financial and economic, institutional and managerial, and behavioral and consumer dimensions. The global non-alcoholic market was worth 1223.93 billion in the year 2023, with an expected growth rate of 7.4% CAGR from 2024 to 30 (Statista). With an increased lookout on health companies, they are contributing to getting new and innovative non-alcoholic beverages in the market. According to a report published by Waitrose Food and Company on October 21, the consumers searching for non-alcoholic beverages is increasing by 22% every year ("Waitrose & Partners"). Another factor that can be taken into account is the changing demographics. Gen Z and millennials are moving towards a health-conscious diet and are helping to eschew the consumption of alcoholic beverages. Asia Pacific dominated the non-alcoholic beverage sector with a market share of 33.5%. The increase in demand for non-alcoholic beverages in developing nations like India, China, Malaysia, and Thailand is expected to be the regional players of the market. The non-alcoholic beverages can be mainly distributed as Carbonated soft drinks (CSDs), ready-to-drink (RTDs), tea, coffee, energy drinks, and sports drinks. Among these categories, CSDs accounted for the largest share in 2023, which was 28.10%. The wide variety of flavors, sizes, and prices available for such drinks makes them the most famous and go-to drinks. North America is expected to grow at a rapid rate, but due to high obesity and high sugar taxes that restrain consumers from consuming high-calorie products like CSDs, their demand is predicted to decrease in the coming years, and functional beverages are said to grow. The top five non-alcoholic beverage companies in 2023 were Nestlé, PepsiCo, Unilever, Keurig Dr. Pepper Inc., and The Coca-Cola Company.

20.3.2 Global Alcoholic Market

Now talking about the global alcoholic market, it was valued at $1624 billion in the year 2021 and is projected to reach $2036.6 billion in the year 2031 with a CAGR of 2.2% from 2021 to 2031 (Bhat and Naik 2017). The alcoholic beverage industry is segmented into various products according to size, category, and flavors. The following types, according to product type, are namely beer, ale, hybrid, wine, whiskey, sparkling wine, fortified wine, spirits, rum, vodka, and others. Other categorization is mainly on the basis of alcoholic content present (high, medium, low), flavor (flavored, unflavored), and packaging (plastic bottle, glass bottle, tins). The countries in the alcoholic beverage market are New Zealand, Vietnam, Australia, etc. According to the report of ICRIER, North America is projected to grow and dominate the global alcoholic beverage market with an increase in consumption of quality alcohol.

20.4 Indian Non-alcoholic Market

The Indian non-alcoholic beverage sector was worth INR 1376 billion in the year 2023 and is expected to reach INR 1986 billion by 2028 with a 7.29% CAGR (Research and Markets 2023). It is difficult to estimate its sales and volume as the government does not record its data separately, but considers it as a part of the food and beverage sector. According to Euromonitor International, the growth of the domestic market has been growing continuously with a rate of 14.5% over the last decade in terms of total sales volume. However, over the last years, the sector has seen a trend of rapid decline in consumption of carbonated soft drinks (CSD), which is the largest segment of this sector. Its future growth is projected to come down to single digits.

The leading food processing companies are diversifying their beverage segment for sustainable growth. Large companies are now selling their manufacturing processes to other companies for a higher reach. For example, in the north region, Varun Beverages Ltd took up the manufacturing of bottles, packaging, and distribution of Pepsi Co. This collaboration was to make the distribution footprints of Pepsi Co. stronger in states such as Haryana, Uttarakhand, Uttar Pradesh, and Haryana. In the year 2010, per capita sales volume was 5.66 l; it increased by 19.75 l between 2010 and 2019. Statista has reported a sales volume of 21.36 l in 2018. India's per capita volume is lower than that of other developing countries like Vietnam (111.89 l) and the Philippines (69.75 l). The average per capita of this sector was USD 8.91 in the year 2019 and USD 9.09 in 2020. In the present, it is at USD 16.57 and is expected to grow annually by 6.72% from 2024 to 2028. Among these, carbonated drinks hold the highest share in the market, second and third are taken by bottled water and Juice, respectively, over the years.

When we talk about the sales volume, bottled water has the highest shares, followed by carbonated drinks and juices. Along with these traditional categories, products like energy drinks, sports drinks, various types of coffee and tea have been introduced in the market and have a lower share than the rest of the products. The market share of energy and sports drinks has been increasing rapidly, with a share of just USD 5.25 million in 2010 to USD 212 million in 2019 to USD 470 million in 2023, with a CAGR of 45-50%. This shows consumer preference shift with an increase in product diversification.

20.5 Indian Alcoholic Market

India is a small player in the context of the global market of the alcoholic beverage sector. India is the third-largest alcoholic beverage industry market in the world, with a whopping 970 crore volume of liters per annum. But in the past few years, its global exports and imports have not been remarkable. In 2019, India's global export share was 0.27% and global imports were 0.75%, and its rank in the same year was 32nd (Mathur 2014).

Although revenue from alcoholic beverages varies by state, it remains one of the top three sources of revenue in all states. In 2017–2018, alcohol revenue accounted for nearly 99% of total revenue in most of the 10 states studied, except Madhya Pradesh and Rajasthan, where it was 77% and 94%, respectively. Between 2013–14 and 2017–18, Telangana had the greatest revenue collection CAGR at 30.39%, followed by Odisha (12.92%) and Assam (12.69%). Maharashtra had the lowest CAGR at 6.29%. In both 2013–14 and 2017–18, Karnataka and Maharashtra collected the most revenue from the alcoholic beverages sector among the top ten states (ICRIER 2019).

20.6 Key Findings

According to the ICRIER 2019 report, revenue by product categories reveals that the category with the highest revenue in 2017–18 was "foreign liquors and spirits," with percentages earned in states like Delhi (95.55%), Karnataka (86.98%), Telangana (79.61%), Odisha (69.18%), and Maharashtra (55.50%). The "country spirits" section generated the most revenue in states like Haryana and Assam, accounting for 98.51% and 81.31% of total revenue, respectively. The money generated from "foreign liquors and spirits" and "country spirits" in Uttar Pradesh and Rajasthan was nearly similar, accounting for 84.93% and 75.6% of the total alcohol revenue, respectively. "Other receipts" was the category in Madhya Pradesh that collected the largest income (67.76%), not alcohol per se. The CAGR of the non alcoholic beverage sector is reported to be 14.5% and 13.7% in terms of the total sales volume and total sales, respectively, between the years 2010 and 2022. But since then, its growth has slowed down due to the high taxes levied and supply chain challenges faced during the pandemic. India has one of the highest tax rates on beverages globally. A growth forecasting model according to the report of ICRIER 2019 with three scenarios which are realistic, optimistic, and pessimistic from 2020 to 2030 established that is GDP grows at 7.88% in realistic sense, 9.76% in optimistic sense, and 6% in pessimistic sense as the sector of non-alcoholic beverage is expected to grow 8.70%, 10.77%, and 6.66%, respectively in realistic, optimistic, and pessimistic scenarios.

Exports and imports of India are low in both the alcoholic and non-alcoholic sectors compared to ASEAN countries. The study found that the non-alcoholic sector contributes significantly to value addition and job creation in India. The total upstream and downstream job creation is estimated to be 6,91,491 in this sector. Even after providing subsidies and Incentives to the beverage sector, it is still not able to grow due to very high taxes levied, and a negative perspective about the industry. There are other issues, such as the supply chain being very fragmented in this industry, and very high logistic costs, which the government is trying to address. Upcoming significant economic growth of the Indian Alcoholic Industry, which is estimated to be $55 billion (USD) as of 2023, with a 7% CAGR to reach $73 billion (USD) by 2027. Spirits hold the largest share, which is 34 billion USD, in the alcoholic industry market along with wine, beer, and imported alcohol.

20.7 Conclusion

The non-alcoholic beverage sector in India holds substantial untapped potential to emerge as a major export-oriented industry. Rationalization of the Goods and Services Tax (GST) structure could significantly enhance efficiency across the supply chain. Key recommendations include the removal of the 12% cess on carbonated soft drinks (CSDs), thereby simplifying the tax burden to the highest GST slab of 28%. For healthier alternatives such as zero-sugar beverages, a reduced GST rate of 18% is advisable. Additionally, a uniform 12% tax rate on bottled water and a reduction in juice taxes, where inverted duty structures exist, could further incentivize consumption and production.

Investment in food clusters and agro-processing zones, especially for fruit- and vegetable-based beverages, can catalyze export-led growth, as demonstrated by success stories in ASEAN nations. India can also draw inspiration from incentive frameworks in other regions, such as the United Arab Emirates. To ensure alignment with health and nutrition goals, tax policies should be harmonized with the Food Safety and Standards Authority of India (FSSAI) regulations. Moreover, the need for detailed data collection and consumer demand analysis is crucial, as consumption patterns differ across product categories and socioeconomic groups. Incorporating international best practices related to taxes and subsidies can further promote the consumption of nutritious products. In sum, a comprehensive impact assessment of tax structures is required, and amendments should be introduced accordingly to position India as a global hub for beverage processing. However, as the sector expands, environmental sustainability emerges as a pressing challenge. Several companies in the beverage industry have historically been criticized for their ecological footprint. For example, Coca-Cola's operations in India have drawn attention for excessive groundwater extraction and water wastage, highlighting the need for stricter environmental accountability (Tiwari and Kapoor 2019). This underscores the importance of adopting sustainable practices across the industry.

Recent developments indicate a positive shift. Diageo India has set ambitious environmental goals, including reducing water usage by 50% by 2030, of which 43% was already achieved by 2023, and cutting carbon emissions by 75% (currently reduced by 56%). The company is also aiming to enhance recyclable packaging to 70% (Sovacool et al. 2021). Similarly, United Breweries Limited (UBL) has adopted multiple green energy initiatives: 19 of its breweries utilize agricultural waste as fuel, and 75% of their power comes from wind energy. With further plans to incorporate solar power and robust water conservation practices, UBL aims for a 50% efficiency improvement in water usage by 2030, having already reduced its carbon footprint by 3351 tonnes in 2016–17.

These initiatives reflect a broader industry trend toward decarbonization and environmental responsibility. As India aspires to lead the global beverage market, industry collaboration, investment in clean technologies, and growing consumer preference for sustainable products will be critical. By aligning economic ambitions with environmental stewardship, India can indeed transform into a sustainable and globally competitive beverage industry hub.

References

Bhat SA, Naik HG (2017) A study on Indian beverages industry. Int J Res Econ Soc Sci 7(8):277–284

Indian Council for Research on International Economic Relations (ICRIER) (2019) Alcoholic beverages. ICRIER. https://icrier.org/pdf/ES/ES_Alcoholic_Beverages.pdf

Indian Council for Research on International Economic Relations (ICRIER) (2022) Contribution of non-alcoholic and alcoholic beverage sector to Indian economy

Jenson M (2018) Contribution of the beverage industry to employment and sustainable agriculture practices in India. Indian Council for Research on International Economic Relations

Kee DMH, Lee JVK, Azlan IYA, Subramaniam JA, Koay JCW, Putri AM, Asthana R, Shrivastava K, Beg A (2023) Sustainability in the food and beverage industry: a comparative study of Malaysia, India, and Indonesia. Int J Tour Hosp Asia Pacific 6(3):1–17

Mathur AK (2014) Alcoholic beverages industry in India: an exploratory study. Eduved Global Manage Res 1(1)

Menon V, Mukherjee E (2021) Executive summary: contribution of non-alcoholic beverage. Indian Council for Research on International Economic Relations (ICRIER)

Ministry of Food Processing Industries (2024) Government of India

Mukherjee A, Mukherjee E (2020) Executive summary: contribution of alcoholic beverage. Indian Council for Research on International Economic Relations (ICRIER)

Mukherjee A, Mukherjee E (2021) Executive summary: contribution of non-alcoholic beverages. Indian Council for Research on International Economic Relations (ICRIER)

Munjal S, Sharma S (eds) (2017) The food & beverage hospitality industry in India: an emergent segment. CRC Press

Rao AP (2016) The economic significance of the alcoholic beverage industry in India. Indian Council for Research on International Economic Relations

Rao A, Guha S (2022) Flavored carbonated beverages: a case study on Bisleri Pop. Adv Manage Technol (AMT) 3(1):38–45

Research and Markets (2023) India non-alcoholic beverages market report 2023–2028: market witnessing increasing shift towards fruit drinks and low-sugar options. GlobeNewswire

Sovacool Benjamin K, Morgan B, Steve G, Jinsoo K, Aoife F, David R (2021) Decarbonizing the food and beverages industry: a critical and systematic review of developments, sociotechnical systems and policy options. Renew Sust Energ Rev 143:45–67

Tiwari P, Kapoor N (2019) Challenges and opportunities in Indian beverage industry. Int J Manage Technol Soc Sci 4(1):54–61

Chapter 21
Biodiversity-Rooted Millets-Based Products: Consumer Behaviour & Market Dynamics for Sustainable Food Engagement

Urishita Gambhir, Peter R. Makgwane, and Mahua Bhattacharjee

Abstract Millet marketing and value chain revolution is required for the revival of millets in India, as these chain inefficiencies are hampering large-scale consumption of millets as well as other components of the value chain, viz., primary processing, storage, and transportation. Given the increasing prevalence of lifestyle diseases, health problems, and increased awareness of fitness and health, urban consumers are turning to millets and millet-based products. Therefore, it is imperative to comprehend consumer perceptions and motivations regarding millet consumption. Influencers that matter include their social network and the next generation of social media. Individuals can access a greater amount of information with only a single button click nowadays. This paper aims to reveal unique preferences, taste, convenience, or price points compared to the national trends, limitations of millet-based products, bottlenecks, strengths, inefficiencies, opportunities for optimization, and unique distribution channels specific to the Delhi market. The paper focuses on filling this gap in knowledge by analysing the various parameters in consumer and retail segments. Overall, it will contribute to a deeper understanding and shed new light on the intricacies of the city's millet-based product market. It will show how products, consumer behaviour, and market linkages all interact within the Delhi context. This information is valuable for producers, retailers, and anyone with a stake in growing the market for millet products in Delhi. Appropriate market tactics are required to create demand for value addition at the farm gate and for more

U. Gambhir (✉)
Amity School of Economics, Amity University, Noida, Uttar Pradesh, India
e-mail: urishita.gambhir@s.amity.edu

P. R. Makgwane
University of South Africa, Pretoria, South Africa
e-mail: makgwpr@unisa.ac.za

M. Bhattacharjee
Research and Livelihood Development, ISRA, PMU Team-NCUI, New Delhi, India
e-mail: mahuabhatt19@gmail.com

M. Thakur, T. Belwal (eds.), *Food Biodiversity and Consumer Engagements*,
Sustainable Development and Biodiversity 46,
https://doi.org/10.1007/978-981-95-7777-4_21

startups operating in the same industry. Therefore, a practical and long-lasting value chain model is required to increase production and consumption. Components of a value chain include farmers, community and self-help groups, food processors, R&D institutions, nutritional evaluation and quality check, and consumers, which need to be organized and monitored at each level and at various scales.

Keywords Marketing chain · Value chain · Retail market · Delhi · Consumers preference

21.1 Introduction

Marketing chains encompass a series of interconnected activities that bring a product to the consumer, where the key components range from production, distribution, promotion, and retailing. Consumer behaviour is influenced by psychological, social, cultural, and economic determinants. The interplay between marketing chain dynamics and consumer behaviour is characterized by a loop such that each influences and shapes the other. Millet economy, in the recent scenario, is considered a sustainable dietary choice, driven by factors like increasing health consciousness and sustainability concerns. Post-covid, consumers were motivated to build and increase their natural immunity, and hence there was an increase in millet consumption by 5%. An attempt is made in this paper to understand the available market of millet-based products in Delhi through mapping the products marketed in e-commerce and retail markets, and to get a quick glimpse of the consumer's specific preference on millets using a mixed-method approach (Thakur and Belwal 2023).

Small-seeded grasses that are typically grown in arid and semiarid settings are referred to as millets, or dryland cereals. They are rich in dietary fiber, extremely nutritious, and free of gluten. Micronutrients such as calcium, iron, phosphorus, and others are abundant in them. Since they have a low Glycaemic Index (GI), their blood sugar does not spike dramatically. In recent years, healthy food has become increasingly popular. Packaging now uses terms like "sustainable," "roasted," and "no added" in place of sweetened beverages and fried foods. For example, a variety of food options have replaced the basics of the world, wheat and rice, with millets. The two most common grains in the world are wheat and rice, yet found that wheat's high glycaemic index and huge carbon footprint are linked to lifestyle diseases and environmental problems. While millets were traditionally eaten in India, they became less popular and nearly forgotten as a result of the 1960s Green Revolution's emphasis on food security. Before the Green Revolution, millets accounted for over 40% of all farmed grains; over time, this percentage has decreased to about 20% due to environmental, nutritional, and agronomic circumstances being compromised. The National Food Security Act, which covers coarse grains, was passed in 2013, and the Initiative for Nutritional Security via Intensive Millet Promotion (INSIMP) was launched in 2012.

There have been changes in dietary patterns with preferential consumption of high-calorie cereals. There has been a tremendous shift in food choices across the globe. Worldwide worry over the rise in the prevalence of diseases, including diabetes, obesity, and cardiovascular issues like heart attacks, coronary artery disease, and arrhythmias, has grown. Millets can help prevent these diseases because of their high protein, vitamin B and C, and mineral content, which includes calcium and iron. As a result, millets are the greatest option for customers who want wholesome, nutrient-dense food that tastes good. The increasing prevalence of non-communicable diseases (NCDs) and malnutrition makes it necessary to talk about the benefits of millets in these contexts.

21.2 Millet Value Chain Activities

Major millets, such as sorghum, finger millet, and pearl millet, and minor millets, such as tiny millet, proso, kodo, foxtail millet, and barnyard millet, are the two typical categories for millets. The primary emphasis of this paper's analysis is the millets as a whole. Due to its high fibre content and prolonged transit time from the stomach to the intestine, pearl millet is good for weight loss. Often referred to as the "King of Millets," sorghum is a high-carb crop in the Gramineae family (Thakur and Tiwari 2019; Thakur 2024). The primary component of sorghum is starch, which digests more slowly than other cereals. The grains of finger millet are high in protein, dietary fibre, calcium, minerals, and polyphenols. They have a high potassium content, which supports healthy kidney and brain function as well as the smooth operation of muscles. A series of sequential and well-coordinated value-adding operations that grow and process a specific variety of millet into edible goods is known as a millet value chain (Reddy 2018). The value of the finished product is impacted by these actions as well as the accompanying facilitators and barriers. A nutrition-focused value chain analysis helps identify where in the food supply and marketing system nutritional quality can be improved or lost, allowing interventions that enhance nutritional outcomes at specific stages. Entry points represent stages where nutrition can be strengthened through actions such as fortification, improved processing, or behaviour change communication, whereas exit points are stages where nutritional value may decline due to factors like inefficient harvesting, inadequate storage, or excessive processing (Jacques 2018).

The separation of glumes and other foreign components during primary processing results in desired, fine edible grains with the best possible quality and customer acceptability, which is the first step in the value addition process for millets. To obtain millet flour, these grains undergo additional secondary processing by being ground. This flour has a shorter shelf life than other grains, lasting only 1–2 months. While research and development are still being done to extend the shelf life of flour, it is crucial to route the flour toward tertiary processing, such as the Ready to Eat and Ready to Cook segments, to reduce waste and increase millets' consumption. According to APEDA, the millets market is expected to increase from its present valuation of over $9 billion to over $12 billion by 2025. However, researchers have found that there are several inefficiencies associated with millet's value chain that are preventing its mass consumption, such as low shelf life, low yield, and high transportation costs (Jacques 2018).

21.3 Consumers and Millet-Based Products

Customers are increasingly asking for snacks and restaurant meals that are nutritious, with no chemicals, and made from locally sourced ingredients, such as traditionally grown millets. Millets have experienced a resurgence in popularity and are now being hailed as superfoods, as evidenced by recent research. As millets gain recognition as highly nutritious and environmentally friendly alternatives to conventional cereals Umanath et al. 2018), it becomes imperative to investigate the underlying factors motivating consumers to choose millet-based products.

Innovative millet processing technologies that provide safe, easy-to-handle, ready-to-cook, and ready-to-eat products and meals at a commercial scale are mainly available in urban areas, where rice and refined wheat flour dominate and are much more accessible and affordable. This has led to an increasing emphasis globally on consuming whole grains, underlining the importance of mainstreaming nutritious smart food crops like millets and promoting them as a staple. This is one of the major objectives of the Smart Food initiative.

The idea of a healthy lifestyle with dietary intake has gained ground rapidly, especially among the more educated and affluent households. They tend to adopt diet charts, and millets constitute a major part of the same. Access, affordability, and perception are the major factors affecting people's consumption patterns. Within the theme of access, the following categories came up: limited availability and lack of trustworthy options, such as organic and branded varieties (Suruchi Singh and Vidya Vemireddy).

The Health Belief Model offers assistance to consumers as well. Individuals who are more inclined to switch to millet eating are those who are predisposed to medical disorders and those who are more health-conscious. Globally, as economies shift from subsistence to more advanced systems, dietary habits are changing. This shift, dubbed the nutrition transition (NT), is typified by modifications to dietary habits and levels of physical exercise. According to studies, the obesity prevalence in the country's urban areas was 12.7%, while it was 5.4% in its rural parts. In metropolitan regions, 42.9% of individuals with a BMI of more than 25 belonged to the upper middle class. The majority of obese and overweight women live in wealthy Indian urban households. This necessitates the need to study the dietary patterns of the affluent urban section of India.

21.4 Marketing Chain and Value Chain

The marketing chain and value chain of millet products are critical components in understanding the journey of these nutritious grains from cultivation to consumption. The marketing chain encompasses various stages, starting from the production of millets by farmers to their distribution and sale to end consumers. This involves processes such as sorting, processing, packaging, and transportation, all of which contribute to shaping the market dynamics of millet products. On the other hand, the value chain of millet products delineates the sequence of activities involved in adding value to these grains at each stage of production and distribution. This includes aspects like improving quality, creating innovative products, and implementing effective marketing strategies to enhance consumer awareness and demand.

Understanding both the marketing and value chains is pivotal for stakeholders in optimizing the production, distribution, and consumption of millet products, thereby fostering their sustainable growth and utilization in addressing food security and nutritional challenges. While the demand for millets for human consumption is probably going to be a crucial factor in the plant's development, specific data on customer preferences and patterns is not readily available. In order to take market-oriented production decisions and profit from expanding markets, producers and other market participants require this kind of information.

Using e-commerce platforms to sell millet-based products can increase the accessibility of these products for consumers, while also providing farmers with better prices and market opportunities. It has been found that developing innovative millet-based products and packaging can increase the marketability of millets, which can benefit farmers by providing them with better prices and market opportunities.

The marketing chain is a subset of the value chain. It focuses on the specific activities within the value chain that are related to getting the product or service into the hands of the customer. Effective marketing chains contribute to value creation by ensuring products reach the right customers efficiently and effectively. By influencing consumer decisions and building brand awareness, marketing activities can enhance the perceived value of a product.

21.5 Millet Retail Market of Delhi: Empirical Research

Delhi, being a bustling metropolis, reflects this trend with a growing availability of millet-based products in both traditional retail markets and the booming e-commerce sector. In retail markets across the city, consumers can find an array of millet varieties, including pearl millet (bajra), foxtail millet (kangni), finger millet (ragi), and sorghum (jowar). These products are commonly available in the form of flour, grains, flakes, and ready-to-eat snacks. Retail outlets ranging from local grocery stores to specialty health food stores stock these items, catering to the increasing demand for healthier alternatives. Simultaneously, the advent of e-commerce platforms has further expanded the accessibility of millet-based products to Delhi's residents. Online marketplaces offer a convenient avenue for consumers to explore a diverse range of millet products from both local and national brands. These platforms provide options for doorstep delivery, enabling consumers to access millet-based goods with ease, irrespective of their location within the city.

Moreover, Delhi's millet market showcases a fusion of traditional and contemporary offerings, with innovative millet-based products such as millet pasta, cookies, and breakfast cereals gaining traction among health-conscious consumers. Overall, the availability of millet-based products in both retail and e-commerce channels reflects a growing awareness and appreciation for these nutritious alternatives in Delhi's culinary landscape.

By analysing the marketing chain, we can zoom in on the specific players and channels involved in bringing these millet products to Delhi consumers. This includes identifying key players, distribution channels, and marketing activities.

Key players are

- Farmers: Small-scale producers cultivating and initially processing millets. Strengthening linkages with FPOs could benefit both farmers and consumers.
- Processors: From small units to large mills, processors transform raw millets into consumer-ready products. Their capacity directly impacts product availability.
- Wholesalers: Wholesalers ensure wider distribution and price stability but can create inefficiencies.
- Retailers: Traditional kirana stores, modern supermarkets, and online platforms influence consumer decisions through product selection, pricing, and promotions.

Distribution channels identified were modern trade, specialty stores, convenience stores, and online retail. In Delhi, millet-based products reach consumers through a multi-pronged distribution network. Traditional kirana stores remain a dominant force, offering loose purchases and credit options but with limited product variety and inconsistent quality. Modern supermarkets are gaining traction, providing a wider range of branded and packaged millet products with consistent quality and promotions. However, higher prices and limited space for niche products can be drawbacks. E-commerce platforms offer a convenient alternative, boasting the most extensive selection, including emerging brands, innovative options, and specialized categories. While offering competitive prices on some products, trust in online vendors and potential delivery delays can pose challenges. This diverse distribution landscape caters to various consumer preferences, but optimizing accessibility and variety requires addressing limitations within each channel.

21.6 Marketing Strategies Adopted in Delhi for Selling Millet-Based Products

Qualitative studies indicate that various strategies are taken to promote these nutritious alternatives, targeting health-conscious consumers seeking sustainable dietary choices. Traditional marketing channels, such as print media, television advertisements, and outdoor campaigns, are often complemented by digital platforms to reach a wider audience. Social media marketing plays a pivotal role, with brands utilizing platforms like Instagram, Facebook, and Twitter to engage with consumers, share educational content, and showcase product offerings. Additionally, collaborations with influencers and health experts further amplify brand visibility and credibility within the target demographic. In-store promotions, sampling sessions, and product demonstrations are also commonly employed to encourage trial and purchase. Moreover, partnerships with retail outlets, including supermarkets, health food stores, and online marketplaces, facilitate product placement and distribution. Through these concerted marketing efforts, millet-based products have carved out a niche in the Delhi market, appealing to health-conscious consumers seeking wholesome and sustainable dietary options.

21.7 Consumers' Choice of Millet-Based Products in Delhi

Based on a 200-random-sample survey, an effort is made to understand the knowledge, perceptions, and factors motivating millet consumption in the capital city of India, Delhi.

21.7.1 Consumption Pattern

Family income: majority (67.4%) of the respondents' families earn above 75,000 INR monthly. The next category earns between 50,000-75,000/month and 25,000-50,000/month, both with the percentage of 11.3% respectively.

Table 21.1 reflects that awareness among people needs to be instigated so that the frequency of consumption can be increased. 40% of the respondents consume millets monthly, and 36.2% rarely consume millets. 16.3% respondents have a weekly consumption.

This Fig. 21.1 shows that it is very clear that the majority of the respondents prefer millets in the form of snacks, be it Jowar, Ragi, or Bajra. 44.3% of the respondents in Delhi prefer millet flour in the raw form. The trend of millet cereals also follows Delhi, with 30.4% respondents. Another famous category that appeals

Table 21.1 Frequency of consumption by family income

Distribution of respondents based on family income	
Frequency of consumption	Percentage amount
Daily	2.10%
Weekly	16.30%
Monthly	40.00%
Rarely	36.20%
Never	5.00%

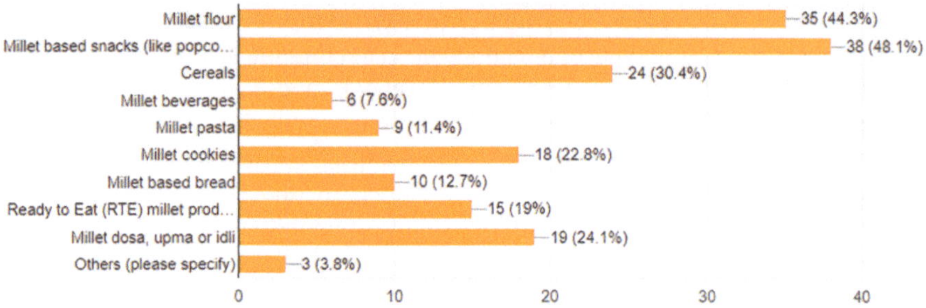

Fig. 21.1 Category-wise consumption of millet-based products

Fig. 21.2 Source of information about millet consumption

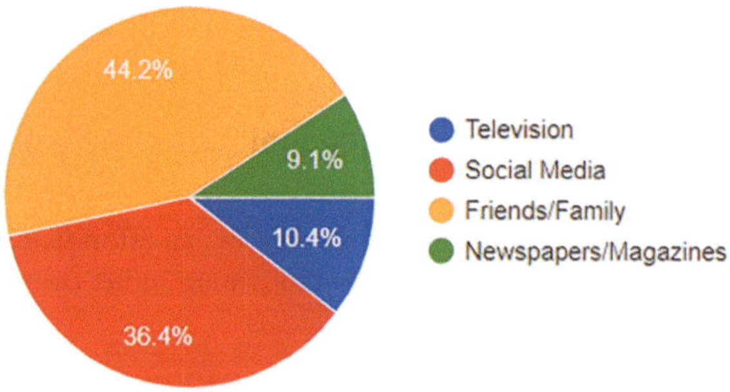

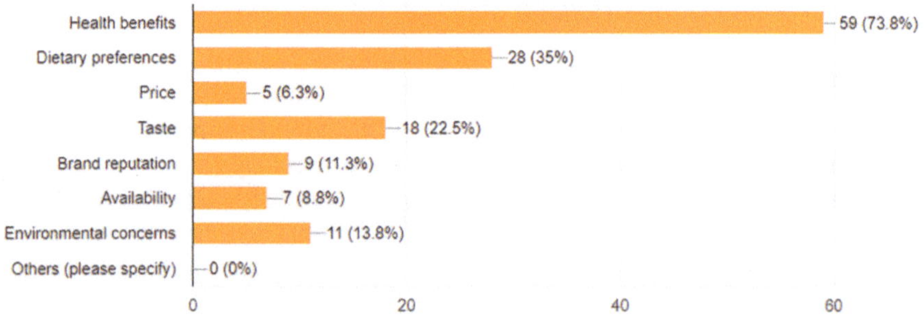

Fig. 21.3 Customers' motivation to consume millets

to kids as well as adults is millet pasta and noodles. Ready-to-eat millet products are also quite popular in Delhi.

Figure 21.2 shows that most people obtain information from friends/family and social media. Very little information is provided by newspapers and television. This indicates that businesses can carve out different creative strategies to promote their products and reach the masses.

Figure 21.3 shows that in the era of health consciousness being promoted among people as a fad, it is essential to ensure maximum nutrient content, as health is the most important factor people consider, among others. Dietary preference is something many dieticians and millennials carefully map out, including their appetite, eating habits, ingredient content, and diet investigation. Health, coupled with taste, is fairly preferred by the respondents.

Overall, millet-based products have a plethora of health benefits for people of all ages. These products have been a part of India's agricultural system as well as food habits, but lost their essence during the Green Revolution. Now, in the culture of self-care and health consciousness, eating right is the key, and millets are a great solution as well as a tastier alternative to other foods. They are natural and can be consumed in a variety, catering to the needs of all age groups. More needs to be done to make people aware, build trustworthy brands and goodwill, work more on taste and prices, and promote through advertising and marketing strategies. Also, focusing on the outlets is important as that will help in easier access to products and increase the frequency of purchase.

21.8 Challenges in the Millet Market

The lack of dedicated supply chains connecting smallholders to market outlets necessitates the improvement of logistical infrastructure and supply chain management techniques. Other gaps in the current value chain include the absence of backward and forward linkages between millets' production and conservation, the lack of efficient machinery for the initial processing of millets, which raises processing costs and results in a low-quality end product, and the lack of appropriate value-added technologies. Processing millet has drawbacks since it lacks gluten, making it more laborious and time-consuming. Additionally, there are not many successful and sustainable business initiatives in the millet space, nor are there enough market strategies to create demand for value addition at the farm gate and among other entrepreneurs in the industry.

Although millet species exhibit extensive genetic diversity and can adapt to a wide range of climatic conditions, their cultivation has declined in recent years. This reduction is primarily due to limited public awareness, a lack of specialized processing equipment, inadequate availability of value-added products, low profitability, poor marketing support, and insufficient policy incentives.

21.9 Conclusion

Millets hold much promise in tackling the dual challenges of health insecurity and environmental sustainability. In order to take advantage of this market potential, marketers can serve as catalysts for the sustainable diet pattern. A step towards a sustainable food economy at the policy level would be to encourage the use of millets and to provide incentives for their cultivation and value addition to make these nutri-cereals available to a wider population.

Recent studies in India have underscored the pivotal role of Farmer-Producer Organizations (FPOs) in enhancing the millet value chain. By establishing primary processing units at the farm gate, FPOs enable farmers to undertake value addition, thereby increasing their share of consumer spending. This approach not only boosts farmers' incomes but also ensures the availability of diverse millet-based products in the market. Additionally, integrating self-help groups and rural women into these processes has been instrumental in promoting inclusive growth and empowering marginalized communities.

To further strengthen the millet value chain, the adoption of improved technologies and the development of robust market linkages are essential. Investments in infrastructure, such as establishing millet food parks and dedicated marketplaces, can provide farmers with better access to consumers and fair pricing. Moreover, declaring a minimum support price (MSP) for millets and ensuring procurement facilities at MSP can safeguard farmers against market volatility, encouraging them to continue millet cultivation.

Consumer awareness campaigns highlighting the nutritional benefits of millets are crucial for driving demand. Incorporating millet-based products into government programs, such as mid-day meals and public distribution systems, can further promote consumption across various socio-economic segments. Collaborations with research institutions to develop climate-resilient millet varieties and innovative processing techniques can enhance both yield and product quality. Collectively, these strategies can create a sustainable and resilient millet value chain in India, addressing health and environmental challenges while supporting farmer livelihoods.

References

Jacques DC (2018) Social capital and transaction costs in millet markets. Heliyon 4

Reddy AA (2018) Analysis of pearl millet market structure and value chain in India. J Agribus Dev Emerg Econ 9(8):123–167

Thakur M (2024) Millets: the multi-cereal paradigm for food sustainability. World sustainability series. Springer, Cham. https://doi.org/10.1007/978-3-031-64237-1_1

Thakur M, Belwal, T. (2023) Bioactive components. Springer, Singapore https://doi.org/10.1007/978-981-19-2366-1_34

Thakur M, Tiwari P (2019) Millets: the underutilized nutritious functional foods. Plant Arch 19(1): 875–883

Umanath M, Balasubramaniam R, Paramasivam R (2018) Millets consumption probability and demand in India: an application of Heckman sample selection model. J Nutr 7(5):63–76

Chapter 22
Biodiversity-Based Livestock Enterprises for Tribal India: Market Chain Innovation and Consumer Engagement

Mahua Bhattacharjee and Peter R. Makgwane

Abstract In the context of developing countries, business profitability and long-term sustainability increasingly depend on adopting an organized approach—both at the enterprise level and throughout the broader market chain. Farmers, particularly those operating in the livestock sector, require targeted support and capacity building to effectively integrate into market structures, thereby enabling them to occupy stronger, more profitable positions within market chains. The efficiency of forward and backward linkages, which connect input, output, and service markets, is largely determined by the enabling environment and the underlying production and distribution processes. This paper aims to assess the differences in expenditure and prospective income between traditional and scientifically informed methods of livestock production and business management. Drawing upon baseline surveys and socio-economic analyses of the pig-rearing sector in Northeast India, the study estimates an optimal combination of livestock stock and capital inputs to support sustainable and profitable enterprise development. The findings reveal that adoption of scientific production techniques and participation in multiple market chains yield an average gross profit that is 45.11% higher than traditional methods reliant on a single market channel. Specifically, in the livestock sector, a circular entrepreneurial model begins with the organization of small producers into functional clusters or self-help groups (SHGs). To support this, market chains and business services must be developed through rigorous outcome analysis, incorporating systematic monitoring and evaluation. In Northeast India, the involvement of women's cooperatives or cluster members can be enhanced through targeted leadership and management training, fostering community-wide participation in critical areas such as production, digitization, waste management, and vaccination programs. Bridging existing gaps

M. Bhattacharjee (✉)
Foundation for Development of Rural Value Chains (FDRVC), New Delhi, India
e-mail: mahuabhatt19@gmail.com

P. R. Makgwane
University of South Africa, Pretoria, South Africa
e-mail: makgwpr@unisa.ac.za

M. Thakur, T. Belwal (eds.), *Food Biodiversity and Consumer Engagements*, Sustainable Development and Biodiversity 46,
https://doi.org/10.1007/978-981-95-7777-4_22

in backward linkages and facilitating integration among producers, resource providers, and market agents remains essential for strengthening market efficiency and ensuring inclusive growth.

Keywords Profitability · Capital stock · Innovation · Cooperative · Entrepreneurship · Livestock sector

22.1 Introduction

22.1.1 Sustainable Food Enterprises and Cooperative Approach

Collective action is essential for establishing a sustainable food system. Sustainable Development Goal (SDG) 2 Zero Hunger encounters significant challenges in fostering effective collaboration among a wide array of supporting stakeholders (Herrero et al. 2009). However, such integration also presents valuable opportunities for innovation within food enterprises. Typically, these enterprises operate at a distance from primary food producers and may lack transparency regarding how premium prices are distributed along the supply chain. The sustainable food framework conceptualizes sustainable food entrepreneurship as a dynamic and cyclical process of continuous adaptation and innovation (Van der Gaast et al. 2022). An illustrative example is provided by the German company Bohlsener Mühle, which historically sourced large quantities of sugar from international suppliers but later engaged in a collaborative initiative with local farmers to cultivate quinoa in northern Germany. This partnership also serves as a strategic response to the challenges posed by climate change. Farmers with limited access to irrigation now work with Bohlsener Mühle to grow quinoa—a resilient crop capable of withstanding both drought and excessive rainfall. Through this collaboration, Bohlsener Mühle actively encourages and supports farmers in experimenting with new crop varieties, ultimately fostering the development of innovative product lines that generate mutual benefits for both the enterprise and the agricultural community, while simultaneously reducing food miles (Weber et al. 2020).

Participatory growth models have gained increasing significance in developing countries as a means of fostering competitive and sustainable economic sectors. Broadly, development approaches can be categorized into two paradigms: growth-centered and people-centered models. In growth-centered theories and strategies, the primary objective is the pursuit of economic growth itself. In contrast, people-centered approaches place human well-being at the core of the development process, positioning economic growth merely as a means to fulfill the needs of individuals and communities rather than an end in itself (Tsito Monaheng 2004). The emergence of community-based or cluster-based development approaches is rooted in the principles of people-centered development. Notably, Michael Porter popularized the concept of industry clusters in his seminal work The Competitive Advantage of Nations. Since then, there has been a marked surge of interest in clusters as dynamic engines of economic growth and innovation. Concepts such as industrial poles, clusters, and localized production systems have become central themes in development discourse, recognized for their capacity to stimulate entrepreneurship, foster knowledge exchange, and drive productivity enhancements (Mac and Belussi 2002).

In India, the monthly per capita consumption expenditure (MPCE) is distributed between food and non-food items at 47% and 53%, respectively, in rural areas, and 40% and 60% in urban areas (Household Consumer Expenditure Survey Report, 2022–23). At the global level, the value added by the agriculture sector has increased by 73%, with meat and primary crop production rising by 44% and 53%, respectively, between 2000 and 2019 (FAO 2018). In this evolving landscape of demand and supply for agricultural commodities, the interplay of factors within the marketing chain has become increasingly crucial for enhancing overall profitability (Bhattacharjee et al. 2023a, b).

The success of livestock enterprises is strongly linked to greater participation in market chains, particularly through clusters or cooperative structures. The long-term sustainability of such enterprises depends on the effective distribution of risk and the cultivation of entrepreneurial motivation, both of which are most efficiently supported within an organized institutional framework. However, in practice, limited leadership capacity and managerial skills often act as significant barriers to the efficient functioning of these markets and hinder market actors from achieving competitive advantages (Bhattacharjee et al. 2023a, b). As a result, one of the fastest-growing sectors within the broader food economy, the livestock sector, remains largely confined to subsistence-level operations. To address this gap, the present study employs baseline survey data and socio-economic analysis from selected regions of Northeast India to identify feasible, profitable combinations of animal stock and capital inputs under both traditional and scientific methods of livestock production.

22.2 Forward and Backward Linkages in the Livestock Sector (Pig Market) with Reference to Women's Participation

The livestock production is majorly concentrated in the hands of women carrying out a wide range of activities varying from care of animals, cleaning of animal sheds, processing livestock products (Kumar and Rai 2011). They constitute about 69% of the workforce engaged in the livestock sector, and in tribal populated states, the share comes to around 90% (Bhattacharjee et al. 2022). When rural women have access to control the livestock products, the household's coping strategies are affected, thereby resulting in a positive impact on overall household well-being and nutrition (IFAD 1999; FAO 2012). A self-sustaining framework is required, especially in developing countries, to sustain their livelihood, where women are found to be the guardians of livestock diversity. Pigs constitute about 2.9% of the livestock sector (Basic Animal Husbandry Statistics 2020). In North East India, more than 80% of the population in Meghalaya, Manipur, Mizoram, Nagaland, and Arunachal Pradesh comes under the social group of the Scheduled Tribe, rear livestock like pigs, and backyard poultry is their primary source of income (Chauhan and Thakor 2006). It holds both social and economic significance in tribal households across the North-Eastern states of India (Ali and Das 2003).

The country's pig meat consumption rose from 362.1 thousand tonnes in 2017 to 364.01 thousand tonnes in 2022 (OECD-FAO Agricultural Outlook 2022). Out of which the North Eastern states consume approximately 70% of pig meat (pork, bacon, ham). The highest consumption is recorded in Assam (50.4%), Meghalaya (45.25%), and Nagaland (32.28%), respectively. Studies by ICMR on nutritional requirements indicate that by 2050, the demand for pig meat would have to be increased by 3 times.

Fulfilling the nutritional demand of this huge livestock instills the need for supportive technical, institutional, and policy initiatives for the value-added chain of pork products. These initiatives will form the basis of the guidelines and framework for the commercialization of pork. The potential of livelihood integration in the pork value chain across the lens of social and gender groups will incorporate all the existing and prospective agents for the innovative value-added chain. This will lead to a roadmap for the improvement of livestock (pig) quality, market linkage, and further investment plans to improve market efficiency by using a holistic approach from the local network to the prospective e-network. It will help to address the marketing and value chain issues of the pig/pork market to attain a model framework for channeling the livestock (pig) rearing towards commercialization and for the economic empowerment of women.

Globally, about 600 million of the world's poorest households keep livestock as a primary source of income (Thornton et al. 2006). In India, about 90% of livestock is owned by small farmers and landless rural households who are extensively dependent on income from the sale of agricultural and animal products. But even in the mixed farming systems, livestock is the main source of income for rural households, which contributes to their employment as well.

Within agriculture, the most prominent activity that has the highest share of women entrepreneurs are livestock (32%) and forestry (1.55%) (Government of India 2021). Therefore, this study tries to explore the role of agricultural entrepreneurship among the tribal women as a catalyst for development, especially for the North Eastern states and for India as a whole. The study tries to develop direct linkages between women's entrepreneurship in the agricultural sector and women's empowerment, which ultimately results in their sustainable development.

22.3 Growth of Meat Market in India: Changes in Production and Consumption

The livestock population in the country is showing an increase of almost 5% over the livestock census 2012, where a 11.19% increase has occurred in urban areas as compared to 5% growth in rural areas. Categorically, data indicate that the highest percentage growth is around sheep production (14.13), followed by goat, whereas the contribution to the livestock sector from piggery is gradually declining.

A linear growth curve shows a steady increase in the production of meat from 2009–10 to 2018–19. The total production has almost doubled from 2009–10 to 2018–19 (Annual Report 2019–2020, Department of Animal Husbandry and Dairying; Fig. 22.1). The meat product has shown growth of 6% —during 2021–23 (Annual Report 2022–23, Department of Animal Husbandry and Dairying). The change in the consumption of eggs, fish, and meat by rural and urban areas is shown by the composition of MPCE since 1999–2000 in Table 22.1.

Figure 22.2 clearly indicates that rural areas spend more on eggs, fish, and meat. In the last decade, a clear 1% gap in expenses was found between rural and urban areas in egg, fish, and meat (All India Household Consumption Expenditure Survey 2022–2023).

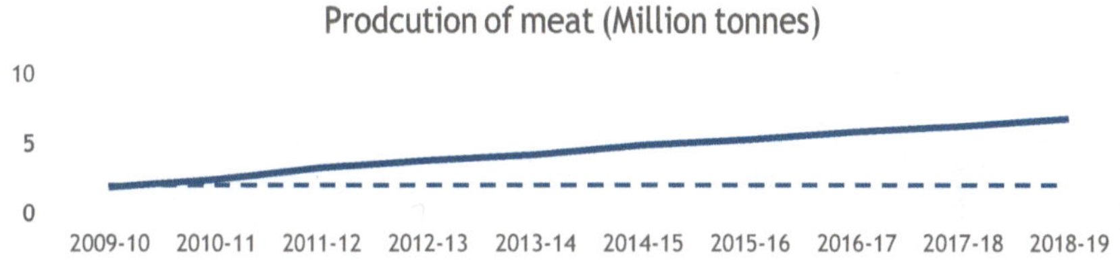

Fig. 22.1 Trend of meat production. (Annual Report, Department of Animal Husbandry and Dairying, 2022–23)

Table 22.1 Wholesale prices of pig (Rs/quintal)

Year	Chennai	Kolkata	Delhi
2019	21,250	19,583	12,325
2020	22,583	21,182	14,573

Agricultural Prices in India, Directorate of Economics and Statistics (2020)

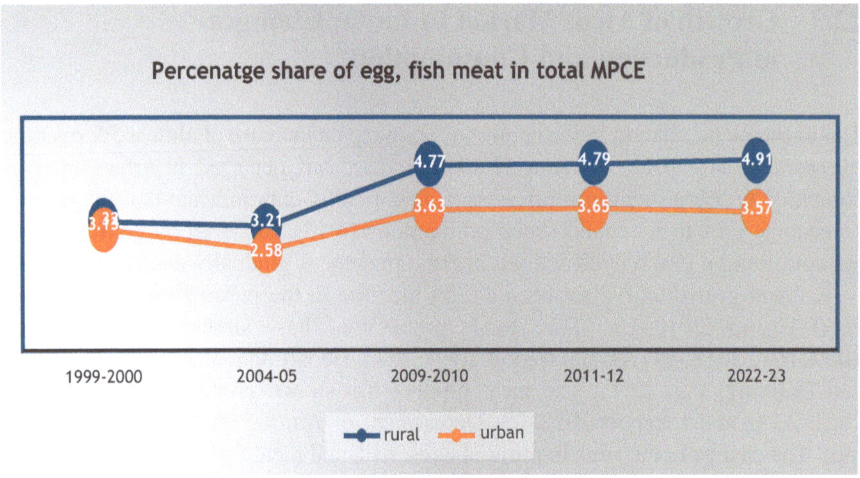

Fig. 22.2 Percentage composition of MPCE for egg, fish, and meat since 1999–2000 in rural and urban areas. (All India Household Consumption Expenditure Survey, 2022–2023)

22.4 Economics of the Livestock Sector in North East (Assam, Meghalaya, and Nagaland)

The growth chart of the pig gives an alarming sign as it shows a negative growth of 12%. Interstate performance is also found to be quite different from 2012 to 2019, where states like Telangana and Andhra Pradesh are showing a bumper increase in pig production from 30 to 50%. The highest decline in the percentage of the pig population is found in UP (70) and Bihar (47). North East states like Assam, Meghalaya, and Mizoram show a 20–30% increase in population within the time period from 2012 to 2019 (20th Livestock Census 2019). The pig/pork price variation in tier 1 cities is given below to understand the gap in the specific livestock market

Many internal and external factors influence the price of meat and meat-based products. Table 22.1 gives a glimpse of price variation in cities from which the major percentage of livestock-based products are traded.

An empirical survey of backyard pig farmers has identified the price determinant in the Northeast region. With the application of the multiple linear regression method, the linear relationship is estimated between the dependent variable (total cost of the small producer) and independent variables (pig stock, feed, vaccinations, transport, and equipment) (Table 22.2). The applied model is given below-

$$Y = \beta_0 + \beta_1 X_1 + \beta_2 X_2 + \beta_3 X_3 + \beta_4 X_4 + \beta_5 X_5 + \beta_6 X_6 + \mu$$

$Y =$ total cost
$X_i =$ variables are the expenditures incurred on

Table 22.2 Cost determinants of pig farming in the North-Eastern region of India

States	Variables (log)	P values	Coefficient	Correlation with total cost
Assam	Stock purchase	0.000	0.068	Positive
	Feed	0.000	0.457	Positive
	Vaccination	0.000	0.058	Positive
	Transportation	0.000	0.031	Positive
	Equipment	0.008	−0.025	Negative
Meghalaya	Stock purchase	0.000	0.431	Positive
	Feed	0.000	0.057	Positive
	Vaccination	0.000	0.054	Positive
	Transportation	0.151	−0.009	Negative
	Equipment	0.184	0.065	Positive
Nagaland	Stock purchase	0.000	0.253	Positive
	Feed	0.000	0.272	Positive
	Vaccination	0.016	0.070	Positive
	Transportation	0.000	0.089	Positive
	Equipment	0.001	0.057	Positive

X_1 = stock purchase, X_2 = feed, X_3 = health related expenditure, X_4 = transportation, X_5 = equipment, X_6 = depreciation (shed maintenance), and μ = random error term

The common key determinants of total cost in Assam, Meghalaya, and Nagaland came out to be the input cost incurred on the purchase of pig stock, feed, vaccinations, and transport. Thus, the three cost-influencing factors for the backyard farmers are the purchase of pig stock, feed cost, and vaccination in Assam and Meghalaya. On the other hand, in addition to the stock purchase of pigs and feed costs, transportation costs are the major cost determinants in Nagaland. The determinants of the price of by state are (a) Assam and (b) Meghalaya—stock purchase, feed and vaccine, and (c) Nagaland—stock purchase, feed, and transportation.

The analysis suggests that to reap the benefits of the economic gain of the pig production system and for the profit maximization of the smallholder pig farmers, the significant factors influencing the cost inputs are stock purchase of pigs, feed cost, vaccinations, and transportation charges. The distribution of the produced animal or meat takes place through a marketing chain, which further decides the income of chain actors.

From Table 22.3, it has been explained that a major percentage of chain actors in Assam and Meghalaya follow a common traditional chain of business as given below, with the exception of Nagaland.

$$\text{Produce} \rightarrow \text{Trader} \rightarrow \text{Retailer} \rightarrow \text{Consumer}$$

With the least intermediaries in the chain of Nagaland, theoretically, it is the most market-efficient chain and should be the most profitable. The study tried to find out the major cause of withdrawal in pig enterprises in Nagaland. It is found that the

Table 22.3 Percentage distribution of participation by chain actors in the pig/pork market chain

States		
Assam	Meghalaya	Nagaland
Chain 1. Producer—trader—wholesaler—retailer—consumer (35%)	Chain 1. Producer—trader—wholesaler—retailer—consumer (30%)	Chain 1. Producer—trader/wholesaler—consumer (20%)
Chain 2. Producer—retailer/butcheries—consumer (15%)	Chain 2. Producer—trader—consumer (15%)	Chain 2. Producer—consumer (70%)
Chain 3. Producer—trader—retailer—consumer (50%)	Chain 3. Producer—trader—slaughterhouse—retailer—consumer (10%)	Chain 3. Producer—trader—retailer—consumer (10%)
	Chain 4. Producer—trader—retailer—consumer (45%)	

increasing cases of sickness in Nagaland are making the business more risk-prone. The restricted supply of local meat and increasing demand from consumers are steadily leading to an increase in the price of pork. Neighboring states and countries have come up actively to cover this gap in the market. Local entrepreneurs are gradually losing the market, mainly due to being unorganized in nature and also due to the non-application of the scientific method of farming, along with poor participation in the vaccination process.

22.5 Sickness and Vaccination

Deworming and vaccination against the ASF are important preventive measures for reducing the vulnerability in the pig business. Among the three states (Fig. 22.3) Assam (74.5%) has the highest percentage of pigs dewormed (treated to be free of worms). Meghalaya and Nagaland share a similar but poor performance on vaccination. Survey indicated that Meghalaya and Nagaland have only 15.3% of pigs, which are dewormed.

During 2020–21, almost half of farmers (56%) reported at least one sudden death among their pigs. Contrary to this, during the same timeline, the majority of farmers (63%) reported the total inflow of pigs, with 7636 additions to the existing herd size through the purchase of piglets, gifts from institutions or others, and new live births on farms. Though the majority of farmers knew Classical swine fever, a very insignificant number of farmers have had their pigs vaccinated (25%), with Assam being the highest (41.2%), followed by Meghalaya (34.5%), and Nagaland (10.9%).

The efficiency of the marketing chain is not sufficient to bring profitability in the livestock sector by comparative advantage. The participatory approach of development is indispensable to distribute vulnerability and to create an environment for the ease of business. The interplay between the functional variables like vaccination, transport cost, and feed cost on one hand and coordination among the research and

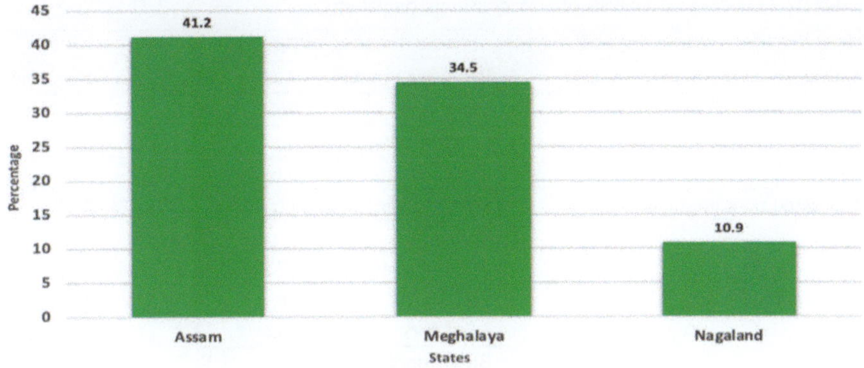

Fig. 22.3 Percentage of pigs vaccinated in Assam, Meghalaya, and Nagaland

administrative units for information on the other plays a key role in transforming the subsistence-based occupation towards entrepreneurial business in the livestock sector.

22.6 Consumers' Choice of Pork in NE Market: Entrepreneurial and Innovative Solution

The pork consumption pattern of the consumers was studied in terms of pork type, processed pork products, pork preparation, willingness to pay more for a safer pork, and family pork consumption in Assam, Meghalaya, and Nagaland (Table 22.4).

In the meat market of the North East, fresh meat is more preferred to processed or any other form. The analysis revealed that 96.4%, 98.2%, and 99.4% of consumers purchased fresh pork in Assam, Meghalaya, and Nagaland, respectively. Processed pork in all three states (Assam—3.6%, Meghalaya—1.8%, and Nagaland—0.6%) is insignificantly preferred. Among the processed foods, Pickles and Nuggets are most preferred, followed by sausage. Pork stew is the most preferred method of pork preparation in Assam, whereas in the other two states, consumers prefer the boiled pork method to prepare the pork.

22.6.1 Changes in Consumption Pattern

The consumers, being health-conscious, are willing to pay more for safer and naturally grown pork (pork with no antibiotics and drugs). Nagaland consumers (73.2%) showed maximum willingness to pay more for safer pork products, followed by Assam (64.0%) and Meghalaya (62.8%). The pork consumption in

Table 22.4 State-wise analysis of pork consumption pattern

Question	Response options	Frequency %		
		Assam	Meghalaya	Nagaland
Type of pork	Fresh pork	96.4	98.2	99.4
	Frozen pork	0	0	0
	Processed pork	3.6	1.8	0.6
Processed pork products	Pork sausage	11.7	21.2	33.5
	Ham	4	5.2	9.4
	Pickle and nuggets	27.5	38.8	47.6
Preparation of pork	Pork stew	38	26.8	5.2
	Fried pork	33.6	32	4.5
	Boiled pork	23.6	34	90.3
	Grilled pork	4.8	4.4	0
Willingness to pay more for safer pork	Yes	64	62.8	73.2
	No	36	37.2	26.8
Pork consumption in your family	Increased	62.8	58.3	65.7
	Decreased	7.2	11.2	4.6
	No change	29.6	29.2	30.7

the family has increased, as reported by 62.8% of consumers in Assam, 58.3% of consumers in Meghalaya, and 65.7% of consumers in Nagaland.

Consumers prefer to buy pork products from supermarkets, butcheries, wholesalers, and producers. The analysis revealed that butcheries/street vendors/small shops are the primary place of purchase for 90% of the pork consumers in Assam. On the contrary, producers directly sell pork products in Meghalaya (65%) and Nagaland (60%) to consumers mostly every month.

22.7 Framework—Combination of Animal Stock and Other Inputs for Profitable Entrepreneurship

The given framework gives the possible outcome with capital investment in animals of Rs. 89,100. The remaining expenditure on feed transport, labor, etc., adds to 1.92 lakh rupees. An entrepreneur has two options to move toward commercialization: The traditional method of production with a single chain of business and the scientific method of production with multiple chains (Fig. 22.4).

The framework gives a holistic insight into the prospective variable cost with respect to the traditional and semi-commercial method of production with an initial animal stock of 4 sows, 1 boar, and 8 piglets. The traditional method is restricted to single chains, whereas multiple chain firms with scientific methods of production and waste management can increase income through three pathways (C1, C2, and C3).

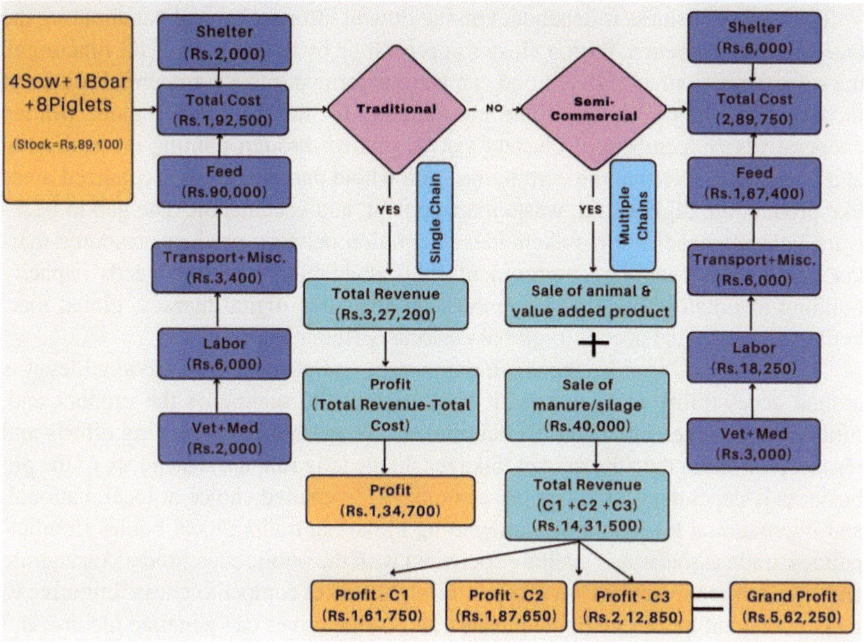

Fig. 22.4 Framework with the possible outcome through capital investment in animals

- C1—Farmer to retailer/wholesaler (selling only piglets).
- C2—Farmer to direct consumer—DTC market (Pork by weight—Rs. 280/kg).
- C3—Farmer to processor/value-added actor (with bio-security measures and following quality control guidelines).

An entrepreneur who deals in multiple market chains has a higher probability of moving towards commercialization through larger profits. Such an entrepreneur can add to their grand profit amount to Rs. 5.62 lakhs (annually).

22.8 Conclusion

A sustainable livestock business is essential for both protecting existing ecosystems and developing better livestock management practices. Research and development (R&D) in social science plays an integral role in creating a sustainable livestock sector. Multiple chains in food-based businesses distribute the risk and increase the scope of competition and profitability, further strengthening the ecosystem for increasing participation. The government is trying to streamline the support mechanism and is working toward digitizing the marketing platforms for entrepreneurial growth in rural and urban areas.

The ease of business is dependent on the flow of information and handholding the budding entrepreneurs. With a cluster approach or by making the SHG functional, the subsistence source of livelihood can be transformed into a commercial source of income for a larger group of the community. In the Northeast region, women cooperatives or members of clusters can be guided through training on leadership and management so that the community as a whole participates in specialized areas like production, digitization, waste management, and vaccination. The gap in backward linkages needs multi-stakeholder integration between producer-resource-market agents. An enabling environment for sustainable livelihood needs capacity building of urban trading/local members on areas like digital literacy, global marketing channels, and access to global customers (India Mart).

The limitation faced by the small enterprises operating at the individual level is around accessibility to resources. It is restricting the scaling of the product and, further, on commercialization. Collaborative linkages allow for pooling efforts and resources to strengthen the market linkages. In the long run, the sustenance of the pig business is dependent upon tapping customers' diversified choice at local, national, and international levels through integrating the urban trading/local bodies (municipalities, trade associations, welfare societies) with the public stakeholders (academic and research organizations) to prosper through market competitiveness. Initiative to make more and more functional clusters and cooperatives can generate income and opportunities to local communities and become drivers for the broad-based local economy development.

Acknowledgement The paper is based on the project titled "Pork Marketing Chain in North East India for Sustainable Livelihood of Tribal Women (Assam, Meghalaya and Nagaland)" funded by National Agricultural Science Fund-Indian Council of Agricultural Research (NASF-ICAR).

References

Ali F, Das I (2003) Tribal situation in North-East India. Stud Tribes Tribals 1(2):141–148

All India Household Consumption Expenditure Survey (2022–2023) Percentage composition of MPCE for egg, fish, and meat since 1999–2000 in rural and urban areas. Ministry of Statistics and Programme Implementation (MoSPI), Government of India

Basic Animal Husbandry Statistics (2020) Basic animal husbandry statistics 2020. Ministry of Fisheries, Animal Husbandry & Dairying, Government of India

Bhattacharjee M, Mehta S, Malik S, Govindasamy K, Singh M, Saha A, Paul P (2022) Economics of marketing chain for small-scale pig producers and bio-security practices: evidence from North-East States (Assam, Meghalaya and Nagaland). Int J Agric Stat Sci 18:22–31

Bhattacharjee M, Mehta S, Singh P, Singh M, Pongener N (2023a) Diversification of livelihood from animal waste and by-products for small and marginal farmers in India. Asian J Dairy Food Res 42:564–569

Bhattacharjee M, Mehta S, Gupta N, Singh P (2023b) Empirical analysis of consumer behavior and preferences for the health security of tribal population in North East India. Int J Food Nutr Sci 12:34–41

Chauhan NM, Thakor R (2006) Participation of the tribal farm women in the decision making. J Ext Educ 15:55–57

Directorate of Economics and Statistics (2020) Agricultural prices in India. Ministry of Agriculture and Farmers Welfare, Government of India

FAO (2012) Invisible guardians: women manage livestock diversity. Food and Agriculture Organization of the United Nations, Rome

FAO (2018) World livestock: transforming the livestock sector through the sustainable development goals. Food and Agriculture Organization of the United Nations

Government of India (2021) Women entrepreneurs in agriculture: trends and insights. Ministry of Agriculture and Farmers Welfare

Herrero M, Thornton PK, Gerber P, Reid RS (2009) Livestock, livelihoods and the environment: understanding the trade-offs. Curr Opin Environ Sustain 24:111–120

IFAD (1999) The role of women in livestock production. Int Fund Agric Dev

Kumar A, Rai DC (2011) Trends and volatility in domestic and international prices of livestock products in India. Indian J Anim Sci 81:61–65

Livestock Census (2019) Livestock Census 2019. Department of Animal Husbandry and Dairying, Ministry of Fisheries, Animal Husbandry and Dairying, Government of India

Mac D, Belussi (2002). Cluster development for pro-poor growth: the UNIDO approach. United Nations Industrial Development Organization (UNIDO)

OECD & FAO (2022) OECD-FAO Agricultural Outlook 2022. Organization for Economic Co-operation and Development & Food and Agriculture Organization of the United Nations

Thornton PK, van de Steeg J, Notenbaert A, Herrero M (2006) The impacts of climate change on livestock and livestock systems in developing countries: a review of what we know and what we need to know. Agric Syst 101(3):113–127

Tsito M (2004) Community development and empowerment. In: Introduction to development studies. Oxford university press

van der Gaast K, van Leeuwen E, Wertheim-Heck S (2022) Food systems in transition: conceptualizing sustainable food entrepreneurship. Int J Agric Sustain 20(5):705–721

Weber H, Wiek A, Lang DJ (2020) Sustainability entrepreneurship to address large distances in international food supply. Bus Strateg Dev 3(3):318–331

Chapter 23
Interpreting Labels for Biodiverse Foods: Consumer Awareness, Comprehension and Purchase Drivers

Vijayata Sengar and Arushi Loiwal

Abstract Poor dietary habits are a leading contributor to deaths worldwide, surpassing other risk factors and driving the rising burden of non-communicable diseases (NCDs) such as obesity and diabetes. High intake of processed and ultra-processed packaged foods, which are often calorie-dense and low in nutrients, has been linked to increased risk of these conditions. Research shows that clear and easy-to-interpret nutrition information, particularly Front-of-Pack Labels (FOPLs), can help consumers make healthier choices by quickly highlighting key nutritional qualities at the point of purchase.

In this context, the current study assessed how well consumers understand and perceive food labels. A consumer survey was conducted in India using snowball (purposive) sampling, gathering data from 425 participants aged 15–49 years. Respondents were asked about their awareness and comprehension of both back-of-pack and front-of-pack nutrition information, as well as their attitudes toward labelling. The findings indicated that, although many participants reported noticing nutrition labels, their actual understanding of the information was limited. A larger proportion (55.3%) felt that back-of-pack labels are most effective for communicating health-related information about processed foods, compared to 44.7% who viewed front-of-pack labels as effective. The results underscore the urgent need for consumer education on how to interpret food labels so that individuals can use this information to make more informed and healthier dietary choices, particularly given the growing consumption of unhealthy foods among young populations.

Keywords Food labels · Back of pack labels · Front of pack labels · Processed packaged foods · Poor diets · Unhealthy dietary patterns · Unhealthy food consumption · Non-communicable diseases

V. Sengar (✉) · A. Loiwal
Department of Foods and Nutrition, Faculty of Family and Community Sciences, The Maharaja Sayajirao University of Baroda, Vadodara, Gujarat, India
e-mail: vijayata.sengar-fn@msubaroda.ac.in

M. Thakur, T. Belwal (eds.), *Food Biodiversity and Consumer Engagements*, Sustainable Development and Biodiversity 46,
https://doi.org/10.1007/978-981-95-7777-4_23

23.1 Introduction

All aspects of producing, preparing, distributing, marketing, and consuming food are together referred to as food systems. They affect livelihoods, diets, food security, and the sustainability of the environment. Food habitats, food supply networks, and individual elements are the parts that make up food systems (food system). Diets are changing everywhere. Consumption of animal foods, sugar, salt, fats and oils, refined grains, and processed foods has increased; consumption of fruits and vegetables has either increased or decreased, depending on where you look; and consumption of coarse grains, staple cereals, and pulses has decreased. These changes are happening at different rates for different populations and places, but the developing world is changing at the fastest rate. For example, in high-income countries, the consumption of sugar, salt, and particularly fat from processed foods has reached a plateau, whereas in middle-income countries, it is rapidly increasing.

A high-quality diet is increasingly being consumed by more individuals, according to the Global Panel on Agriculture and Food Systems for Nutrition. Low-quality diets are still common in most countries, despite advances in nutrition. Current estimations indicate that the primary cause of the global disease burden is inadequate diets that are deficient in essential nutrients and excessive in harmful substances (Gain Global). Dietary changes and their impact on nutrition are referred to as the "nutrition shift," and they are strongly linked to the increased prevalence of non-communicable diseases, including diabetes and heart disease, as well as overweight and obesity (Hawkes et al. 2017). Due to the nutrition transition, consumption of unhealthy diets has increased. Unhealthy diets are responsible for more deaths worldwide and are the leading cause of non-communicable diseases (NCDs). Excessive consumption of processed packaged foods (PPFs) is associated with an increased risk of obesity and related NCDs. Effective nutrition labelling, including simple-to-use Front of Pack Labelling, has been identified as one of the strategies that countries should use to address the growing global concern of unhealthy dietary patterns. Dietary label nutrition information can help consumers make educated dietary decisions. Consumers must read, comprehend, and interpret food labels to choose nutritious foods (Hawkes et al. 2017). Labelling can also be a sustainable strategy for transforming food systems. Eco-labels and nutrition labels are useful resources for encouraging the consumption of more sustainably produced and nutritious foods (Mhurchu and Gorton 2007).

Therefore, nutrition information on food labels needs to be accurate and presented in a way that customers can easily understand. Studies have shown that back-of-pack labels are difficult for laymen to comprehend and thus leave the consumer confused. Front-of-Pack Labels have been successfully implemented in a few countries and can be one of the effective strategies to address NCDs. Thus, the present study was carried out to assess the consumers' understanding and perceptions regarding food labels, i.e., Back of Pack Labels (BoPLs) and Front of Pack Labels (FoPLs).

23.2 Methods and Materials

This cross-sectional study has been planned to assess consumers' understanding and perceptions of food labels for processed packaged foods. Snowball sampling (purposive sampling) was used as the sampling technique. Data were collected online with 425 study participants (sample size calculation shown below) in the age group of 15–49 years residing in India. The sample size of 310 was as per the sample size calculations given below:

23.3 Results

The present study dealt with assessing consumers' understanding and perceptions regarding food labels for processed packaged foods.

23.3.1 Perceptions Regarding Back of Pack Labels (BOPL)

Participants' perceptions were analysed to see how much they could understand from BOPLs. It was observed that although they reported information from BOPL, the extent of information understood was low. Around 82.3% reported that BOPL informed them that PPF is high in fat, whereas only 29.4% reported that BOPL completely informed them that PPF is high in fat (Figs. 23.1 and 23.2). Participants' perceptions were analysed to see how much they could understand from BOPLs. It was observed that although they reported information from BOPL, the extent of information understood was low. Around 82.3% reported that BOPL informed them that PPF is high in fat, whereas only 29.4% reported that BOPL completely informed them that PPF is high in fat (Figs. 23.3 and 23.4).

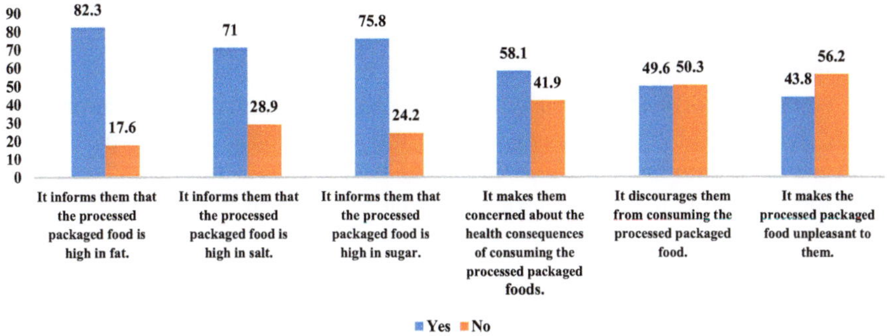

Fig. 23.1 Perceptions regarding Back of Pack Label ($N = 425$)

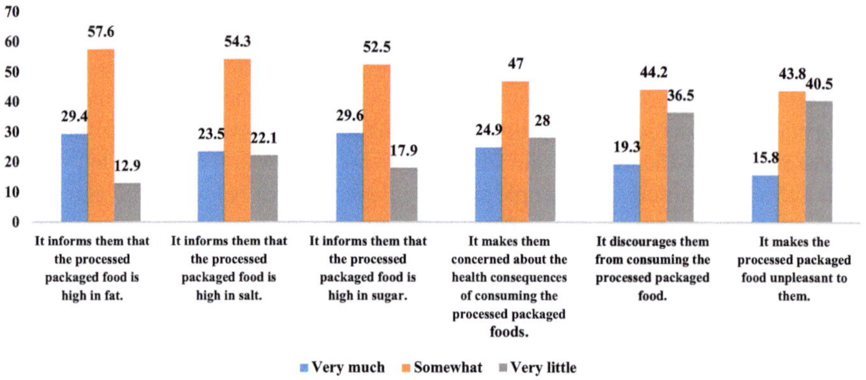

Fig. 23.2 Perceptions regarding the extent of information provided by the Back of Pack Label ($N = 425$)

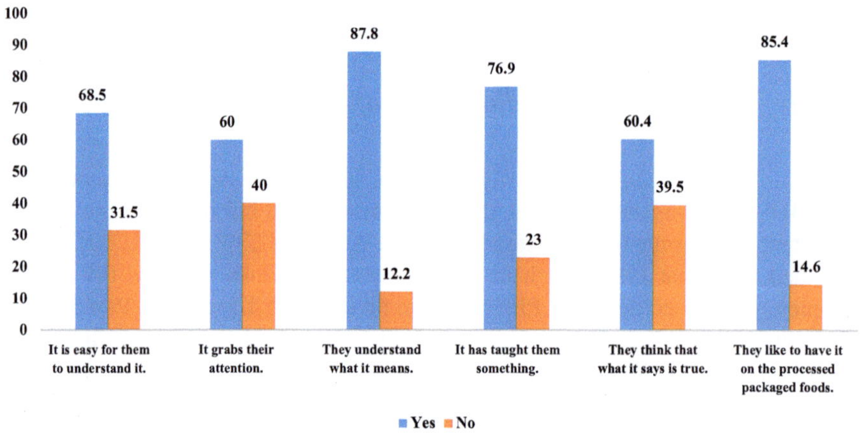

Fig. 23.3 Perceptions regarding Back of Pack Label ($N = 425$)

23.3.2 Perceptions Regarding Front of Pack Labels (FOPL)

Participants' perceptions were analysed to see how much they could understand from FOPLs. It was observed that although they reported receiving information from FOPL, the extent of information understood was low. Around 57.1% reported that FOPL would inform them that the PPF is high in fat, whereas only 23.5% reported that FOPL would completely inform them that PPF is high in fat (Figs. 23.5 and 23.6). Around 78.1% reported that FOPL would be easy for them to understand, whereas only 46.1% reported that they would completely be able to understand the FOPL (Figs. 23.7 and 23.8). Around 55.3% of the study participants considered the back of pack label as the best way of informing them about the health effects of consuming processed packaged foods, while only 44.7% considered the front of pack label effective for the same (Fig. 23.9).

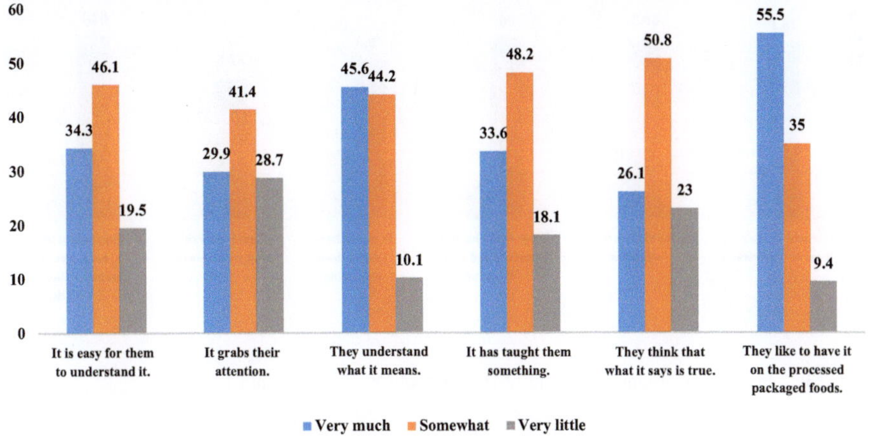

Fig. 23.4 Perceptions regarding the extent of information provided by Back of Pack Label ($N = 425$)

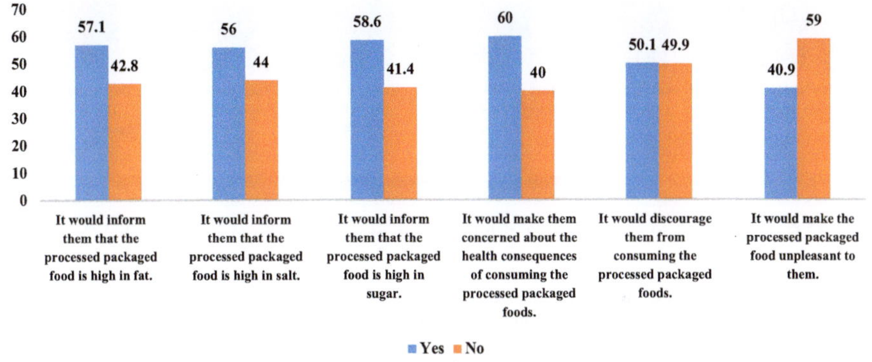

Fig. 23.5 Perceptions regarding Front of Pack Label ($N = 425$)

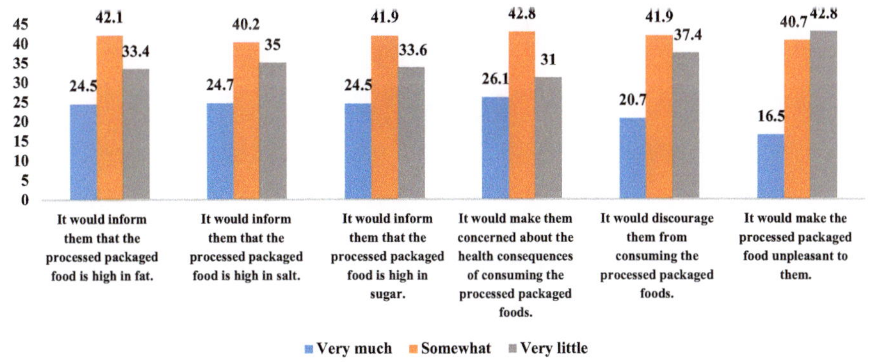

Fig. 23.6 Perceptions regarding extent of information provided by Front of Pack Label ($N = 425$)

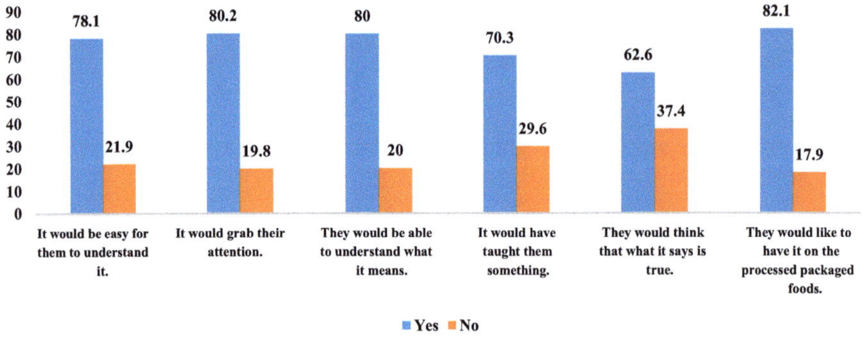

Fig. 23.7 Perceptions regarding Front of Pack Label ($N = 425$) (contd.)

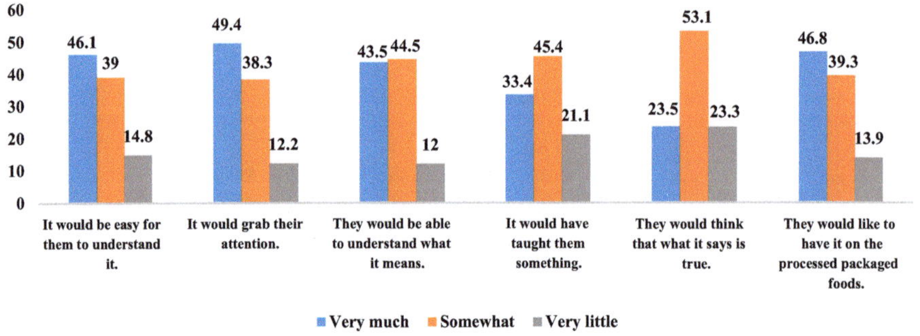

Fig. 23.8 Perceptions regarding extent of information provided by Front of Pack Label ($N = 425$)

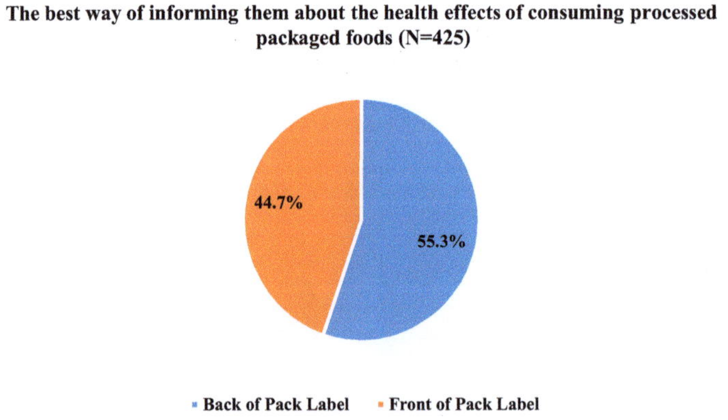

Fig. 23.9 Comparison between Back of Pack Label and Front of Pack Label-study participants' responses ($N = 425$)

23.4 Discussion

Nutrition transition over decades has led to an increased consumption of processed and ultra-processed foods. This can be detrimental in terms of the growing incidence of non-communicable diseases. WHO has identified unhealthy diets as one of the leading causes of NCDs among populations. Food Labelling can be used as a strategy to address this matter. The present study was conducted to assess consumers' understanding and perceptions regarding food labels (Potter et al. 2023). To understand this current scenario, a consumer survey was conducted. The results showed that around 55.3% of the study participants considered the back of pack label as the best way of informing them about the health effects of consuming processed packaged foods, while only 44.7% considered the front of pack label effective for the same.

This is because most of the processed packaged foods that are available in the Indian market do not have front of pack labels along with back of pack labels. In one of the research studies, it was reported that simpler front-of-pack labelling helped participants take much less time to make a healthier choice. Therefore, in a retail environment where judgments are made quickly, simpler front-of-pack labelling designs seem more appropriate (Feunekes et al. 2008).

23.5 Conclusion

The consumption of unhealthy foods continues to rise, particularly among young people, highlighting an urgent need for improved nutrition literacy. Educating consumers on how to interpret food labels can enhance their ability to use this information when choosing foods, ultimately supporting healthier eating behaviours and better diet-related health outcomes.

23.6 Recommendations

Future long-term studies need to be conducted on consumer education and behaviour change using strategic communication.

References

Feunekes GI, Gortemaker IA, Willems AA, Lion R, van den Kommer M (2008) Front-of-pack nutrition labelling: testing effectiveness of different nutrition labelling formats front-of-pack in four European countries. Appetite 50(1):57-70. https://doi.org/10.1016/j.appet.2007.05.009. PMID: 17629351

Hawkes C, Harris J, Gillespie S (2017) Changing diets: urbanization and the nutrition transition. In: 2017 global food policy report. International Food Policy Research Institute (IFPRI), pp 34–41

Mhurchu CN, Gorton D (2007) Nutrition labels and claims in New Zealand and Australia: a review of use and understanding. Aust N Z J Public Health 31(2):105–112. https://doi.org/10.1111/j.1753-6405.2007.00026.x

Potter C, Pechey R, Cook B, Bateman P, Stewart C, Frie K, Clark M, Piernas C, Rayner M, Jebb SA (2023) Effects of environmental impact and nutrition labelling on food purchasing: an experimental online supermarket study. Appetite 180:106312. https://doi.org/10.1016/j.appet.2022.106312

Part IV
Mindful Consumption: Psychology Behind Sustainable Eating

Chapter 24
From Stress to Snacking: Psychological Pathways and Consumer Behavior in Today's Food Environment

Siddharth Soni, Sushma Kumari, Manoj Kumar, and Shivanjali Shrivastava

Abstract This chapter explores the intricate relationship between psychological stress and eating behaviors, an area gaining increasing attention in both academic and public health spheres. Stress, whether short-term or long-lasting, can profoundly influence dietary habits, leading to patterns such as emotional eating, overeating, or, conversely, reduced appetite. Drawing from foundational psychological frameworks like the affect regulation model and the psychosomatic theory, the discussion highlights how stress can affect food preferences, portion sizes, and meal routines. The chapter also delves into the biological processes involved, particularly the function of the hypothalamic-pituitary-adrenal (HPA) axis and the role of cortisol in enhancing cravings for calorie-dense foods high in sugar and fat. Beyond the biological aspects, it examines key psychosocial elements such as individual coping mechanisms, body image, gender differences, and cultural background, which shape how people respond to stress in relation to eating. Adopting an interdisciplinary perspective that blends insights from psychology, neuroscience, and nutritional science, the chapter offers a comprehensive view of the stress-eating dynamic. It concludes by presenting practical, evidence-based strategies for psychological intervention, including stress reduction and cognitive-behavioral approaches designed to support healthier eating habits and improved mental well-being. Such an integrated framework is essential for informing both therapeutic practices and public health policies.

Keywords Stress · Eating habits · Psychological factors · Emotions

---

S. Soni (✉)
School of Psychology and Counselling, O.P. Jindal Global University, Sonipat, Haryana, India

S. Kumari
Berlin School of Business and Innovation, Berlin, Germany
e-mail: Sushma.kumari@berlinsbi.com

M. Kumar · S. Shrivastava
Amity Centre of Behavioural Science, Amity University, Noida, India
e-mail: mmathur@amity.edu; sshrivastava3@amity.edu

M. Thakur, T. Belwal (eds.), *Food Biodiversity and Consumer Engagements*, Sustainable Development and Biodiversity 46,
https://doi.org/10.1007/978-981-95-7777-4_24

24.1 Introduction

In India, the interplay between stress and eating habits is a significant concern, influenced by a combination of traditional dietary practices, modern lifestyle changes, and socio-economic factors. Rapid urbanization, high work pressures, and changing family structures contribute to elevated stress levels across various demographics. This stress often manifests in altered eating behaviors, where individuals may resort to unhealthy comfort foods or experience a loss of appetite. Traditional Indian diets, which are rich in carbohydrates and spices, may be supplemented by processed and high-calorie foods due to increased availability and convenience. Stress-induced eating patterns, such as emotional eating or binge eating, are becoming more prevalent, leading to health issues like obesity, diabetes, and cardiovascular diseases. Additionally, the pressure to conform to societal standards of appearance and success exacerbates stress, influencing eating habits in both detrimental and complex ways. The Indian context highlights the need for culturally sensitive approaches to stress management and dietary guidance, aiming to integrate traditional practices with modern health strategies to address these interconnected challenges effectively.

Stress and eating habits are deeply intertwined, affecting physical health, mental well-being, and overall quality of life. Understanding how these factors interact is crucial for comprehending how individuals cope with daily challenges and maintain their health. Recognizing the connection between stress and eating behaviors is key to developing effective strategies that promote healthy eating habits and manage stress. This chapter explores the definitions of stress and eating habits, examines their interrelationship, and highlights the importance of studying their psychological impact.

24.1.1 Definition of Stress

Stress is a complex response involving both psychological and physiological reactions to perceived threats or challenges. It manifests in various forms, each with distinct characteristics:

Acute Stress: Short-term stress that arises from immediate threats or pressures, such as a looming deadline or a sudden argument. It triggers the body's "fight-or-flight" response, releasing hormones like cortisol and adrenaline to prepare for immediate action.

Chronic Stress: Long-term stress resulting from ongoing situations or conditions, such as long-term financial difficulties, a demanding job, or a troubled relationship. Chronic stress keeps the body in a state of heightened alertness, which can have detrimental effects on health over time.

Episodic Acute Stress: Characterized by frequent episodes of acute stress, often experienced by individuals with a chaotic lifestyle who are constantly dealing with crises and deadlines. This repeated activation of the stress response can lead to significant health issues.

Stress triggers a cascade of physiological responses, including the release of cortisol and adrenaline. These hormones help the body respond quickly to threats but can cause harm if the stress response is activated too frequently or for prolonged periods. Chronic activation of the stress response can lead to various health problems, including cardiovascular disease, digestive issues, and weakened immune function.

24.1.2 Understanding Stress

24.1.2.1 Biological Mechanisms of Stress

The biological mechanisms of stress involve a sophisticated interplay between the brain and various physiological systems, primarily mediated through the hypothalamic-pituitary-adrenal (HPA) axis and the autonomic nervous system (ANS). When a person encounters a stressor, the hypothalamus in the brain initiates the stress response by secreting corticotropin-releasing hormone (CRH). This hormone then stimulates the pituitary gland to release adrenocorticotropic hormone (ACTH), which prompts the adrenal glands to produce and release cortisol, a key glucocorticoid hormone. Cortisol plays a critical role in the stress response by increasing blood glucose levels to provide energy, suppressing non-essential functions such as immune activity, and modulating brain function to enhance cognitive performance and alertness (Sapolsky 2004; McEwen 2008).

Simultaneously, the sympathetic nervous system (SNS), a component of the autonomic nervous system, is activated. This leads to the release of catecholamines, including adrenaline (epinephrine) and noradrenaline (norepinephrine), from the adrenal medulla. These chemicals prepare the body for immediate physical action by elevating heart rate, increasing blood pressure, enhancing respiratory rate, and redistributing blood flow to major muscle groups. This physiological response is often referred to as the "fight-or-flight" response, equipping the individual to either confront or evade the perceived threat (Everly and Lating 2013).

Once the stressor is removed, the parasympathetic nervous system (PNS) activates to counterbalance the SNS and restore the body to a state of equilibrium. This process, known as the "rest-and-digest" response, reduces heart rate, promotes digestion, and facilitates relaxation. Chronic activation of the stress response, however, can lead to HPA axis dysregulation and sustained elevated cortisol levels, which are associated with negative health outcomes. Persistent high cortisol levels can impair immune function, contribute to cardiovascular diseases, exacerbate metabolic disorders such as diabetes, and affect brain regions involved in memory and emotional regulation, such as the hippocampus and amygdala (Chrousos 2009; Kiecolt-Glaser et al. 2003).

The biological mechanisms of stress underscore the critical importance of effective stress management strategies to mitigate its adverse effects on health and well-being.

24.1.3 Psychological Responses to Stress

Psychological responses to stress include a range of emotional and cognitive reactions that individuals experience when facing perceived threats or challenges. These responses are shaped by how individuals appraise stressors, their coping strategies, and their psychological resilience. Understanding these responses is crucial for effectively addressing mental health issues and developing strategies to manage stress.

Emotional Responses: The emotional responses to stress can vary based on the nature of the stressor and the individual's coping mechanisms. Key emotional responses include:

Anxiety, Depression, Irritability: Stress often triggers a range of emotional and psychological responses that can significantly impact an individual's well-being. One common outcome is anxiety, which manifests as feelings of worry, fear, and apprehension. Anxiety can be acute or chronic, with prolonged stress potentially leading to generalized anxiety disorder (GAD) or panic disorder, conditions characterized by persistent and excessive worry that disrupts daily functioning (American Psychological Association 2020). Additionally, extended or severe stress can contribute to depressive symptoms such as persistent sadness, loss of interest in activities, and feelings of hopelessness. Chronic stress is a known risk factor for major depressive disorder (MDD) and may also worsen preexisting depression (Kiecolt-Glaser et al. 2003). Stress can further lead to increased irritability and mood swings, resulting in heightened frustration and anger. These emotional fluctuations often cause conflicts with others, strain relationships, and may isolate individuals, thereby intensifying feelings of loneliness (Everly and Lating 2013). Together, these stress-induced emotional challenges highlight the profound effect stress can have on mental health and social functioning.

Cognitive Responses: Cognitive responses to stress involve how individuals perceive and interpret stressors and their effects:

Cognitive Distortions and Mental Blocks: Stress often triggers negative thinking patterns such as catastrophizing, where individuals expect the worst outcomes, and rumination, which involves repeatedly dwelling on distressing events. These cognitive distortions can exacerbate stress and impair problem-solving abilities (Beck 1979). Additionally, stress negatively affects cognitive functions, including concentration, memory, and decision-making. Persistent stress can make it difficult to focus and retain information, which can reduce performance in work or academic environments (Sapolsky 2004). High levels of stress also narrow one's focus and reduce cognitive flexibility, further decreasing the ability to generate effective solutions to problems (McEwen 2008).

Behavioral Responses: Behavioral responses to stress include the observable actions and behaviors resulting from psychological stress:

Avoidance, Emotional Eating, and Substance Use: Stress often leads individuals to engage in avoidance behaviors as a way to escape stressful situations. This can include procrastination, withdrawing from social interactions, and neglecting responsibilities. While avoidance may offer short-term relief, it typically results in increased stress and negative consequences over time (Carver 1997). Additionally, stress can alter eating patterns, causing emotional eating where food is used as a coping mechanism. This behavior may contribute to unhealthy eating habits, weight gain, and related health issues (Macht 2008). Some individuals also turn to substances such as alcohol, tobacco, or recreational drugs to manage their stress. Although these substances might provide temporary relief, they often lead to further health complications and risk of dependency (Torres and Nowson 2007).

Coping Mechanisms: Effective coping mechanisms are vital for managing stress and reducing its negative effects. Coping strategies can be classified into:

Problem-Focused Coping: This strategy involves directly addressing the source of stress through problem-solving and taking actionable steps. It is typically associated with positive outcomes and improved well-being (Lazarus 1984).

Emotion-Focused Coping: This approach includes managing emotional responses to stress through techniques such as seeking social support, engaging in relaxation practices, and practicing mindfulness. Emotion-focused coping can alleviate emotional distress and enhance resilience (Gross 2002).

Recognizing and understanding psychological responses to stress and utilizing effective coping strategies are crucial for maintaining mental health and improving quality of life. Addressing these responses through therapeutic interventions and support systems can help individuals navigate stress more effectively and promote psychological well-being.

24.1.4 Eating Habits and Their Importance

Eating habits refer to the patterns and behaviors associated with food consumption and play a crucial role in determining nutritional status and overall health. Key aspects include:

Frequency, Choice, and Quantity: Eating habits refer to the patterns and behaviors related to food consumption, playing a crucial role in determining nutritional status and overall health. Key aspects include the frequency and timing of meals, which help regulate metabolism and maintain energy levels throughout the day. Food choices, particularly consuming a balanced diet rich in fruits, vegetables, whole grains, and lean proteins, support optimal health. Additionally, the quantity of food consumed in one sitting is important, as appropriate portion sizes help maintain a healthy weight and prevent overeating. Healthy eating habits are essential not only for physical health but also for mental well-being and the prevention of chronic diseases such as obesity, diabetes, and heart disease. Conversely, poor eating habits can lead to nutritional deficiencies, weight gain, and an increased risk of chronic illnesses.

24.1.5 The Link Between Stress and Eating Habits

Stress significantly influences eating behaviors, often disrupting healthy eating patterns in multiple ways. Some individuals respond to stress by overeating, particularly consuming high-calorie, high-sugar, and high-fat "comfort" foods a behavior known as stress eating or emotional eating. While these comfort foods may provide temporary relief, they can lead to weight gain and long-term health problems. Conversely, others experience a loss of appetite and eat less during stressful periods, which can cause nutritional deficiencies, weakened immune function, and unintended weight loss. The body's stress response also impacts hormones like cortisol and insulin, which regulate appetite and cravings. Elevated cortisol levels tend to increase appetite and cravings for energy-dense foods, while chronic stress-induced insulin resistance affects how the body processes and stores nutrients, further influencing eating behaviors.

24.1.6 Psychological Factors and Eating Behavior

Several psychological factors mediate the relationship between stress and eating habits, including:

Emotional, Mindful, and Restrained Eating: Several psychological factors mediate the relationship between stress and eating habits. Emotional eating involves using food as a coping mechanism for emotional distress, where individuals may seek comfort in food during stressful times, often resulting in overeating and unhealthy food choices. Mindless eating refers to consuming food without attention to hunger cues or portion sizes, frequently as a distraction from stress, which can also lead to overeating and poor dietary decisions. Additionally, restrained eating, where individuals deliberately restrict their food intake to control weight, can backfire during stressful periods, triggering binge eating episodes and a loss of control over food consumption.

24.2 Importance of Studying the Psychological Impact

Studying the relationship between stress and eating behaviors helps identify how stress-related eating habits contribute to health problems. This knowledge guides the development of interventions to mitigate these effects and promote better health outcomes. Knowledge of the psychological mechanisms underlying stress and eating behaviors can inform the design of effective behavioral strategies. These strategies can help individuals develop healthier eating habits and better stress management techniques. Addressing both stress and eating habits holistically can improve overall well-being and quality of life. Interventions that consider the psychological, physiological, and behavioral aspects of stress and eating are likely to be more effective in promoting lasting change. By exploring the psychological aspects of stress and eating habits, we can develop comprehensive strategies to help individuals manage stress and adopt healthier eating behaviors. This approach not only improves physical health but also enhances mental well-being, contributing to a better quality of life overall.

24.2.1 Eating Habits: A Psychological Perspective

Eating habits are deeply influenced by psychological factors and play a significant role in overall health and well-being. Understanding these habits from a psychological perspective involves examining their definitions, underlying psychological influences, and the role of emotions in eating behavior. Eating habits can significantly impact physical health, influencing weight management, energy levels, and the risk of chronic diseases such as diabetes, heart disease, and obesity. Healthy eating habits are associated with balanced nutrition, supporting overall health and reducing the risk of nutritional deficiencies and related illnesses.

24.2.2 Importance of Eating Habits

Healthy eating habits are essential for maintaining physical well-being by helping to regulate body weight, support heart health, and supply necessary nutrients for proper bodily functions. In contrast, unhealthy eating patterns can lead to obesity, nutritional deficiencies, and a higher likelihood of developing chronic illnesses (Skerrett and Willett 2010). Nutrition also has a significant effect on mental health; consuming nutrient-dense foods can improve mood, enhance cognitive abilities, and promote emotional balance, while poor nutrition may contribute to mood disorders and cognitive decline (Jacka et al. 2017). Additionally, well-balanced dietary habits enhance overall quality of life by supporting physical health, fostering mental resilience, and encouraging positive social experiences involving food (Gordon et al. 2014).

24.2.3 Psychological Factors Influencing Eating Habits

Psychological factors play a crucial role in shaping eating habits. These factors include:

Stress: Stress can significantly impact eating behaviors, leading to either overeating or undereating. Stress-induced overeating often involves consumption of high-calorie comfort foods, while stress-induced undereating can result in poor nutrient intake and weight loss (Macht 2008).

Self-Esteem: Individuals with low self-esteem may engage in unhealthy eating patterns as a form of self-punishment or due to negative body image. Conversely, high self-esteem is often associated with healthier eating habits and better body image (Stice et al. 2004).

Habits and Routines: Long-standing eating habits and routines can shape current eating behaviors. For example, habitual snacking or regular fast food consumption can become entrenched behaviors that are difficult to change (Fletcher et al. 2008).

Socio-cultural Influences: Cultural norms and social environments influence eating habits by dictating food choices, portion sizes, and mealtime practices. Social influences, such as peer pressure or family traditions, can also impact individual eating behaviors (HIGGINS 2007).

24.2.4 The Role of Emotions in Eating Behavior

Emotions play a crucial role in shaping eating behaviors through several key factors. Emotional eating involves using food, often comfort foods, to cope with feelings such as sadness, boredom, or stress, which can result in unhealthy eating patterns and weight gain (Macht 2008). For some individuals, food serves as a reward or incentive, reinforcing certain eating behaviors and potentially leading to overeating and poor dietary habits. Additionally, mood states like depression, anxiety, and stress influence appetite and food choices depression may increase cravings for high-calorie foods, whereas anxiety can reduce appetite. The consumption of comfort foods during emotional distress may provide short-term relief but often contributes to unhealthy eating patterns and long-term health problems.

Understanding the interplay between psychological factors, eating habits, and emotional influences can help in developing strategies to promote healthier eating behaviors and improve overall well-being.

24.2.5 *The Interplay Between Stress and Eating Habits*

24.2.5.1 How Stress Affects Eating Patterns

Stress has a profound impact on eating behaviors, leading to notable deviations from typical dietary habits. The effects of stress on eating can manifest as either stress-induced overeating or stress-induced undereating, both of which can have significant health consequences.

24.2.6 *Stress-Induced Overeating*

Stress-induced overeating occurs when individuals consume more food than usual, often gravitating towards high-calorie, sugary, and fatty foods in response to stress. Several mechanisms contribute to this behavior:

Emotional Eating and Hormonal Change: Stress can significantly influence eating patterns, particularly through emotional eating and hormonal changes. Emotional eating involves using food, especially high-calorie comfort foods, as a way to cope with negative emotions such as anxiety, sadness, or frustration. While this may offer temporary emotional relief, it often results in unhealthy eating habits and potential weight gain (Macht 2008). Additionally, the body's stress response activates the release of hormones like cortisol and adrenaline. Elevated cortisol levels are closely linked to increased appetite and cravings for energy-dense foods, making individuals more likely to reach for sugary and fatty options during stressful times.

Altered Reward Processing and Behavioral Changes: Stress can significantly influence how the brain perceives and responds to food rewards. It heightens the reward value of high-calorie, palatable foods, making individuals more inclined to consume them for comfort and emotional relief (Tian et al. 2012). This altered reward processing increases the likelihood of seeking out sugary and fatty foods during stressful times. Additionally, stress often triggers behavioral changes, such as frequent snacking or increased consumption of unhealthy foods. These habits, driven by a desire for immediate gratification and distraction from stress, can contribute to weight gain and elevate the risk of related health issues (Macht 2008; Torres and Nowson 2007).

24.2.7 *Stress-Induced Undereating*

Stress-induced undereating happens when stress causes a decrease in appetite and food intake. This response can be attributed to several factors:

Suppressed Appetite and Digestive Disruptions: Acute stress often leads to a temporary suppression of appetite, primarily due to the activation of the sympathetic nervous system and the release of adrenaline. In some cases, this response can be intense enough to cause noticeable weight loss and nutritional deficiencies (Kiecolt-Glaser et al. 2003). Additionally, stress can disrupt normal digestive functions, resulting in symptoms such as nausea, stomach cramps, or indigestion. When stress becomes chronic, these gastrointestinal issues can persist, further interfering with appetite and overall nutritional health (Sapolsky 2004).

Cognitive and Psychological Impacts on Eating Behavior: Elevated stress levels can disrupt cognitive functioning, particularly concentration and memory, making it challenging to maintain consistent eating routines. This cognitive distraction often leads to skipped meals or reduced food intake (McEwen 2008). Additionally, psychological conditions commonly associated with stress, such as anxiety and depression, can further suppress appetite and diminish interest in food, contributing to undereating and unintended weight loss.

24.2.8 Impact of Stress on Specific Eating Disorders

Stress significantly affects various eating disorders, influencing their development, progression, and management. This section explores how stress impacts four specific eating disorders: Binge Eating Disorder, Anorexia Nervosa, Bulimia Nervosa, and Emotional Eating.

Binge Eating Disorder: Binge eating disorder (BED) is characterized by episodes of consuming large quantities of food in a short period, often accompanied by feelings of loss of control and distress. Stress plays a critical role in the onset and exacerbation of BED. Stress is a critical factor in triggering binge eating episodes. Many individuals cope with stress by consuming large quantities of high-calorie, palatable foods, leading to emotional and uncontrolled eating behaviors (Hudson et al. 2007). On a neurobiological level, stress activates brain regions involved in reward and pleasure, such as the nucleus accumbens, thereby enhancing the appeal of food and reinforcing binge eating tendencies (Stice et al. 2011). Psychologically, individuals with BED often struggle with heightened anxiety and depression, which are exacerbated by stress and further drive emotional eating as a coping strategy (Carter et al. 2015).

Anorexia Nervosa: Anorexia Nervosa is an eating disorder marked by an intense fear of gaining weight and a distorted body image, leading to severe restriction of food intake. Stressful life events and chronic psychological stress can significantly intensify restrictive eating behaviors commonly observed in anorexia nervosa. For many individuals, limiting food intake becomes a way to regain a sense of control amidst external chaos or internal emotional turmoil. This stress is often internalized through traits like self-criticism and perfectionism, further reinforcing disordered eating patterns (Keel and Mitchell 1997). Additionally, chronic stress impacts hormonal systems such as the hypothalamic-pituitary-adrenal (HPA) axis, disrupting appetite regulation and contributing to the persistence of restrictive eating behaviors (Kaye et al. 2001).

Bulimia Nervosa: Bulimia nervosa involves recurrent episodes of binge eating followed by compensatory behaviors such as vomiting, laxative abuse, or excessive exercise. Stress has a profound impact on this disorder. Stress plays a significant role in the onset and maintenance of bulimia nervosa by influencing behavioral, emotional, and biological processes. Individuals may resort to binge eating as a coping mechanism during stressful periods, followed by purging behaviors to offset the perceived consequences of overeating (Garg et al. 2007). Stress also heightens negative emotions and distorted body image, intensifying feelings of guilt and shame that can drive the cycle of bingeing and purging (Fairburn et al. 2003). Biologically, stress affects neurotransmitter systems involved in mood and appetite regulation, and such disruptions may be amplified under stress, further contributing to the persistence of bulimic behaviors (Johnson et al. 2008).

Emotional Eating: Emotional eating involves consuming food in response to emotional states rather than physical hunger. Stress is a key contributor to emotional eating, where individuals seek comfort in food to alleviate negative emotions, often resulting in cravings for high-calorie foods and unhealthy eating habits (Macht 2008). This behavior is closely linked to difficulties in emotion regulation and stress management, with stressed individuals frequently turning to food as a coping mechanism and distraction. Over time, chronic stress and emotional eating can create a vicious cycle of poor dietary habits and weight gain, which in turn heightens stress levels and may increase the risk of developing more serious eating disorders or other health problems (Oliver et al. 2000).

24.3 Stress Management and Healthy Eating

Effective management of stress and the promotion of healthy eating habits are integral to overall well-being. This section explores various techniques for managing stress and strategies for fostering healthy eating practices, highlighting their importance and practical applications.

Techniques for Managing Stress: Effective stress management is crucial for maintaining mental and physical health. Several techniques can help individuals reduce stress and enhance their quality of life.

Mindfulness: Mindfulness involves paying attention to the present moment without judgment. It can help individuals become more aware of their stress responses and reduce overall stress levels. Research has shown that mindfulness practices can lower stress hormones, improve emotional regulation, and enhance psychological resilience.

Meditation: Meditation is a practice that focuses the mind to achieve a state of relaxation and mental clarity. Techniques such as guided imagery, progressive muscle relaxation, and breath awareness are commonly used. Regular meditation can reduce symptoms of anxiety, depression, and stress, and has been linked to improved cognitive function and emotional well-being (Goyal et al. 2014).

Cognitive-Behavioral Therapy (CBT): CBT is a structured, goal-oriented psychotherapy that focuses on identifying and changing negative thought patterns and behaviors. It is effective for managing stress by helping individuals reframe their perceptions of stressors and develop coping strategies. CBT can reduce stress, anxiety, and depressive symptoms by altering maladaptive thinking and behavior patterns (Beck 1979). Common CBT techniques include cognitive restructuring (challenging and changing negative thoughts), behavioral activation (engaging in rewarding activities), and problem-solving skills. By addressing both cognitive and behavioral aspects, CBT helps individuals manage stress more effectively and improve their overall mental health (Hofmann et al. 2012).

Physical Activity and Exercise: Benefits, regular physical activity is one of the most effective ways to manage stress. Exercise promotes the release of endorphins, which are natural mood elevators, and helps reduce the levels of stress hormones like cortisol. Physical activity also improves sleep quality, enhances cognitive function, and increases overall energy levels (Craft and Perna 2004). Engaging in moderate-intensity exercise, such as brisk walking, cycling, or swimming, for at least 30 min most days of the week can significantly reduce stress. Additionally, incorporating activities like yoga or tai chi can combine physical exercise with relaxation techniques, further enhancing stress reduction (Salmon 2001).

24.3.1 Strategies for Promoting Healthy Eating Habits

Promoting healthy eating habits involves a combination of education, behavioral strategies, and practical approaches to nutrition.

Nutritional Counselling: Purpose, nutritional counselling provides individuals with personalized guidance on dietary choices and eating behaviors. It helps identify specific nutritional needs and challenges, and supports the development of tailored dietary plans to achieve health goals (Micha et al. 2017). Counselors work with clients to set realistic dietary goals, address barriers to healthy eating, and create actionable plans. Nutritional counselling often involves educating individuals about the benefits of balanced nutrition, portion control, and the importance of a varied diet (Rosen 2005).

Developing a Balanced Diet: A balanced diet includes a variety of foods from all food groups to ensure adequate intake of essential nutrients. This typically involves consuming fruits, vegetables, whole grains, lean proteins, and healthy fats. Emphasizing portion control and moderation helps prevent overeating and supports weight management (Academy of Nutrition and Dietetics 2016).

To develop a balanced diet, individuals should plan meals ahead of time, make healthy food choices when shopping, and cook at home to control ingredients and portion sizes. Incorporating a range of nutrient-dense foods and limiting the intake of processed foods, sugars, and unhealthy fats are key strategies for maintaining a balanced diet (Harvard T.H. Chan School of Public Health 2020).

Behavioral Interventions: Behavioral interventions focus on changing eating behaviors and habits through techniques such as self-monitoring, goal setting, and reinforcement. These interventions can help individuals develop healthier eating patterns and manage weight more effectively (Stice et al. 2006). Engaging in group programs or support groups can provide additional motivation and accountability. Behavioral interventions often include strategies for dealing with emotional eating, managing cravings, and establishing healthier routines around food (Butryn et al. 2011).

24.4 Conclusion

The relationship between stress and eating habits is a multifaceted issue with profound implications for health and well-being. In India, the intersection of traditional dietary practices, modern lifestyle changes, and socio-economic factors highlights the complex nature of this relationship. Rapid urbanization and increased work pressures contribute to elevated stress levels, which in turn influence eating behaviors in significant ways. Stress can manifest in both overeating and undereating, each with distinct health consequences. Stress-induced overeating often leads to the consumption of high-calorie comfort foods, which can contribute to obesity, diabetes, and cardiovascular diseases. Conversely, stress-induced undereating can result in nutritional deficiencies, weight loss, and weakened immune function. The biological mechanisms underlying stress, including hormonal changes and altered reward processing, play a critical role in these behaviors.

Psychological responses to stress, such as emotional eating, mindless eating, and restrained eating, further complicate the interplay between stress and eating habits. Emotional states, including anxiety and depression, significantly impact food choices and eating patterns. Understanding these psychological factors is crucial for developing effective interventions. Addressing the psychological impact of stress on eating habits requires a holistic approach that combines stress management techniques and strategies for promoting healthy eating. Techniques such as mindfulness, cognitive-behavioral therapy, and physical activity can help manage stress effectively. Concurrently, nutritional counselling, balanced diet planning, and behavioral interventions can support healthier eating behaviors. Ultimately, integrating culturally sensitive approaches with modern health strategies can help individuals in India and beyond navigate the challenges of stress and eating habits. By addressing both psychological and physiological aspects, we can develop comprehensive strategies that improve overall well-being, prevent chronic diseases, and enhance quality of life. Understanding and managing the complex interplay between stress and eating habits is essential for fostering a healthier, more balanced approach to both mental and physical health.

References

Academy of Nutrition and Dietetics. (2016). Nutrition basics. Academy of Nutrition and Dietetics

American Psychological Association (2020) Stress effects on the body. https://www.apa.org/topics/stress/body

Beck AT (1979) Cognitive therapy and the emotional disorders. Penguin

Butryn ML, Webb V, Wadden TA (2011) Behavioral interventions for obesity. In: Stern JS (ed) Obesity: a reference handbook. ABC-CLIO, pp 139–155

Carter JC, Kenny TE, Hudson JI (2015) Cognitive-behavioral therapy for binge eating disorder. Curr Psychiatry Rep 17(8):Article 55. https://doi.org/10.1007/s11920-015-0596-4

Carver CS (1997) You want to measure coping but your protocol's too long: consider the brief COPE. Int J Behav Med 4(1):92–100. https://doi.org/10.1207/s15327558ijbm0401_6

Chrousos GP (2009) Stress and disorders of the stress system. Nat Rev Endocrinol 5(7):374–381. https://doi.org/10.1038/nrendo.2009.106

Craft LL, Perna FM (2004) The benefits of exercise for the clinically depressed. Prim Care Companion J Clin Psychiatry 6(3):104–111. https://doi.org/10.4088/pcc.v06n0301

Everly GS, Lating JM (2013) A clinical guide to the treatment of the human stress response, 4th edn. Springer. https://doi.org/10.1007/978-1-4614-5538-7

Fairburn CG, Cooper Z, Shafran R (2003) Enhanced cognitive behavior therapy for adults with bulimia nervosa: a randomized controlled trial. Behav Res Ther 41(5):639–656. https://doi.org/10.1016/S0005-7967(02)00088-8

Fletcher B, Hanson J, Page N, Pine K (2008) The impact of habit formation on diet and exercise behaviors. Am J Lifestyle Med 2(3):165–170. https://doi.org/10.1177/1559827608314149

Garg M, Williams N, Frank L (2007) Stress and eating: a review of the literature. Appetite 48(1): 42–46. https://doi.org/10.1016/j.appet.2006.07.002

Gordon R, Dylan S, Richards M (2014) Quality of life and eating habits: the role of nutrition in health and well-being. J Nutr Health Food Eng 1(2):45–50. https://doi.org/10.15406/jnhfe.2014.01.00009

Goyal M, Singh S, Sibinga EMS, Gould NF, Rowland-Seymour A, Sharma R et al (2014) Meditation programs for psychological stress and well-being: a systematic review and meta-analysis. JAMA Intern Med 174(3):357–368. https://doi.org/10.1001/jamainternmed.2013.13018

Gross JJ (2002) Emotion regulation: affective, cognitive, and social consequences. Psychophysiology 39(3):281–291. https://doi.org/10.1017/S0048577201393198

Harvard T.H. Chan School of Public Health (2020) Healthy eating plate. https://www.hsph.harvard.edu/nutritionsource/healthy-eating-plate/

Higgins ET (2007) The self in social psychology: the role of self and social identity in health. Soc Personal Psychol Compass 1(1):1–17. https://doi.org/10.1111/j.1751-9004.2007.00001.x

Hofmann SG, Asnaani A, Vonk IJ, Sawyer AT, Fang A (2012) The efficacy of cognitive behavioral therapy: a review of meta-analyses. Cogn Ther Res 36:427–440. https://doi.org/10.1007/s10608-012-9476-1

Hudson JI, Hiripi E, Pope HG Jr, Kessler RC (2007) The prevalence and correlates of binge eating disorder in the community. Int J Eat Disord 40(5):472–479. https://doi.org/10.1002/eat.20322

Jacka FN, O'Neil A, Opie R, Itsiopoulos C, Cotton S, Mohebbi M et al (2017) The association between diet and depression: a review of the literature. Aust N Z J Psychiatry 51(3):272–279. https://doi.org/10.1177/0004867416686780

Johnson JL, Larson JJ, Kaye WH (2008) Neurobiological factors in bulimia nervosa. Curr Opin Psychiatry 21(6):659–665. https://doi.org/10.1097/YCO.0b013e3283119259

Kaye WH, Bulik CM, Thornton L, Barbarich N, Masters K (2001) Stress and eating disorders: a review. J Clin Psychiatry 62(Suppl 10):52–58

Keel PK, Mitchell JE (1997) Outcome of anorexia nervosa: a review of the literature. J Clin Psychiatry 58(Suppl 4):161–165

Kiecolt-Glaser JK, Preacher KJ, MacCallum RC, Atkinson C, Malarkey WB, Glaser R (2003) Chronic stress and age-related increases in the proinflammatory cytokine IL-6. Proc Natl Acad Sci 100(15):9090–9095. https://doi.org/10.1073/pnas.1531903100

Lazarus RS (1984) Stress, appraisal, and coping. Springer

Macht M (2008) How emotions affect eating: a five-way model. Appetite 50(1):1–11. https://doi.org/10.1016/j.appet.2007.07.002

McEwen BS (2008) Central effects of stress hormones in health and disease: understanding the protective and damaging effects of stress and stress mediators. Eur J Pharmacol 583(2–3):174–185. https://doi.org/10.1016/j.ejphar.2007.11.071

Micha R, Peñalvo JL, Cudhea F, Imamura F, Rehm CD, Mozaffarian D (2017) Association between dietary factors and mortality from heart disease, stroke, and type 2 diabetes in the United States. JAMA 317(9):912–924. https://doi.org/10.1001/jama.2017.0947

Oliver G, Wardle J, Gibson EL (2000) Stress and emotional eating. Appetite 35(2):199–209. https://doi.org/10.1006/appe.2000.0369

Rosen JC (2005) Nutritional counseling and management. In: Preedy JC, Watson RR, Patel VR (eds) Nutrition in clinical practice. Springer, pp 347–358. https://doi.org/10.1007/0-387-24642-1_25

Salmon P (2001) Effects of physical exercise on anxiety, depression, and sensitivity to stress: a unifying theory. Clin Psychol Rev 21(1):33–61. https://doi.org/10.1016/S0272-7358(99)00032-X

Sapolsky RM (2004) Why zebras don't get ulcers: the acclaimed guide to stress, stress-related diseases, and coping. Holt Paperbacks

Skerrett PJ, Willett WC (2010) Dietary fat and cardiovascular disease: a summary of the evidence. Nutr Rev 68(3):104–114. https://doi.org/10.1111/j.1753-4887.2010.00287.x

Stice E, Fisher M, Lowe MR (2004) Relation of dieting to binge eating, purging, and body dissatisfaction: a longitudinal study. J Abnorm Psychol 113(2):150–155. https://doi.org/10.1037/0021-843X.113.2.150

Stice E, Presnell K, Shaw H (2006) Mechanisms of action of behavioral weight control programs. In: Bray GA, Bouchard C (eds) Handbook of obesity. CRC Press, pp 521–551

Stice E, Spoor S, Bohon C, Small DM (2011) The role of reward sensitivity in the development of obesity: a review. Behav Res Ther 49(11):847–856. https://doi.org/10.1016/j.brat.2011.08.001

Tian H, Wang J, Chen J (2012) Stress, eating, and the brain's reward system: insights into stress-induced food intake. Front Psychol 3:Article 219. https://doi.org/10.3389/fpsyg.2012.00219

Torres SJ, Nowson CA (2007) Relationship between stress, eating behavior, and obesity. Nutrition 23(11–12):887–894. https://doi.org/10.1016/j.nut.2007.08.008

Chapter 25
Culinary Psychology: Exploring the Interplay Between Food Practices and Psychological Well-Being

Manoj Kumar, Sonakshi Ruhela, Siddharth Soni, and Shivanjali Shrivastava

Abstract This chapter aims to highlight the significance of meal choices and food intake for behavioral responses and psychological well-being. Furthermore, by educating the public on the connection between diet and mental health, negative consequences like depression can be prevented by encouraging healthy eating choices. Globally, there are substantial changes in the way people eat. Along with the spread of the range and severity of mental illnesses that result in violent acts, there are also notable changes in mood, including melancholy, happiness, and violence. Over the past 10 years, as food options have gotten more varied and simpler to get, food intake and meal choices have changed. These changes could have both positive and negative effects. The relationship between food intake and psychological well-being is explored in this chapter. This chapter aims to prove or disprove the idea that psychological health is influenced by eating, or vice versa. Different diets can affect a person's brain and mental health, which can affect behavioral and psychological responses. Numerous studies have shown that food insecurity negatively affects psychological health and well-being, which can result in violence, marital discord, hopelessness, and loss of happiness. Flavonoids and herbal extracts, for instance, may help treat mood disorders and enhance gut flora. Guidance for mood and cognitive function therapies may come from an understanding of the communication between the gut and brain axis. This book chapter focuses on lowering the cost of treating and preventing diseases and improving the psychological health of the general public, as the majority of diseases and behaviors have a strong dietary component.

M. Kumar (✉) · S. Shrivastava
Amity Centre of Behavioural Science, Amity University, Noida, India
e-mail: mmathur@amity.edu; sshrivastava3@amity.edu

S. Ruhela
Rochester Institute of Technology, Dubai, United Arab Emirates
e-mail: Sorcad@rit.edu

S. Soni
School of Psychology and Counselling, O.P. Jindal Global University, Sonipat, Haryana, India

M. Thakur, T. Belwal (eds.), *Food Biodiversity and Consumer Engagements*, Sustainable Development and Biodiversity 46,
https://doi.org/10.1007/978-981-95-7777-4_25

Keywords Food choice · Behavioral responses · Psychological Well-being · Gut–Brain axis · Nutrients · Depression · Mental health

25.1 Introduction

The onset of adolescence triggers significant changes, notably a rapid acceleration in growth. This surge in growth is fueled by hormonal activity impacting every organ in the body, underscoring the significance of healthy eating. Nutritional necessities increase steadily during childhood, peak during adolescence, and then stabilize or decrease as individuals transition into adulthood. The adage "You are what you eat" holds, with compelling evidence supporting a direct association between diet and mental well-being. While navigating research on "nutrition and mental health" can be overwhelming, the good news is that the connection between food and mental well-being is simpler than often perceived. It largely revolves around a return to basics, emphasizing simple, nutritious foods for brain health and fostering beneficial eating habits.

Food encompasses any solid or liquid substance that, upon ingestion, digestion, and absorption by the body, furnishes crucial nutrients to sustain health. It constitutes a fundamental requirement for life, supplying energy for bodily functions, facilitating progression and tissue repair, stimulating the body against illness, and regulating physiological processes.

Nutrition is the discipline encompassing the study of foods, their constituent nutrients, and other substances they harbor, along with their physiological processes inside the body, including absorption, ingestion, digestion, metabolism, and excretion. Whereas this description encapsulates its physiological aspects, nutrition extends beyond mere bodily functions, incorporating social, psychological, and economic dimensions as well.

Nutrients are the essential components found in food that must be provided to the body in appropriate quantities. These encompass proteins, carbohydrates, minerals, fats, water, vitamins and fiber. Maintaining health necessitates a diverse array of nutrients. Many foods contain multiple nutrients; for instance, milk contains proteins, fats, and more. Nutrients are categorized into macronutrients and micronutrients based on the daily intake required. The following diagram illustrates the differentiation between macronutrients and micronutrients (Thakur and Modi 2020).

According to the World Health Organization (WHO), "Health is the state of complete physical, mental, and social well-being, not merely the absence of diseases or infirmity." This definition, established in 1948, remains unchanged. While physical health is readily understood, mental health encompasses emotional and psychological well-being, empowering individuals to utilize their emotional and cognitive capacities, engage effectively in society, and fulfill the everyday demands of life. Simply lacking a recognized mental disorder does not guarantee good mental health. A person's mental health can be evaluated by assessing their ability to function effectively and successfully. Indicators of good mental health include feeling competent and capable, managing normal levels of stress, maintaining fulfilling relationships, leading an independent life, and demonstrating resilience in overcoming challenging circumstances (Thakur et al. 2024a, b).

A sustainable approach also emphasizes the importance of minimizing environmental impact while maximizing nutritional value. This involves adopting practices such as organic farming, reducing food waste, and promoting local and seasonal foods. By focusing on nutrient-dense foods, such as fruits, vegetables, whole grains, and lean proteins, we can ensure a balanced diet that supports both personal health and the well-being of the planet. Sustainable diets not only help preserve natural resources but also enhance food security and reduce greenhouse gas emissions, contributing to a healthier future for all (Thakur et al. 2024a, b).

The relationship between nutrition and mental health is mutually influential: our dietary choices impact our mental well-being, and conversely, our mental health status can influence what and how we eat. The American Society for Nutrition (ASN) and the American Psychiatric Association (APA) are collaborating to emphasize the interconnected nature of nutrition and mental health. To gain insights into public perceptions regarding the association between diet and mental well-being, the APA surveyed between March 16 and 17, 2023, polling a sample of 2200 adults. The findings revealed that:

- Two-thirds (66%) of surveyed adults expressed confidence in their understanding of the connection between diet and mental health.
- A significant majority (81%) of adults indicated a willingness to modify their diet in a manner that positively influences their mental health.
- Furthermore, four in ten respondents (43%) expressed a high level of willingness to adjust their diet specifically to enhance mental well-being.

An expanding body of research underscores the mental health advantages of maintaining a healthy diet. Numerous studies have highlighted the efficacy of dietary interventions in alleviating depression, with particular emphasis on the benefits of adhering to the Mediterranean Diet. A comprehensive review published in The American Journal of Clinical Nutrition in 2019 revealed that heightened consumption of fruits and vegetables yields positive effects on psychological well-being (Tuck et al. 2019). Additionally, daily intake of vegetables demonstrates therapeutic potential in mitigating symptoms of depression among individuals diagnosed with clinical depression. A balanced diet rich in vitamins, minerals, healthy fats, and fiber sourced from fruits, vegetables, whole grains, nuts, and seeds holds the capacity to mitigate inflammation and modulate neurotransmitter activity, thereby ameliorating depressive symptoms.

Increased consumption of healthy foods such as vegetables and fruits is associated with improved mood and mental well-being. Over the past decade, there has been a rise in the intake of fruits, vegetables, salads, and dairy products (Liu et al. 2016). Conversely, the proliferation of fast-food establishments in public spaces, including workplaces and neighborhoods, has led to increased accessibility and consumption of fast food across all age groups (James et al. 2017). Scholars have examined consumers' perceptions regarding the impact of food choices on happiness and mood fluctuations. It is common for individuals to alter their food preferences or choices based on transient psychological states, such as turning to "comfort foods" during times of sadness or experiencing changes in appetite in response to stress. However, maintaining a balanced diet poses challenges, particularly for individuals with mental disorders, due to economic constraints, environmental factors, and side effects of psychiatric medications (Spencer and Deak 2017). Evidence supporting the positive effects of dietary changes on depressive symptoms could incentivize resources to promote healthy living behaviors and prevent mental illness, thus reducing societal costs associated with mental health expenditures (Ljungberg et al. 2020). Sedentary lifestyle, smoking, substance abuse, and unhealthy eating habits have all been linked to an elevated risk of depression, with stronger associations observed when these factors are combined (Berk et al. 2013). Psychological factors influence food choices in both the short and long term, with specific nutrients known to impact biological processes underlying cognition, emotion, and behavior. Gender and personality play significant roles in spicy food consumption, with preferences for spicy foods associated with higher levels of anger. Studies suggest that eating spicy foods may reflect a thrill-seeking behavior with a social component, particularly among American college students, where a preference for spicy meals correlates positively with high sensation-seeking traits (Byrnes and Hayes 2013).

25.2 Background

This narrative review explores the significant impact of food consumption and meal choices on brain function and mental health, encompassing psychological and behavioral responses such as mood regulation, interpersonal relationships, and aggression. It underscores the necessity for further research due to global shifts in dietary behaviors, fluctuations in mood, rising rates of mental illness, and instances of violence. The proliferation of diverse and accessible food options in recent years may have both advantageous and detrimental effects on mental well-being and behavior. The adverse consequences of food insecurity on psychological welfare, including feelings of despair, diminished happiness, marital discord, and violent behavior, underscore the urgency of comprehending the nexus between food and mental health. Examining the influence of various dietary patterns and food accessibility on these outcomes can yield valuable insights. The review stresses the importance of elucidating the interplay between mental health and food consumption, behavior, and societal outcomes. Additionally, it highlights the potential use of herbal extracts and flavonoids as interventions to enhance gut microbiota and address mood disorders, underscoring the need for further investigation. Understanding the communication within the gut-brain axis and the impact of dietary components on this communication can inform interventions targeting mood and cognitive function. Addressing these issues can mitigate treatment expenses and curb crime and delinquency at the community level. Proactive measures through mindful food choices are imperative. Continued research in this area is essential for developing interventions that encourage positive mental health results and potentially prevent harmful behavioral consequences, including criminal behavior.

25.3 Determinants of Food Choices

- *Biological Factors*: Biological factors such as hunger, appetite, and taste preferences significantly impact food choices. Different nutrients elicit varying levels of satisfaction upon consumption, with fat being the least satiating, followed by carbohydrates and then protein. Palatability, influenced by taste, smell, texture, and appearance, often drives food consumption for pleasure rather than solely for nutritional purposes.
- *Economic Factors*: The cost of food and individual income are key economic determinants of food choices. Low-income groups often face challenges in maintaining balanced diets due to affordability issues. Interestingly, an increase in income does not always lead to healthier dietary choices, and concerns about food waste may deter individuals from trying new foods.
- *Physical Factors*: Physical factors such as access to food outlets, education, skills, and time constraints also shape food choices. The proximity of shops and availability of food options, along with transportation links, influence accessibility. While education on healthy eating is essential, a lack of knowledge and cooking skills can impede the adoption of healthier diets. Time constraints often result in the selection of convenient foods despite their higher costs.
- *Social Factors*: Social class, culture, and social context exert significant influence on food choices. Dietary habits often vary across different social classes, with higher classes typically exhibiting healthier dietary patterns. Cultural influences affect both food choices and preparation methods, which may undergo changes with migration. Additionally, social context, including interactions with peers and family, directly and indirectly shapes food choices.
- *Psychological Factors*: Psychological factors, such as mood, stress, and guilt, also impact food choices. Stress can lead to both overeating and undereating, while mood influences cravings and food preferences. Feelings of guilt often follow indulgence or restriction attempts, subsequently influencing future food decisions.
- *Attitudes, Beliefs, and Knowledge*: Attitudes, beliefs, and knowledge about food, alongside optimistic bias, play a crucial role in shaping food choices. Various factors, including perceived quality, price, taste, healthiness, and family preferences, contribute to food decision-making. These factors may vary based on individual characteristics such as gender, age, and education level.

25.4 Effect of Foods on Mood

The effect of different foods on mood can vary widely based on several factors, including nutritional content, individual metabolism, and personal preferences. Here are some general effects that certain foods can have on mood:

25.4.1 Chocolate and Sweets

Research indicates that consuming sweets and sugary meals can elevate mood, with women showing a greater susceptibility to this effect than men (Thakur and Belwal 2023). Chocolate, in particular, is renowned for evoking feelings of happiness and alleviating tension (Meier et al. 2017). This mood-lifting effect can be attributed to psychoactive chemicals present in chocolate, such as anandamides, which target opioid receptors in the central nervous system, akin to the effects of cannabis. Additionally, compounds like tyramine and phenylethylamine act similarly to amphetamine (Parker et al. 2006).

However, it's essential to approach chocolate consumption with moderation. Research suggests that excessive chocolate intake may lead to heightened anxiety over time, while moderate-to-low consumption is associated with improved mood and increased happiness (Moreno-Dominguez et al. 2012). Consuming high-cacao breakfast foods, such as high-cacao chocolate or biscuits, can induce feelings of satisfaction and reduce the likelihood of overeating later in the morning.

Despite its potential mood-enhancing effects, some individuals participating in weight-loss programs may experience negative emotions after consuming chocolate. Thus, it's important to strike a balance and consume chocolate in moderation to avoid adverse outcomes such as anxiety and guilt.

25.4.2 Vegetables and Fruits

The consumption of fruits and vegetables is associated with numerous benefits for mood, cognitive function, and psychological well-being in women. These foods are rich in essential vitamins and minerals, contributing to the daily recommended intake necessary for maintaining good health and preventing disease. Research indicates that individuals who regularly consume a variety of fruits and vegetables, encompassing a range of colors, exhibit higher levels of happiness, imagination, and curiosity (Conner et al. 2015).

Furthermore, epidemiological studies suggest that the intake of flavonoids found in fruits and vegetables may decrease the risk of developing depression. Flavonoids have been shown to enhance cognitive processes associated with executive function, thereby reducing depressive cognitive patterns and elevating mood (AIAmmar et al. 2020). Therefore, incorporating a diverse array of fruits and vegetables into one's diet can contribute to improved mental and emotional well-being.

25.4.3 *Caffeine*

Caffeine is often regarded as a ubiquitous stimulant that can influence daily mood and mental health. As a central nervous system stimulant, it temporarily increases activity, providing a boost in alertness and reducing fatigue. However, chronic caffeine users may experience withdrawal symptoms when faced with a temporary shortage of caffeine. Research has shown that consuming caffeine in the morning can significantly alleviate morning anxiety and enhance performance (Dominguez et al. 2021; Souissi et al. 2012).

Regular consumption of caffeine, such as in coffee, has been associated with a reduced likelihood of experiencing depression (Lucas et al. 2011). However, excessive caffeine consumption can lead to negative side effects, including insomnia, headaches, restlessness, and stress (Rogers 2007). Some researchers have suggested reducing caffeine intake due to a lack of significant mood elevation benefits (James and Gregg 2004). Additionally, individuals aged 45 and older may be more susceptible to the stress-inducing effects of caffeine (James and Gregg 2004). While caffeine consumption can have both positive and negative effects on mood, performance, and behavior, moderation is key to minimizing potential adverse outcomes (Malone et al. 2015).

25.4.4 *Oilseeds and Seafood*

It has been demonstrated that omega-3 fatty acids, such as alpha-linolenic acid (ALA), docosahexaenoic acid (DHA), and eicosapentaenoic acid (EPA), have positive benefits on behavior, mood, neuroticism, and impulse control. The main sources of these healthy acids are oilseeds, fish, and other marine foods. Numerous mental health problems, including schizophrenia, substance abuse, major depressive disorder, bipolar disorder, and attention deficit disorder, have been shown to benefit from omega-3 fatty acids (Stahl et al. 2008). Mood problems have been linked to deficiencies in EPA and DHA, both of which are necessary for brain development and function. Depression has been associated with low blood levels of these fatty acids (Pawels and Volterrani 2008). Thus, eating foods high in omega-3 fatty acids, including fish and oilseeds, can help one maintain mental health and general well-being. In contrast, a lack of these fatty acids may be linked to mood disorders and decreased depression.

25.4.5 *Spicy Foods*

Despite the potential discomfort, irritability, and even pain caused by consuming foods containing capsaicin, such as hot peppers, many individuals worldwide, particularly men, enjoy doing so. Sociological observations suggest that the ability of spicy foods to elicit these unpleasant physiological responses has led to an association between these meals and qualities commonly associated with masculinity in various cultural contexts (Tornwall et al. 2012).

Research has explored the association between a preference for spicy food and endogenous testosterone, a hormone often linked to stereotypically masculine preferences and behaviors. Additionally, various factors, including genetic predispositions, psychological factors, physiological responses, and social influences, contribute to individuals' preferences for and consumption of foods containing capsaicin (Erdost et al. 2007).

Consuming spicy foods has been associated with traits such as power, bravery, and masculinity, with personality types and gender playing significant roles in this association. Additionally, preferences for spicy foods have been linked to advanced levels of anger and sensation-seeking behavior. Hormonal factors, such as testosterone, have also been implicated in dietary preferences, including behaviors related to aggression, social dominance, novelty seeking, and daring behavior.

25.4.6 Food and Depression

Depression stands as one of the most prevalent psychiatric conditions affecting mood, emotion, and behavior across diverse populations, regardless of age, background, or cultural context. Emotional eating, often stemming from depression, can profoundly impact dietary choices. Despite the reduction in physically demanding jobs due to technological advancements, psychological stress has increased in today's society. Managing daily affairs, including family and work responsibilities, can be challenging for individuals grappling with mental illness, and seeking treatment may be hindered by feelings of guilt and shame (Berk et al. 2013).

Three clinical recommendations were derived from a narrative review that summarized the body of data on lifestyle choices in the treatment of depression. First, smoking, substance abuse, poor eating, and physical inactivity were found to be lifestyle factors that increase the risk of depression; their co-occurrence further reinforced this association (Berk et al. 2013). Second, it has been demonstrated that physical activity reduces depressive symptoms. Finally, despite the paucity of evidence, suggesting dietary modifications and offering assistance in quitting smoking have demonstrated potential as depression treatments (Ljungberg et al. 2020).

Emotional eating, which is frequently brought on by depression, involves consuming sweet and high-fat foods, which can momentarily elevate mood by stimulating the reward regions of the brain (Konttinen et al. 2010). According to research, those who are depressed and emotional eaters typically eat more sweets, whereas people who are not depressed tend to eat fewer fruits and vegetables. Therefore, the relationship between grief and emotional eating has a big impact on dietary decisions, frequently leading to the intake of harmful foods.

25.4.7 *Food and Happiness*

Certain eating habits, such as calorie tracking, dieting, and restricting particular foods, have historically been categorized as harmful by researchers, the media, and regulators. According to Cornil and Chandon (2016), health intervention programs frequently encourage consumer behavior that puts health advantages ahead of the momentary pleasure of decadent foods. Restrictive eating habits and diets, however, are frequently unproductive and may even be a contributing factor to eating disorders and long-term weight gain.

According to Renner et al. (2012), a new viewpoint proposes moving away from considering food only as energy and toward its function in fostering happiness and human eating habits. In this regard, Block (2011) argues for a paradigm change from "food as health" to "food as well-being." Making "good" dietary choices, such as eating more fruits and vegetables, can improve mental and physical health and may be a long-term investment in general well-being, according to a number of studies.

In a nationally representative panel study of over 12,000 Australian people, for instance, Mujcic et al. (2016) discovered that eating fruits and vegetables over 2 years predicted gains in happiness, life satisfaction, and general well-being. This review questions the widely held notion that eating "unhealthy" meals that are heavy in calories, fat, or sugar makes you happier.

25.5 Healthy Lifestyle and Physical Activity

Several studies have highlighted a negative relationship between physical activity and dietary fat intake among both men and women. For instance, data from the Behavioral Risk Factor Surveillance System revealed that inactive men consumed a higher percentage of their calories from fat (37.1%) compared to men who engaged in regular and intense exercise (30.2%) (Simoes et al. 1995). Similarly, inactive women consumed a higher percentage of their calories from fat (30.3%) compared to active women (24.6%).

However, findings from studies comparing marathon runners or team athletes with the general population have yielded mixed results, with some studies supporting and others opposing the inverse association between physical activity and dietary fat intake. Additionally, randomized trials of exercise have shown little impact on dietary fat intake, although occasional increases in carbohydrate consumption have been observed (Wood et al. 1985). These discrepancies suggest that individuals who choose to be more physically active may also opt for healthier dietary patterns, whereas increased exercise may not necessarily influence diet choices.

Despite this, increased physical activity consistently emerges as a crucial factor in the long-term maintenance of weight loss. Studies, including randomized controlled trials and cross-sectional comparisons, have consistently shown that individuals who maintain weight loss over the long term are more likely to engage in regular physical activity (Klem et al. 1997; Pronk and Wing 1994). The National Weight Control Registry, which includes over 750 successful weight losers, further supports this notion. Participants in this registry, who have lost an average of 60 pounds and maintained it for about 5 years, reported expending a significant amount of energy through physical activity, with approximately 25% coming from high-intensity activities (Contento et al. 1995).

25.6 The Connection Between Nutrition and Psychological Well-Being

The significance of diet in promoting both physical and psychological well-being cannot be emphasized enough. Take, for example, the Mediterranean diet, which has been linked to a notable 32% decrease in the risk of depression. In today's fast-paced society, we often face numerous stressors that can deeply affect our mental health. Despite these challenges, we are constantly striving to make nutritious food choices to support our physical well-being. This underscores the profound connection between nutrition and mental health. It's imperative to further explore how our dietary habits can impact our emotional and psychological well-being. By understanding this relationship more deeply, we can better prioritize our mental health alongside our physical health through mindful food choices.

Fatty Acid

Incorporating essential fatty acids, such as omega-3 and omega-6, into our diet plays a pivotal role in maintaining mental well-being. These fatty acids are vital for our body's functions but cannot be produced internally, making their intake through diet or supplements essential. Diets that are abundant in essential fatty acids often emphasize foods like fish, nuts, seeds, and plant-based oils such as flaxseed, soybean, and canola oil. By including these nutrient-rich sources in our meals, we can ensure adequate intake of essential fatty acids, thereby supporting our mental health and overall well-being.

Vitamin D

Vitamin D, often referred to as the "sunshine vitamin" for its synthesis in the body upon exposure to sunlight, serves a crucial role beyond just promoting a sun-kissed complexion. This nutrient is essential for enhancing mental health, with insufficient levels associated with conditions like depression, anxiety, and mood disorders.

Vitamin D plays a key role in neurotransmitter production in the brain, including serotonin and dopamine, which are vital for mood regulation. To ensure sufficient vitamin D intake, it's important to incorporate foods rich in these nutrients into your diet. Excellent sources include fatty fish like salmon and mackerel, fortified foods such as milk and cereal, cheese, egg yolks, and beef liver.

For individuals managing sobriety and depression, recognizing the benefits of a vitamin D-rich diet is crucial. Such dietary choices can help reduce cravings and improve overall mood, supporting efforts toward recovery.

In addition to vitamin D, essential fatty acids are also essential for mental health, playing a pivotal role in brain function. Research indicates that inadequate intake of essential fatty acids can contribute to various mental health issues, including depression, anxiety, and ADHD. Studies also suggest that increasing omega-3 consumption may alleviate symptoms in conditions like schizophrenia and bipolar disorder. Therefore, prioritizing a diet rich in essential fatty acids can be beneficial for mental well-being.

25.7 Super Food Improving Mental Health

When discussing superfoods, the focus often centers on their physical health benefits, but some also offer specific advantages for mental well-being. These foods contain unique nutrients that optimize brain function and contribute to feelings of happiness and relaxation.

- *Blueberries*: Despite their small size, blueberries are rich in antioxidants that combat brain aging and enhance memory. They can be easily incorporated into your diet by adding them to morning cereal, blending them into a smoothie, or enjoying them as a standalone snack.
- *Avocados*: Avocados are packed with healthy fats that support brain health and cognitive function. Adding avocado to your meals, such as spreading it on whole-grain toast for breakfast, can help keep your brain sharp and focused.
- *Protein sources*: Lean meats, fish, eggs, and beans are crucial for brain health as they contain tryptophan, an amino acid that aids in serotonin production. Serotonin is a neurotransmitter that promotes feelings of well-being and happiness. Including these protein sources in your diet can support mood regulation and overall mental health.

While dieting alone cannot cure mental health issues, it plays a significant role in supporting brain function and mood regulation. A healthy diet, incorporating top superfoods like blueberries, avocados, and protein sources, can complement other treatment strategies for individuals dealing with depression and other mental health conditions.

25.8 Gut–Brain Axis

The gut–brain axis emphasizes the connection between our digestive system and mental health. While "trust your gut" is commonly associated with decision-making, it also reflects how gut health affects overall well-being, including mental health.

This axis functions as a two-way communication system between the brain and the digestive tract, with ongoing interaction. The gut houses millions of bacteria crucial for regulating mood and mental well-being. A balanced and healthy gut microbiota can positively impact mood, while an imbalance may lead to anxiety, stress, and depression.

Taking proactive measures to maintain gut health through a balanced diet can significantly influence mental well-being. By ensuring a balanced and healthy gut microbiota, we can support our mental health and overall wellness. This underscores the importance of the gut–brain axis in shaping our mental and emotional state.

25.9 The Link Between Nutrition and Mental Health

Recognizing the connection between nutrition and mental health is essential. Adding foods such as avocados and blueberries to your diet can have a positive effect on mental well-being, along with consuming healthy fats and essential vitamins. While a healthy diet isn't a standalone solution for mental illness, it is a crucial component of a holistic treatment approach.

If you or someone you care about is struggling with addiction or looking to improve mental health, our facility offers comprehensive mental health care services. Visit our Mental Health page to explore the programs we provide and take the first step toward recovery and well-being. Don't hesitate to reach out and seek the support you need.

25.10 Discussion

The discussion elaborates on the relationship between food and psychological well-being and the importance of dietary choices in maintaining overall well-being. From the onset of adolescence, when rapid growth occurs and nutrient needs peak, to adulthood, where dietary habits play a crucial role in mental health, the impact of nutrition is profound. Research highlights the bidirectional nature of this relationship, with diet influencing mental health and mental health status affecting dietary habits.

Various studies have proved the significant role of specific nutrients in promoting psychological well-being. For instance, omega-3 fatty acids, found in fish and oilseeds, have been linked to improved behavior, mood regulation, and cognitive function (Stahl et al. 2008). Similarly, vitamin D, essential for neurotransmitter production, has been associated with depression, anxiety, and mood disorders. Moreover, antioxidants in foods like blueberries combat brain aging and enhance memory, contributing to overall mental well-being.

Furthermore, the gut–brain axis serves as a critical link between diet and mental health. The balance of gut microbiota influences mood regulation, with an imbalance potentially leading to feelings of anxiety, stress, and depression. Thus, maintaining gut health through a balanced diet is essential for supporting psychological well-being.

While diet alone cannot cure mental illness, it forms an integral part of a holistic treatment plan. Incorporating nutrient-rich foods like avocados, blueberries, and lean proteins into one's diet can positively impact mental health (Gibson 2006). This underscores the importance of dietary interventions alongside other therapeutic approaches in managing mental health conditions.

25.11 Conclusion

The relationship between food and psychological well-being is undeniable. As evidenced by numerous studies, dietary choices play a significant role in influencing mood, cognitive function, and overall mental well-being. From omega-3 fatty acids to antioxidants and essential vitamins, specific nutrients are crucial for promoting mental health.

Understanding the bidirectional relationship between diet and mental health is essential for developing effective treatment strategies. While diet alone cannot replace medical interventions, it complements other therapeutic approaches, forming a crucial component of holistic care. By incorporating nutrient-rich foods into one's diet and maintaining gut health, individuals can support their psychological well-being and enhance their overall quality of life.

In light of these findings, it is imperative to raise awareness about the importance of nutrition in mental health care. Providing access to comprehensive mental health care services that integrate dietary interventions can empower individuals to take proactive steps towards recovery and well-being. Together, by acknowledging the significance of nutrition in mental health and implementing evidence-based interventions, we can work towards promoting mental wellness in our communities.

References

AlAmmar WA, Albeesh FH, Khattab RY (2020) Food and mood: the corresponsive effect. Curr Nutr Rep 9(3):296–308

Berk M, Sarris J, Coulson CE, Jacka FN (2013) Lifestyle management of unipolar depression. Acta Psychiatr Scand 127:38–54

Block LG, Grier SA, Childers TL, Davis B, Ebert JE, Kumanyika S et al (2011) From nutrients to nurturance: A conceptual introduction to food well-being. J Public Policy Mark 30(1):5–13

Byrnes NK, Hayes JE (2013) Personality factors predict spicy food liking and intake. Food Qual Prefer 28(1):213–221

Conner TS, Brookie KL, Richardson AC, Polak MA (2015) On carrots and curiosity: eating fruit and vegetables is associated with greater flourishing in daily life. Br J Health Psychol 20(2):413–427

Contento I, Balch GI, Bronner YL, Lytle LA, Maloney SK, Olson CM, Swadener SS (1995) The effectiveness of nutrition education and implications for nutrition education policy, programs, and research: a review of research

Cornil Y, Chandon P (2016) Pleasure as an ally of healthy eating? Contrasting visceral and Epicurean eating pleasure and their association with portion size preferences and wellbeing. Appetite 104:52–59

Domínguez R, Veiga-Herreros P, Sánchez-Oliver AJ, Montoya JJ, Ramos-Álvarez JJ, Miguel-Tobal F et al (2021) Acute effects of caffeine intake on psychological responses and high-intensity exercise performance. Int J Environ Res Public Health 18(2):584

Erdost H, Ozfiliz N, Ozguden C, Gunes N, Onen S, Ilhan T, Ozer A (2007) Expression of a capsaicin receptor (VR1) in the testes of mice after an application of capsaicin. Bull Vet Inst Pulawy 51(4):649

Gibson EL (2006) Emotional influences on food choice: sensory, physiological and psychological pathways. Physiol Behav 89(1):53–61

James JE, Gregg ME (2004) Effects of dietary caffeine on mood when rested and sleep restricted. Hum Psychopharmacol Clin Exp 19(5):333–341

James P, Seward MW, James O'Malley A, Subramanian SV, Block JP (2017) Changes in the food environment over time: examining 40 years of data in the Framingham Heart Study. Int J Behav Nutr Phys Act 14:1–9

Klem ML, Wing RR, McGuire MT, Seagle HM, Hill JO (1997) A descriptive study of individuals successful at long-term maintenance of substantial weight loss. Am J Clin Nutr 66(2):239–246

Konttinen H, Männistö S, Sarlio-Lähteenkorva S, Silventoinen K, Haukkala A (2010) Emotional eating, depressive symptoms and self-reported food consumption. A population-based study. Appetite 54(3):473–479

Liu X, Yan Y, Li F, Zhang D (2016) Fruit and vegetable consumption and the risk of depression: a meta-analysis. Nutrition 32(3):296–302

Ljungberg T, Bondza E, Lethin C (2020) Evidence of the importance of dietary habits regarding depressive symptoms and depression. Int J Environ Res Public Health 17(5):1616

Lucas M, Mirzaei F, Pan A, Okereke OI, Willett WC, O'Reilly ÉJ et al (2011) Coffee, caffeine, and risk of depression among women. Arch Intern Med 171(17):1571–1578

Malone RM, Giles K, Maloney NG, Fyfe CL, Lorenzo-Arribas A, O'Connor DB, Johnstone AM (2015) Effects of stress and mood on caffeine consumption in shift and non-shift workers. Proc Nutr Soc 74(OCE1):E135

Meier BP, Noll SW, Molokwu OJ (2017) The sweet life: the effect of mindful chocolate consumption on mood. Appetite 108:21–27

Moreno-Dominguez S, Rodríguez-Ruiz S, Martín M, Warren CS (2012) Experimental effects of chocolate deprivation on cravings, mood, and consumption in high and low chocolate-cravers. Appetite 58(1):111–116

Mujcic R, Oswald J, A. (2016) Evolution of well-being and happiness after increases in consumption of fruit and vegetables. Am J Public Health 106(8):1504–1510

Parker G, Parker I, Brotchie H (2006) Mood state effects of chocolate. J Affect Disord 92(2–3):149–159

Pawels EK, Volterrani D (2008) Fatty acid facts, part I. Essential fatty acids as treatment for depression, or food for mood? Drug News Perspect 21(8):446–451

Pronk NP, Wing RR (1994) Physical activity and long-term maintenance of weight loss. Obes Res 2(6):587–599

Renner B, Sproesser G, Strohbach S, Schupp HT (2012) Why we eat what we eat. The Eating Motivation Survey (TEMS). Appetite 59(1):117–128

Rogers PJ (2007) Caffeine, mood and mental performance in everyday life. Nutr Bull 32:84–89

Simoes EJ, Byers T, Coates RJ, Serdula MK, Mokdad AH, Heath GW (1995) The association between leisure-time physical activity and dietary fat in American adults. Am J Public Health 85(2):240–244

Souissi M, Abedelmalek S, Chtourou H, Atheymen R, Hakim A, Sahnoun Z (2012) Effects of morning caffeine' ingestion on mood states, simple reaction time, and short-term maximal performance on elite judoists. Asian J Sports Med 3(3):161

Spencer RL, Deak T (2017) A users guide to HPA axis research. Physiol Behav 178:43–65

Stahl LA, Begg DP, Weisinger RS, Sinclair AJ (2008) The role of omega-3 fatty acids in mood disorders. Curr Opin Investig Drugs 9(1):57–64

Thakur M, Belwal, T. (2023) Bioactive components. Springer, Singapore https://doi.org/10.1007/978-981-19-2366-1_34

Thakur M, Modi VK (2020) Emerging technologies in food science – focus on the developing world. Springer Nature

Thakur M, Ferrante C, Sanches Silva A (2024a) Editorial: sustainable development in the agro-food and nutrition sector. Front Nutr 11:1419937. https://doi.org/10.3389/fnut.2024.1419937

Thakur M, Modi VK, Feng S (2024b) Editorial: food sustainability: challenges and opportunities for the future. Front Nutr 10:1341427. https://doi.org/10.3389/fnut.2023.1341427

Törnwall O, Silventoinen K, Kaprio J, Tuorila H (2012) Why do some like it hot? Genetic and environmental contributions to the pleasantness of oral pungency. Physiol Behav 107(3):381–389

Tuck NJ, Farrow C, Thomas JM (2019) Assessing the effects of vegetable consumption on the psychological health of healthy adults: a systematic review of prospective research. Am J Clin Nutr 110(1):196–211

Wood PD, Terry RB, Haskell WL (1985) Metabolism of substrates: diet, lipoprotein metabolism, and exercise. Fed Proc 44(2):358–363

Chapter 26
Mindful Consumption Amid Cravings: Behavioral Dynamics from Urge to Sustainable Eating

Shivanjali Shrivastava, Sonakshi Ruhela, Manoj Kumar, and Siddharth Soni

Abstract This chapter dissects the complex interplay between the physiological and psychological forces that drive our desires for specific foods, and the subsequent actions we take in response to those cravings. Hunger, a basic biological need, plays a role, but our cravings extend far beyond simple caloric requirements. We explore the influence of the brain's reward system, which is activated by the pleasurable sensations associated with particular foods. This hedonic aspect, along with learned associations and emotional states, can trigger powerful cravings. However, the journey from craving to consumption doesn't end there. This chapter delves deeper, investigating the psychological mechanisms that bridge the gap between desire and action. We examine how emotions, both positive and negative, can influence our food choices. Stress, for instance, might lead to comfort food cravings, while feelings of happiness might motivate us to indulge in celebratory treats. Additionally, established habits play a significant role. Repeatedly reaching for sugary snacks in the afternoon can create a powerful, automated behavior that's difficult to break. Finally, we explore the often-overlooked influence of environmental cues. The sight, smell, and even the packaging of food can trigger cravings and influence our decisions, especially in environments saturated with tempting options. By elucidating these intricate dynamics, we gain valuable insights that can be harnessed to promote healthy eating patterns. Understanding how cravings arise and how they translate into actions empowers individuals to make informed dietary decisions. This knowledge can help us identify and manage triggers, develop coping mechanisms for emotional eating, and establish mindful eating habits that prioritize long-term

S. Shrivastava (✉) · M. Kumar
Amity Centre of Behavioural Science, Amity University, Noida, India
e-mail: sshrivastava3@amity.edu; mmathur@amity.edu

S. Ruhela
Rochester Institute of Technology, Dubai, UAE
e-mail: sorcad@rit.edu

S. Soni
School of Psychology and Counselling, O.P. Jindal Global University, Sonipat, Haryana, India

M. Thakur, T. Belwal (eds.), *Food Biodiversity and Consumer Engagements*,
Sustainable Development and Biodiversity 46,
https://doi.org/10.1007/978-981-95-7777-4_26

health over fleeting cravings. This chapter holds particular significance for researchers in nutrition, psychology, and public health, offering a framework for understanding and influencing food choices.

Keywords Food cravings · Behavioral dynamics · Eating decisions · Biological influences · Environmental cues

26.1 Introduction

Understanding the factors that influence our eating behaviors is crucial for developing effective strategies to promote healthier eating patterns. This chapter dissects the complex interplay between physiological and psychological forces driving our cravings and the actions we take in response. While hunger is a basic biological need, our cravings often extend beyond mere caloric requirements. We delve into the brain's reward system, emotional states, learned associations, and environmental cues to uncover the mechanisms that lead from craving to consumption. Eating behavior is influenced by a multitude of factors that extend beyond basic biological needs. These factors include the brain's reward system, which is activated by the pleasurable sensations associated with particular foods (Volkow et al. 2011). The hedonic aspect of eating is powerful, often driving individuals to consume foods rich in sugar, fat, and salt, which activate the brain's reward circuits and release dopamine, a neurotransmitter associated with pleasure and reward. This response can lead to a preference for certain foods, even in the absence of hunger, highlighting the complexity of our eating behaviors.

Emotional states also play a significant role in food cravings and consumption. Stress, for instance, can lead to cravings for comfort foods, often high in sugar and fat, as a form of self-soothing (Adam and Epel 2007). High levels of cortisol, the hormone released during stress, can increase appetite and drive cravings for energy-dense foods (McEwen 2007). Conversely, positive emotions such as happiness can motivate indulgence in celebratory treats (Andermann and Anker 2015). Emotional eating can become a coping mechanism, creating a cycle where negative emotions trigger overeating, leading to guilt and more negative emotions (Konttinen et al. 2010). Learned associations are another critical component in the development of food cravings. Habits formed through repeated behaviors in specific contexts become automatic responses, making them difficult to change. For instance, regularly eating a snack while watching television can become a deeply ingrained habit. The cues and routines associated with these habits are crucial in understanding habit formation and breaking unhealthy patterns. Environmental cues further complicate our eating behaviors. The sight, smell, and even the packaging of food can trigger cravings and influence our decisions, especially in environments saturated with tempting options (Wansink 2004). Marketing strategies, such as attractive packaging and strategic placement in stores, are designed to entice consumers and increase sales (Cohen and Babey 2012). The availability and convenience of unhealthy food options, often more accessible than healthier alternatives, further complicate our food choices.

Recognizing and modifying these environmental factors can help mitigate their impact on our eating behavior. Simple changes, such as rearranging the kitchen to place healthier foods at eye level, can make healthier choices more accessible and appealing (Wansink 2004). Research indicates that small modifications to food environments, such as reducing portion sizes and improving food labeling, can lead to healthier eating behaviors (Marteau et al. 2011).

Understanding the intricate dynamics of food cravings and eating behaviors offers valuable insights into the complex interplay of physiological, psychological, and environmental factors. This knowledge empowers individuals to make healthier dietary choices and provides a foundation for developing effective strategies to combat unhealthy eating habits. Continued research and practical application of this knowledge can help address the growing public health challenges related to diet-related diseases, such as obesity, diabetes, and cardiovascular disease.

Future research should focus on further exploring the interactions among biological, psychological, and environmental factors influencing eating behaviors. Investigating the effectiveness of different intervention strategies and identifying new approaches for promoting healthy eating can contribute to improved public health outcomes. Additionally, exploring the impact of emerging technologies, such as mobile health apps and digital interventions, on eating behaviors and health outcomes can provide new opportunities for advancing public health strategies.

26.2 The Role of Hunger

Hunger is a fundamental biological drive that influences food intake, but its role in our eating behaviors is multifaceted. This section explores the physiological signals and hormones involved in hunger, as well as the complex interplay of factors that drive food consumption beyond mere caloric needs.

Physiological Signals and Hunger Hormones Understanding the biological underpinnings of hunger involves examining the key hormones and signals that regulate appetite and satiety.

Ghrelin and Appetite Regulation Ghrelin, often referred to as the 'hunger hormone,' is secreted by the stomach and plays a critical role in meal initiation. Research shows that ghrelin levels increase before meals and decrease after eating, signaling the brain to increase appetite and prepare for food intake (Cummings et al. 2001). Elevated ghrelin levels are associated with increased hunger and food consumption, and chronic elevation can contribute to weight gain and obesity (Schellekens et al. 2012).

Leptin and Energy Balance Leptin, known as the "satiety hormone," is produced by adipose tissue and helps regulate long-term energy balance by inhibiting hunger.

It acts on the hypothalamus to suppress appetite and promote energy expenditure (Flier 2004). Individuals with higher body fat levels have higher leptin levels, but leptin resistance can occur in obesity, leading to continued overeating despite sufficient energy stores (Myers et al. 2010).

26.3 Beyond Caloric Needs

While hunger is a primary driver of food intake, various other factors contribute to eating behaviors that extend beyond simple caloric requirements.

Emotional Eating Emotional states such as stress, anxiety, and depression can trigger emotional eating, where individuals consume food for comfort rather than hunger. Studies indicate that emotional eating is linked to higher body mass index (BMI) and an increased risk of obesity (Konttinen et al. 2010). Stress, in particular, has been shown to increase the consumption of high-calorie, palatable foods (Adam and Epel 2007).

Environmental Cues The availability and visibility of food significantly influence eating behavior. Research has demonstrated that people are more likely to eat food that is readily accessible and visible, regardless of their hunger levels (Wansink 2004). Marketing strategies, such as attractive packaging and strategic placement in stores, are designed to entice consumers and increase sales (Cohen and Babey 2012).

The Impact of Food Advertising Food advertising, especially for unhealthy food products, profoundly impacts dietary choices, particularly among children and adolescents. Exposure to food advertisements increases the likelihood of choosing and consuming advertised products, contributing to poor dietary habits and obesity (Harris et al. 2009). Regulatory measures to limit the advertising of unhealthy foods to children have been suggested as a strategy to combat obesity (Hawkes 2007).

Social and Cultural Influences Social and cultural factors play a significant role in shaping eating behaviors and food preferences. Cultural norms and traditions influence the types of foods consumed and the contexts in which eating occurs. Social interactions, such as family meals and social gatherings, can encourage overeating and the consumption of energy-dense foods (Story et al. 2008).

The Role of Social Norms Social norms regarding portion sizes and acceptable eating behaviors influence individual food intake. Research shows that people tend to eat more when dining in groups and are influenced by the eating habits of their peers. Interventions that promote healthier social norms around eating can help encourage better dietary habits.

26.4 Importance of Understanding the Dynamics of Eating Behavior

The dynamics of eating behavior are complex and multifaceted, influenced by an interplay of physiological signals, psychological states, and environmental factors. Physiologically, hunger and satiety signals regulate our basic need for energy intake, but cravings often extend beyond these biological imperatives. The brain's reward system, which is activated by the pleasurable sensations associated with eating, plays a significant role in driving cravings. This system involves the release of neurotransmitters such as dopamine, which reinforce the desire for certain foods, particularly those high in sugar and fat (Berthoud 2011; Rolls 2017). Psychologically, emotions and learned associations significantly impact our food choices. For instance, stress can trigger cravings for comfort foods, while positive emotions might lead to celebratory eating. Understanding these dynamics is essential for addressing eating-related health issues, as it allows for the development of more effective and sustainable dietary interventions (Epel et al. 2001; Gibson 2006).

26.5 Physiological and Psychological Aspects of Cravings

Cravings are driven by a complex interplay of physiological and psychological factors. On the physiological side, hunger is a fundamental biological drive that prompts us to consume food to meet our energy needs. However, cravings often involve more intricate processes. The brain's reward system is central to this experience, with certain foods activating the release of dopamine, a neurotransmitter associated with pleasure and reward. This activation reinforces the behavior of seeking out and consuming these foods, creating a cycle of craving and consumption. Neuroimaging studies have shown that high-calorie, palatable foods activate brain regions similar to those involved in drug addiction, such as the striatum and prefrontal cortex (Berridge 2009; Kringelbach 2004).

Hormonal signals also play a crucial role in regulating hunger and cravings. Ghrelin, often referred to as the "hunger hormone," increases appetite and promotes food intake, while leptin, known as the "satiety hormone," signals fullness and reduces appetite. Dysregulation of these hormones can lead to increased cravings and overeating. For example, individuals with obesity often exhibit leptin resistance, where the brain does not respond appropriately to leptin signals, leading to persistent hunger and increased food intake (Flint et al. 2000).

Psychologically, cravings are significantly influenced by emotional states. Stress, for instance, can lead to a preference for comfort foods that are high in sugar and fat, providing temporary relief from negative emotions. This phenomenon, known as emotional eating, is well-documented and has been linked to poor dietary habits and weight gain (Epel et al. 2001; Gibson 2006). On the other hand, positive emotions can also drive eating behavior, with feelings of happiness and celebration often leading to the consumption of indulgent foods.

Learned associations from past experiences further complicate these dynamics. Childhood experiences with food, such as receiving sweets as rewards, can create lasting preferences and cravings in adulthood. Cognitive factors, such as expectations and beliefs about the rewarding properties of certain foods, also play a role in driving cravings. When we anticipate that a particular food will bring pleasure or comfort, we are more likely to crave and seek out that food (Robinson and Berridge 1993).

In addition to these internal factors, environmental cues such as the sight and smell of food can trigger cravings. For example, walking past a bakery and smelling freshly baked bread can stimulate a craving, even if you are not physically hungry. Marketing and packaging also play a role in influencing our food choices by creating appealing images and associations that entice us to indulge (Cohen and Babey 2012). Understanding the physiological and psychological aspects of cravings is crucial for developing effective interventions to promote healthier eating habits. By identifying the triggers and underlying mechanisms of cravings, we can develop strategies to manage them, such as mindfulness techniques, stress management, and cognitive-behavioral approaches. These strategies can help individuals make more informed and conscious food choices, ultimately leading to better health outcomes.

26.6 The Biological Basis of Hunger and Cravings

26.6.1 Hunger as a Basic Biological Need

Hunger is an essential biological drive that ensures survival by motivating individuals to seek and consume food. This drive is regulated by a complex interplay of physiological signals that monitor the body's energy needs and nutritional status. The hypothalamus, a small region at the base of the brain, plays a central role in regulating hunger. It integrates signals from the body, such as hormone levels and nutrient status, to maintain energy homeostasis (Schwartz et al. 2000).

Ghrelin, often referred to as the "hunger hormone," is produced in the stomach and stimulates appetite by acting on the hypothalamus. Ghrelin levels increase before meals and decrease after eating, thus playing a critical role in meal initiation (Cummings et al. 2001). Conversely, leptin, a hormone produced by adipose tissue, signals satiety and reduces appetite. High levels of leptin are associated with increased fat stores, providing feedback to the brain to reduce food intake and increase energy expenditure (Friedman and Halaas 1998).

These hormonal signals are part of a feedback loop that helps regulate body weight and energy balance. Disruptions in this system, such as leptin resistance, can lead to overeating and obesity. Leptin resistance, a condition where the brain does not respond appropriately to leptin signals, is commonly observed in individuals with obesity, contributing to persistent hunger and increased food intake despite adequate or excessive energy stores.

26.6.2 Differentiating Hunger from Cravings

While hunger and cravings are often used interchangeably, they represent distinct physiological and psychological states. Hunger is driven by the body's need for energy and nutrients, characterized by physiological signals such as stomach contractions and changes in blood glucose levels. Cravings, on the other hand, are intense desires for specific foods, often driven by psychological and emotional factors rather than a true physiological need for nutrition. Cravings can occur even when the body does not require energy-dense food. They are typically focused on foods high in sugar, fat, and salt, which are associated with high palatability and hedonic value. These foods activate the brain's reward system, creating a strong desire to consume them for pleasure rather than sustenance (Pelchat 2002). Emotional states can significantly influence cravings. For example, stress and negative emotions can trigger cravings for comfort foods that provide temporary relief and pleasure. This phenomenon, known as emotional eating, is well-documented and contributes to unhealthy eating patterns and weight gain. Positive emotions, such as happiness and celebration, can also lead to cravings for indulgent foods, highlighting the complex interplay between emotions and food desires.

26.6.3 Role of the Brain's Reward System in Food Cravings

The brain's reward system plays a crucial role in the experience of food cravings. This system, which includes structures such as the nucleus accumbens, ventral tegmental area (VTA), and prefrontal cortex, is activated by the pleasurable sensations associated with eating, particularly foods high in sugar and fat (Kelley and Berridge 2002). The release of dopamine, a neurotransmitter associated with pleasure and reward, reinforces the behavior of seeking and consuming these foods.

Neuroimaging studies have shown that high-calorie, palatable foods activate brain regions involved in reward processing, similar to the activation patterns observed in response to addictive substances like drugs and alcohol. This similarity has led to the concept of "food addiction," where the brain's response to certain foods mimics the addictive response to drugs, driving compulsive eating behaviors (Gearhardt et al. 2011).

The prefrontal cortex, which is involved in decision-making and self-control, also plays a role in regulating food cravings. Individuals with reduced activity in this region may have difficulty resisting cravings and controlling their eating behaviors, contributing to overeating and weight gain (Goldstone et al. 2009).

26.6.4 Influence of Pleasurable Sensations Associated with Specific Foods

The pleasurable sensations associated with eating certain foods are a major driver of food cravings. Foods high in sugar, fat, and salt are particularly effective at activating the brain's reward system, creating strong sensory experiences that reinforce the desire to consume these foods (Berthoud 2011). This hedonic aspect of eating often overrides physiological signals of hunger and satiety, leading to overeating and weight gain.

Palatable foods trigger the release of dopamine in the brain's reward pathways, creating feelings of pleasure and satisfaction. This response not only reinforces the behavior of seeking and consuming these foods but also creates a memory of the pleasurable experience, making it more likely that individuals will crave these foods in the future (Berridge 2009). The sensory properties of food, such as taste, texture, and aroma, play a significant role in this process. For example, the combination of sweet and fatty flavors in desserts and snacks provides a particularly powerful sensory experience that can drive strong cravings. Environmental cues, such as the sight and smell of food, can also trigger cravings by activating sensory memories and associations (Siep et al. 2009).

Understanding the influence of pleasurable sensations on food cravings is crucial for developing strategies to manage cravings and promote healthier eating habits. Techniques such as mindful eating, which encourages individuals to focus on the sensory experience of eating without judgment, can help reduce the impact of hedonic drives on eating behavior and improve dietary choices (Kristeller and Wolever 2011).

26.7 Psychological Influences on Cravings

26.7.1 Hedonic Eating and the Brain's Reward Circuitry

Hedonic eating refers to the consumption of food for pleasure rather than hunger. This behavior is largely driven by the brain's reward circuitry, which includes regions such as the nucleus accumbens, ventral tegmental area (VTA), and prefrontal cortex. These areas are critical for processing rewards and reinforcing behaviors that provide pleasure, including the consumption of highly palatable foods (Kelley and Berridge 2002). Dopamine, a neurotransmitter associated with pleasure and reward, plays a central role in this process. When we consume foods that are rich in sugar, fat, or salt, dopamine is released in the brain's reward pathways, creating feelings of pleasure and satisfaction. This response not only reinforces the behavior of seeking and consuming these foods but also strengthens the memory of the pleasurable experience, making it more likely that we will crave these foods in the future.

Neuroimaging studies have shown that the consumption of palatable foods activates the same brain regions as addictive substances, suggesting that similar mechanisms may underlie both food cravings and drug addiction. For instance, high-calorie foods trigger significant activity in the striatum, a key component of the reward system, highlighting the powerful impact of these foods on the brain. This activation can create a cycle of craving and consumption that is difficult to break, particularly in environments where palatable foods are readily available.

26.7.2 Learned Associations and Their Impact on Food Desires

Learned associations play a significant role in shaping our food desires and cravings. From an early age, we form associations between specific foods and various experiences or emotions. For example, if a child is consistently rewarded with sweets for good behavior, they may develop a strong preference for sugary foods and associate them with positive feelings (Birch 1999). These learned associations are reinforced over time and can persist into adulthood, influencing our food choices and cravings. The concept of "conditioned taste preference" explains how repeated exposure to certain flavors, particularly those associated with positive outcomes, can lead to long-lasting food preferences. Conversely, negative experiences associated with specific foods can result in aversions, demonstrating the powerful impact of learning on our eating behavior. Advertising and marketing also play a crucial role in shaping learned associations. The portrayal of foods in advertisements often links them to positive emotions, social acceptance, and desirable outcomes, reinforcing cravings and desires for these products. Studies have shown that exposure to food advertisements can increase the likelihood of craving and consuming the advertised products, particularly among children and adolescents (Harris et al. 2009).

26.7.3 Emotional States as Triggers for Cravings

Emotional states are powerful triggers for food cravings, influencing both the type and amount of food consumed. Emotional eating, or the consumption of food in response to emotions rather than hunger, is a well-documented phenomenon. Negative emotions such as stress, anxiety, sadness, and boredom are particularly potent triggers for cravings, often leading to the consumption of high-calorie, palatable foods that provide temporary relief and comfort. Stress, for instance, activates the hypothalamic-pituitary-adrenal (HPA) axis, leading to the release of cortisol, a hormone that can increase appetite and cravings for energy-dense foods. This stress-induced eating is thought to be a coping mechanism, providing a temporary distraction from the source of stress and triggering the release of pleasure-inducing neurotransmitters such as dopamine (Epel et al. 2001).

Positive emotions, such as happiness and excitement, can also trigger cravings, though the mechanisms may differ. Celebratory eating, where individuals indulge in special foods to mark positive events or achievements, is a common practice across cultures. These foods are often rich in sugar and fat, reinforcing the association between positive emotions and indulgent eating.

26.7.4 Influence of Stress on Comfort Food Cravings

Stress is one of the most significant psychological triggers for comfort food cravings. The relationship between stress and eating behavior is complex, involving both physiological and psychological mechanisms. When individuals experience stress, the body responds by activating the HPA axis, leading to the release of cortisol. Elevated cortisol levels have been linked to increased appetite and cravings for high-energy foods, particularly those high in sugar and fat (Adam and Epel 2007). Comfort foods are typically those that are associated with positive emotions and past experiences, providing a sense of nostalgia and emotional comfort. The consumption of these foods during stressful times can serve as a coping mechanism, offering temporary relief from negative emotions and activating the brain's reward system (Dallman et al. 2003). This stress-induced eating behavior is particularly problematic as it can lead to the development of unhealthy eating patterns and contribute to weight gain and metabolic disorders (Torres and Nowson 2007).

26.7.5 Happiness and Celebratory Eating

Happiness and celebratory occasions are also powerful triggers for specific eating behaviors. During moments of joy and celebration, individuals are more likely to indulge in foods that they might otherwise avoid. This behavior is often culturally reinforced, with certain foods being associated with holidays, birthdays, and other special events. These foods are typically rich and indulgent, contributing to the formation of strong hedonic associations. The act of celebratory eating is often social, involving shared meals and communal experiences. This social aspect can amplify the pleasure derived from eating, as the positive emotions associated with social interactions enhance the sensory enjoyment of the food. The presence of others can also create a sense of permission to indulge, further reinforcing the behavior (Macht and Dettmer 2006).

Understanding the psychological influences on cravings, including the impact of hedonic eating, learned associations, emotional states, stress, and celebratory occasions, is crucial for developing effective strategies to manage cravings and promote healthier eating habits. By addressing the underlying psychological triggers and developing coping mechanisms, individuals can gain better control over their eating behaviors and make more informed food choices.

26.8 Environmental and Social Influences on Food Choices

26.8.1 Impact of Environmental Cues on Cravings

Environmental cues are significant triggers for food cravings and consumption. These cues include visual, olfactory, and situational stimuli that can activate cravings, even in the absence of actual hunger. The sight and smell of food are powerful environmental triggers that can prompt eating behavior by stimulating the brain's reward system. For instance, food advertisements, displays in grocery stores, and even the aroma of cooking food can provoke cravings and influence food choices (Burger and Stice 2011).

The concept of "food environments" refers to the accessibility and availability of various foods in a person's surroundings, which can affect dietary patterns and eating behaviors. Studies have shown that environments saturated with high-calorie, palatable foods increase the likelihood of overeating and unhealthy food choices (Powell et al. 2007). For example, individuals who live in neighborhoods with a high density of fast-food outlets are more likely to consume fast food and experience higher rates of obesity.

Additionally, the packaging and presentation of food products can influence perceptions and cravings. Research indicates that larger portion sizes and attractive packaging can lead to increased consumption, as people often eat more when food is presented in larger quantities or more appealing formats (Wansink 2004). This effect is partly due to cognitive biases, where individuals rely on external cues rather than internal signals of hunger and satiety to regulate their eating behavior.

Sensory Triggers Environmental cues play a significant role in triggering cravings and influencing eating decisions. The sight, smell, and even the packaging of food can stimulate our desire to eat. For instance, research shows that people are more likely to crave and consume foods that are highly visible and easily accessible.

Marketing and Availability Marketing strategies, such as attractive packaging and strategic placement in stores, are designed to entice consumers and increase sales. The availability and convenience of unhealthy food options, often more accessible than healthier alternatives, further complicate our food choices. Studies have found that placing healthier foods at eye level and making them more convenient can significantly increase their consumption.

Modifying Environmental Factors In environments saturated with tempting options, these cues can overpower our intentions to eat healthily. Recognizing and modifying these environmental factors can help mitigate their impact on our eating behavior. Simple changes, such as rearranging the kitchen to place healthier foods at eye level, can make healthier choices more accessible and appealing. Research indicates that small modifications to food environments, such as reducing portion sizes and improving food labeling, can lead to healthier eating behaviors.

26.8.2 Social Influences on Eating Behavior

Social factors play a crucial role in shaping eating behavior and food choices. Social influences include peer pressure, family eating habits, cultural norms, and social gatherings, all of which can impact what, when, and how much people eat. Social interactions around food can either encourage healthy eating or contribute to unhealthy eating patterns (Herman and Polivy 2008).

Family eating patterns are often established early in life and can have a long-lasting impact on dietary habits. Family meals provide opportunities for modeling eating behaviors and socialization, which can influence children's food preferences and eating practices. For example, children who regularly eat meals with their families are more likely to consume healthier diets and exhibit lower rates of obesity. Cultural norms and traditions also shape food choices and eating behaviors. Different cultures have specific foods, rituals, and practices associated with meals and celebrations. These cultural influences can affect food preferences, portion sizes, and the frequency of certain foods in the diet (Rozin 2005). For instance, traditional holiday meals often feature rich, indulgent foods, which can contribute to overeating and deviations from typical eating patterns.

Social gatherings and celebrations frequently involve large amounts of food, and the presence of others can influence individual eating behaviors. People are more likely to eat more when dining with others, particularly in social contexts that encourage eating as a form of enjoyment or celebration. The social context can create an environment where eating becomes a communal activity, reinforcing the consumption of larger portions and a greater variety of foods.

26.8.3 Influence of Food Marketing and Advertising

Food marketing and advertising significantly influence food choices and eating behaviors. The marketing strategies used by food companies are designed to attract attention, create desire, and promote consumption of their products. These strategies include colorful packaging, appealing advertising campaigns, and sponsorship of popular events, all of which can enhance the appeal of certain foods and drive cravings.

Advertisements often link foods to positive emotions, social status, and lifestyle aspirations, creating strong associations that can influence food preferences and consumption patterns. Research has shown that exposure to food advertisements can lead to increased consumption of the advertised products, especially among children and adolescents who are particularly susceptible to marketing messages.

The use of endorsements by celebrities and influencers in food advertising also plays a role in shaping food choices. Celebrity endorsements can increase the perceived desirability of a product and encourage consumption, particularly if the celebrity is perceived as attractive or influential. The impact of these endorsements is amplified when they align with social trends and cultural values, making the advertised foods more appealing to target audiences.

26.8.4 Role of Portion Sizes and Eating Situations

Portion sizes and eating situations can significantly impact food intake and overall eating behavior. Larger portion sizes are associated with increased consumption, as people often eat more when presented with larger quantities of food (Herman and Polivy 2008). The phenomenon of portion size influence is driven by external cues, where individuals use the amount of food served as a guide for how much to eat, rather than relying on internal hunger signals. Eating situations, such as dining out at restaurants or attending social events, can also affect food choices and portion sizes. Research indicates that people tend to consume more when dining out due to larger portion sizes, higher calorie content, and the social context of eating. Additionally, special occasions and celebrations often involve indulgent foods and larger quantities, contributing to temporary deviations from regular eating habits (Herman and Polivy 2008).

Understanding the environmental and social influences on food choices is crucial for developing strategies to promote healthier eating habits. By addressing factors such as food environments, social influences, marketing practices, and portion sizes, individuals and public health professionals can work to create environments that support healthier dietary choices and reduce the prevalence of obesity and related health issues.

26.9 Habit Formation and Automated Behaviors

Context-Dependent Repetition Habit formation is a powerful force in eating behavior. Repeatedly choosing certain foods in specific contexts can create automated behaviors that are difficult to change. Habits are formed through a process known as “context-dependent repetition,” where repeated behaviors in the same context become automatic responses. Research suggests that it takes an average of 66 days to form a new habit, though this can vary depending on the complexity of the behavior.

Identifying Cues and Routines For example, consistently reaching for sugary snacks in the afternoon can become a deeply ingrained habit. These habits are maintained by cues in our environment and the routine behaviors that follow. Understanding the cues and routines that drive these habits is essential for developing interventions to promote healthier choices.

Breaking and Forming New Habits Breaking unhealthy habits and establishing new, healthier ones requires conscious effort and consistent practice. Studies have shown that implementing "if-then" planning, where individuals plan specific responses to triggers (e.g., "If I feel like snacking in the afternoon, then I will eat a piece of fruit"), can significantly improve habit change success rates.

26.10 Bridging the Gap: From Cravings to Consumption

Understanding the transition from cravings to consumption involves examining the psychological mechanisms at play. This section delves into how sensory cues, emotional states, and learned behaviors drive the urge to eat specific foods and explores mindfulness and cognitive-behavioral strategies for managing these cravings.

26.10.1 Psychological Mechanisms

Cravings are complex phenomena influenced by multiple psychological factors. Sensory cues, such as the sight, smell, and taste of food, can trigger intense desires to eat. These sensory experiences are often linked to memories and learned associations, making certain foods particularly irresistible (Hill et al. 1984). Emotional states significantly impact cravings and eating behaviors. For instance, stress and negative emotions can lead to emotional eating, where individuals consume food to cope with their feelings. Approximately 40% of individuals report eating more during stressful times, often choosing high-calorie, high-sugar foods that provide temporary comfort (Groesz et al. 2012). Conversely, positive emotions, such as happiness and celebration, can also prompt indulgence in favorite foods.

Learned behaviors and habits further reinforce cravings. Repeatedly consuming certain foods in specific contexts can create automatic responses. For example, consistently eating sugary snacks in the afternoon can develop into a habitual behavior that triggers cravings at the same time each day (Neal et al. 2006).

26.10.2 Mindfulness Practices

Mindfulness involves paying attention to the present moment without judgment. In the context of eating, mindfulness practices help individuals become more aware of their hunger and fullness cues, reducing the likelihood of mindless eating. Research has shown that mindfulness-based interventions can significantly reduce binge eating episodes and increase awareness of eating patterns. For instance, a study by Kristeller and Wolever (2011) found that participants who underwent mindfulness-based eating awareness training experienced a 75% reduction in binge eating episodes. Practicing mindfulness can involve simple techniques such as eating slowly, savoring each bite, and avoiding distractions like television or smartphones during meals. By focusing on the sensory experience of eating, individuals can better recognize when they are satisfied and avoid overeating (Bays 2009).

26.10.3 Cognitive-Behavioral Strategies

Cognitive-behavioral strategies (CBT) focus on identifying and challenging unhelpful thoughts and behaviors related to food. CBT can help individuals reframe their thoughts about food and develop healthier eating patterns. For example, a person might learn to replace the thought "I need to eat this dessert to feel better" with "I can find other ways to cope with my emotions."

CBT techniques include keeping a food diary to track eating habits and triggers, setting realistic and achievable goals, and developing coping strategies for dealing with cravings and emotional eating. Studies have shown that CBT is effective in reducing the frequency of binge eating episodes and improving overall eating behaviors. A meta-analysis by Vocks et al. (2010) reported that CBT resulted in significant reductions in binge eating and improvements in psychological well-being.

26.11 Sustainable Solutions for Healthy Eating

Long-Term Strategies for Healthy Eating Adopting sustainable solutions for managing food cravings and promoting healthy eating is essential for achieving long-term success. Unlike short-term diets and quick fixes that often fail to address the root causes of unhealthy eating behaviors, sustainable strategies focus on making gradual, manageable changes that can be maintained over time. This involves incorporating a balanced diet that includes a variety of nutrient-dense foods while allowing for occasional indulgences to prevent feelings of deprivation. For example, the Mediterranean diet, which emphasizes fruits, vegetables, whole grains, and healthy fats, has been shown to support long-term health and reduce the risk of chronic diseases.

Building Resilience Against Cravings Developing resilience against cravings is a key component of sustainable eating strategies. This involves cultivating habits and skills that help individuals manage cravings effectively without resorting to restrictive diets. Mindfulness practices, such as mindful eating, enhance awareness of hunger and fullness cues, making it easier to differentiate between physical hunger and emotional cravings. Cognitive-behavioral strategies, such as identifying and challenging unhelpful thoughts about food, also play a crucial role in managing cravings. Evidence indicates that individuals who use these techniques are more successful in maintaining healthy eating patterns and achieving long-term weight management (Forman et al. 2013).

Environmental and Policy Changes Creating a supportive environment for sustainable healthy eating habits is critical. This includes increasing access to nutritious foods and making them more affordable, particularly in underserved areas. Policy changes, such as taxing sugary beverages and restricting the marketing of unhealthy foods to children, can help shift societal norms towards healthier eating behaviors. Schools and workplaces can contribute by offering healthy meal options and fostering environments that encourage physical activity and well-being (Swinburn et al. 2011).

Community Engagement Engaging communities in the promotion of healthy eating is vital for sustainability. Community-based interventions that involve local stakeholders, such as schools, healthcare providers, and community organizations, can help build a supportive network for healthy behaviors. Programs that offer education on nutrition and cooking skills, support local food production, and create social connections around healthy eating can empower individuals to make lasting changes. By addressing social determinants of health and involving communities in the process, we can create a more equitable and sustainable approach to promoting healthy eating.

26.12 Implications for Research and Public Health

Designing Effective Interventions The insights from this chapter are invaluable for researchers in the fields of nutrition, psychology, and public health. By understanding the intricate behavioral dynamics of eating, we can design more effective interventions aimed at promoting healthier eating patterns. For example, interventions can be tailored to address the specific psychological and emotional triggers of cravings, such as stress, boredom, or social influences. Personalized approaches that consider an individual's unique physiological and psychological profile can be more effective than one-size-fits-all solutions. Moreover, integrating technology, such as mobile health applications and wearable devices, can provide real-time feedback and support to individuals trying to manage their eating behaviors. These tools can help track food intake, monitor emotional states, and provide immediate interventions when cravings arise. Such data-driven approaches can enhance the efficacy of dietary interventions and support long-term behavioral change.

Public Health Initiatives The framework presented in this chapter provides a robust basis for further research and practical applications across various settings, from individual counseling to broad public health campaigns. Public health initiatives can leverage this understanding to create environments that foster healthy eating behaviors. For instance, implementing policies that reduce the availability of unhealthy food options in public spaces, such as schools, workplaces, and recreational areas, can discourage unhealthy eating habits. Additionally, increasing access to nutritious foods through initiatives like urban farming, farmers' markets, and subsidies for healthy foods can make it easier for individuals to make healthier choices (Swinburn et al. 2011). Public health campaigns can also focus on altering the social norms around food consumption, promoting the idea that healthy eating is desirable and achievable. By addressing both the availability of healthy options and the social context of eating, public health initiatives can more effectively encourage healthy eating patterns on a community-wide scale.

Educational Programs Educational programs play a crucial role in promoting healthier eating behaviors by teaching individuals about the factors influencing their food choices and providing strategies for managing cravings. School-based interventions, in particular, have shown promise in improving children's dietary habits and reducing obesity rates. These programs can incorporate nutrition education, teaching children about the benefits of healthy eating and the risks associated with poor dietary choices. Additionally, behavioral strategies such as goal setting, self-monitoring, and problem-solving can be integrated into the curriculum to help children develop skills for managing cravings and making healthier food choices (Katz et al. 2011). Beyond schools, workplace wellness programs can educate employees about the importance of nutrition and provide resources for healthy eating. Community workshops and public seminars can also raise awareness and empower individuals with the knowledge and tools needed to make informed dietary decisions. By spreading knowledge and fostering skills for healthier eating, educational programs can contribute significantly to the prevention of diet-related diseases and the promotion of overall public health.

26.13 Conclusion

By elucidating the intricate dynamics of food cravings and eating behaviors, this chapter offers valuable insights into the complex interplay of physiological, psychological, and environmental factors. Understanding these elements empowers individuals to make healthier dietary choices and provides a foundation for developing effective strategies to combat unhealthy eating habits. Recognizing the role of hormones like ghrelin and leptin in regulating hunger and satiety allows us to appreciate the biological underpinnings of our cravings. Furthermore, understanding how emotional states and learned associations influence eating behaviors sheds light on the psychological mechanisms that drive food consumption beyond mere caloric needs. By becoming aware of these factors, individuals can develop more mindful eating practices, such as paying closer attention to hunger and fullness cues and avoiding distractions during meals. Additionally, knowledge about the influence of environmental cues, such as food advertising and availability, empowers individuals to create a food environment that supports healthy eating. This might involve making healthier food choices more accessible at home and work, reducing exposure to unhealthy food advertisements, and advocating for policy changes that promote public health. Ultimately, by understanding the multifaceted nature of food cravings, individuals can take proactive steps to cultivate healthier eating habits that align with their long-term health goals.

26.14 Future Research and Applications

Through continued research and practical application, we can influence food choices and promote better health outcomes. This knowledge is crucial for addressing the growing public health challenges related to diet-related diseases, such as obesity, diabetes, and cardiovascular disease. Future research should focus on exploring the long-term effects of interventions aimed at modifying eating behaviors and managing cravings. Longitudinal studies can provide valuable insights into how these interventions impact dietary habits, weight management, and overall health over time. Additionally, investigating the role of genetic and epigenetic factors in eating behaviors can lead to personalized approaches to nutrition and behavioral interventions. By understanding individual differences in susceptibility to cravings and eating patterns, researchers can develop targeted strategies that are more effective for specific populations. Practical applications of this research include the development of comprehensive public health campaigns that address the psychological and environmental factors influencing food choices. For example, community-based programs that combine nutrition education with behavioral strategies can help individuals build sustainable, healthy eating habits. Furthermore, integrating mindfulness and cognitive-behavioral strategies into clinical practice can provide additional support for individuals struggling with emotional eating and other eating-related issues. By harnessing these insights, we can move toward a future where healthy eating is the norm rather than the exception. This requires a concerted effort from researchers, policymakers, healthcare providers, and individuals to create a supportive environment that promotes nutritious food choices and fosters a culture of health and well-being. Through collaborative efforts and a commitment to ongoing research, we can make significant strides in combating diet-related diseases and improving the quality of life for people worldwide.

References

Adam TC, Epel ES (2007) Stress, eating and the reward system. Physiol Behav 91:449–458

Andermann ML, Anker JJ (2015) Food intake and reward: A review of the literature. Progress in Neuro-Psychopharmacology & Biological Psychiatry 60:8–21

Bays J (2009) Mindful eating and its role in preventing and treating obesity. Gastroenterol Nurs 32: 22–26

Berridge KC (2009) "Liking" and "wanting" food rewards: Brain mechanisms and roles in eating disorders. Physiol Behav 97:537–550

Berthoud HR (2011) The neurobiology of appetite and weight control: Possible targets for new interventions. J Clin Invest 121:2163–2173

Birch LL (1999) Development of food preferences. Annu Rev Nutr 19:41–62

Burger KS, Stice E (2011) Variability in the response to food cues as a function of weight status: A meta-analysis. Health Psychol 30:235–242

Cohen DA, Babey SH (2012) Contextual factors influencing eating behaviors: The case of fast food. Am J Clin Nutr 95:788–792

Cummings DE, Purnell JQ, Frayo RS, Schmidova K, Wisse BE, Weigle DS, Tschöp MH (2001) A preprandial rise in plasma ghrelin levels suggests a role in appetite regulation. J Clin Endocrinol Metab 86:5992–5995

Dallman MF, Pecoraro NC, la Fleur SE (2003) Chronic stress and obesity: A new view of "comfort food.". Proc Natl Acad Sci U S A 100:11696–11701

Epel ES, McEwen BS, Ickovics JR (2001) Stress and body shape: Stress-induced obesity and the biological mechanisms of stress-induced eating. Horm Behav 40:330–338

Flier JS (2004) Obesity wars: Molecular progress confronts an expanding epidemic. Cell 116:337–350

Flint A, Raben A, Blundell JE, Astrup A (2000) Rebound weight gain after a period of weight loss is related to high ghrelin and low leptin concentrations. Int J Obes 24:269–275

Forman EM, Butryn ML, Hoffman KL (2013) Comparison of a standard behavioral weight loss program with a mindfulness-based eating awareness program: A randomized controlled trial. Behav Ther 44:484–495

Friedman JM, Halaas JL (1998) Leptin and the regulation of body weight in mammals. Nature 395: 763–770

Gearhardt AN, Corbin WR, Brownell KD (2011) Preliminary validation of the Yale Food Addiction Scale. Appetite 52:430–436

Gibson EL (2006) Emotional influences on food choice: Sensory, physiological and psychological pathways. Physiol Behav 89:53–61

Goldstone AP, Preedy VR, Beck K (2009) The neural and endocrine basis of the response to food. Int J Obes 33:S10–S18

Groesz LM, McCoy S, Karl J (2012) Stress and body weight: A review of the evidence. Obes Rev 13:429–443

Harris JL, Bargh JA, Brownell KD (2009) Priming effects of television food advertising on eating behavior. Health Psychol 28:404–413

Hawkes C (2007) The role of food marketing in the obesity epidemic. Int J Environ Res Public Health 4:89–114

Herman CP, Polivy J (2008) External cues in the control of food intake in humans: The sensory-normative distinction. Physiol Behav 94:722–728

Hill AJ, Maggi RS, Blundell JE (1984) The effect of sensory experience on appetite: A review of the evidence. Appetite 5:93–104

Katz DL, O'Connell M, Yares M (2011) School-based interventions to reduce obesity: A systematic review. Prev Med 52:83–89

Kelley AE, Berridge KC (2002) The neuroscience of food intake and reward. Nat Rev Neurosci 3: 453–462

Konttinen H, Männistö S, Sarlio-Lähteenkorva S (2010) Emotional eating and obesity: The role of depression and gender. Eur J Clin Nutr 64:1424–1430
Kringelbach ML (2004) Food for thought: Hedonic experience across the senses. Curr Opin Neurol 17:538–543
Kristeller JL, Wolever RQ (2011) Mindfulness-based eating awareness training for treating binge eating disorder: A randomized controlled trial. J Consult Clin Psychol 79:625–636
Macht M, Dettmer D (2006) Everyday mood and emotions after eating a meal: The role of hunger and food intake. Appetite 47:460–464
Marteau TM, Hollands GJ, Fletcher PC (2011) Changing diet and physical activity to reduce obesity: A systematic review and meta-analysis of randomized controlled trials. Int J Obes 35: 1395–1404
McEwen BS (2007) Physiology and neurobiology of stress and adaptation: Central role of the brain. Physiol Rev 87:873–904
Myers MG, Leibel RL, Seeley RJ, Schwartz MW (2010) Obesity and leptin resistance: Distinguishing cause from effect. Trends Endocrinol Metab 21:643–651
Neal DT, Wood W, Quinn JM (2006) Habits—A repeat performance. Curr Dir Psychol Sci 15:198–202
Pelchat ML (2002) The role of food cravings in eating behavior. Appetite 39:255–260
Powell LM, Szczpka S, Chaloupka FJ (2007) The role of food environment in food choices and eating behavior. Economic Research Service Report 14:67–89
Robinson TE, Berridge KC (1993) The neural basis of food cravings: A motivational systems approach. Biol Psychiatry 34:491–505
Rolls ET (2017) The neurophysiology of feeding and appetite. Curr Opin Neurobiol 46:49–57
Rozin P (2005) The role of culture in food preference and consumption. Food Qual Prefer 16:181–185
Schellekens H, Finger BC, Dinan TG (2012) Ghrelin and obesity: Role of the gut-brain axis. Neuropeptides 46:263–274
Schwartz MW, Seeley RJ, Campfield LA (2000) Identification of targets for interventions in the control of food intake. J Clin Invest 106:261–263
Siep N, Huitema BE, Jansen A (2009) The impact of environmental cues on food intake and food choices. Appetite 53:212–220
Story M, Kaphingst KA, Robinson-O'Brien R, Glanz K (2008) Creating healthy food environments: Policy and environmental approaches. Annu Rev Public Health 29:253–272
Swinburn BA, Sacks G, Hall KD (2011) The global obesity pandemic: Shaped by global drivers and local environments. Lancet 378:804–814
Torres SJ, Nowson CA (2007) Relationship between stress, eating behavior, and obesity. Nutrition 23:887–894
Vocks S, Lehr A, Wolkenstein L (2010) Cognitive-behavioral therapy for binge eating disorder: A meta-analysis. Behav Res Ther 48:205–213
Volkow ND, Wang GJ, Baler RD (2011) Reward, dopamine and the control of food intake: Implications for obesity. Trends Cogn Sci 15:37–46
Wansink B (2004) Environmental factors that increase the food intake and consumption volume of unknowing consumers. Annu Rev Nutr 24:455–479

MIX
Papier aus verantwortungsvollen Quellen
Paper from responsible sources
FSC® C105338

If you have any concerns about our products,
you can contact us on
ProductSafety@springernature.com

In case Publisher is established outside the EU,
the EU authorized representative is:
Springer Nature Customer Service Center GmbH
Europaplatz 3, 69115 Heidelberg, Germany

Printed by Libri Plureos GmbH
in Hamburg, Germany